Big Data Analytics: Systems, Algorithms, Applications

C. S. R. Prabhu · Aneesh Sreevallabh Chivukula ·
Aditya Mogadala · Rohit Ghosh ·
L. M. Jenila Livingston

Big Data Analytics: Systems, Algorithms, Applications

Springer

C. S. R. Prabhu
National Informatics Centre
New Delhi, Delhi, India

Aditya Mogadala
Saarland University
Saarbrücken, Saarland, Germany

L. M. Jenila Livingston
School of Computing Science
and Engineering
Vellore Institute of Technology
Chennai, Tamil Nadu, India

Aneesh Sreevallabh Chivukula
Advanced Analytics Institute
University of Technology, Sydney
Ultimo, NSW, Australia

Rohit Ghosh
Qure.ai
Goregaon East, Mumbai, Maharashtra, India

ISBN 978-981-15-0093-0 ISBN 978-981-15-0094-7 (eBook)
https://doi.org/10.1007/978-981-15-0094-7

This Springer imprint is published by the registered company Springer Nature Singapore Pte Ltd.
The registered company address is: 152 Beach Road, #21-01/04 Gateway East, Singapore 189721, Singapore

Foreword

Big Data phenomenon has emerged globally as the next wave of technology, which will influence in a big way and contribute to better quality of life in all its aspects. The advent of Internet of things (IoT) and its associated Fog Computing paradigm is only accentuating and amplifying the Big Data phenomenon.

This book by C. S. R. Prabhu and his co-authors is coming up at the right time. This book fills in the timely need for a comprehensive text covering all dimensions of Big Data Analytics: systems, algorithms, applications and case studies along with emerging research horizons. In each of these dimensions, this book presents a comprehensive picture to the reader in a lucid and appealing manner. This book can be used effectively for the benefit of students of undergraduate and post-graduate levels in IT, computer science and management disciplines, as well as research scholars in these areas. It also helps IT professionals and practitioners who need to learn and understand the subject of Big Data Analytics.

I wish this book all the best in its success with the global student community as well as the professionals.

Dr. Rajkumar Buyya
Redmond Barry Distinguished
Professor, Director, Cloud Computing
and Distributed Systems (CLOUDS)
Lab, School of Computing and
Information Systems, The University
of Melbourne, Melbourne, Australia

Preface

The present-day Information Age has produced an overwhelming deluge of digital data arriving from unstructured sources such as online transactions, mobile phones, social networks and emails popularly known as Big Data. In addition, with the advent of Internet of things (IoT) devices and sensors, the sizes of data that will flow into the Big Data scenario have multiplied many folds. This Internet-scale computing has also necessitated the ability to analyze and make sense of the data deluge that comes with it to help intelligent decision making and real-time actions to be taken based on real-time analytics techniques.

The Big Data phenomenon has been impacting all sectors of business and industry, resulting in an upcoming new information ecosystem. The term 'Big Data' refers to not only the massive volumes and variety of data itself, but also the set of technologies surrounding it, to perform the capture, storage, retrieval, management, processing and analysis of the data for the purposes of solving complex problems in life and in society as well, by unlocking the value from that data more economically. In this book, we provide a comprehensive survey of the big data origin, nature, scope, structure, composition and its ecosystem with references to technologies such as Hadoop, Spark, R and its applications. Other essential big data concepts including NoSQL databases for storage, machine learning paradigms for computing, analytics models connecting the algorithms are all aptly covered. This book also surveys emerging research trends in large-scale pattern recognition, programming processes for data mining and ubiquitous computing and application domains for commercial products and services. Further, this book expands into the detailed and precise description of applications of Big Data Analytics into the technological domains of Internet of things (IoT), Fog Computing and Social Semantic Web mining and then into the business domains of banking and finance, insurance and capital market before delving into the issues of security and privacy associated with Big Data Analytics. At the end of each chapter, pedagogical questions on the comprehension of the chapter contents are added.

This book also describes the data engineering and data mining life cycles involved in the context of machine learning paradigms for unstructured and structured data. The relevant developments in big data stacks are discussed with a focus on open-source technologies. We also discuss the algorithms and models used in data mining tasks such as search, filtering, association, clustering, classification, regression, forecasting, optimization, validation and visualization. These techniques are applicable to various categories of content generated in data streams, sequences, graphs and multimedia in transactional, in-memory and analytic databases. Big Data Analytics techniques comprising descriptive and predictive analytics with an emphasis on feature engineering and model fitting are covered. For feature engineering steps, we cover feature construction, selection and extraction along with preprocessing and post-processing techniques. For model fitting, we discuss the model evaluation techniques such as statistical significance tests, cross-validation curves, learning curves, sufficient statistics and sensitivity analyses. Finally, we present the latest developments and innovations in generative learning and discriminative learning for large-scale pattern recognition. These techniques comprise incremental, online learning for linear/nonlinear and convex/multi-objective optimization models, feature learning or deep learning, evolutionary learning for scalability and optimization meta-heuristics.

Machine learning algorithms for big data cover broad areas of learning such a supervised, unsupervised and semi-supervised and reinforcement techniques. In particular, supervised learning subsection details several classification and regression techniques to classify and forecast, while unsupervised learning techniques cover clustering approaches that are based on linear algebra fundamentals. Similarly, semi-supervised methods presented in the chapter cover approaches that help to scale to big data by learning from largely un-annotated information. We also present reinforcement learning approaches which are aimed to perform collective learning and support distributed scenarios.

The additional unique features of this book are about 15 real-life experiences as case studies which have been provided in the above-mentioned application domains. The case studies provide, in brief, the experiences of the different contexts of deployment and application of the techniques of Big Data Analytics in the diverse contexts of private and public sector enterprises. These case studies span product companies such as Google, Facebook, Microsoft, consultancy companies such as Kaggle and also application domains at power utility companies such as Opower, banking and finance companies such as Deutsche Bank. They help the readers to understand the successful deployment of analytical techniques that maximize a company's functional effectiveness, diversity in business and customer relationship management, in addition to improving the financial benefits. All these companies handle real-life Big Data ecosystems in their respective businesses to achieve tangible results and benefits. For example, Google not only harnesses, for profit, the big data ecosystem arising out of its huge number of users with billions of web searches and emails by offering customized advertisement services, but also is offering to other companies to store and analyze the big datasets in cloud platforms. Google has also developed an IoT sensor-based autonomous Google car with real-time analytics for driverless navigation. Facebook, the largest social network in the world, deployed big data techniques for personalized search and advertisement. So LinkedIn also deploys big data techniques for effective service delivery.

Microsoft also aspires to enter the big data business scenario by offering services of Big Data Analytics to business enterprises on its Azure cloud services. Nokia deploys its Big Data Analytics services on the huge buyer and subscriber base of its mobile phones, including the mobility of its buyers and subscribers. Opower, a power utility company, has deployed Big Data Analytics techniques on its customer data to achieve substantial benefits on power savings. Deutsche Bank has deployed big data techniques for achieving substantial savings and better customer relationship management (CRM). Delta Airlines improved its revenues and customer relationship management (CRM) by deploying Big Data Analytics techniques. A Chinese city traffic management was achieved successfully by adopting big data methods.

Thus, this book provides a complete survey of techniques and technologies in Big Data Analytics. This book will act as basic textbook introducing niche technologies to undergraduate and postgraduate computer science students. It can also act as a reference book for professionals interested to pursue leadership-level career opportunities in data and decision sciences by focusing on the concepts for problem solving and solutions for competitive intelligence. To the best of our knowledge, big data applications are discussed in a plethora of books. But, there is no textbook covering a similar mix of technical topics. For further clarification, we provide references to white papers and research papers on specific topics.

New Delhi, India — C. S. R. Prabhu
Ultimo, Australia — Aneesh Sreevallabh Chivukula
Saarbrücken, Germany — Aditya Mogadala
Mumbai, India — Rohit Ghosh
Chennai, India — L. M. Jenila Livingston

Acknowledgements

The authors humbly acknowledge the contributions of the following individuals toward the successful completion of this book.

Mr. P. V. N. Balaram Murthy, Ms. J. Jyothi, Mr. B. Rajgopal, Dr. G. Rekha, Dr. V. G. Prasuna, Dr. P. S. Geetha, Dr. J. V. Srinivasa Murthy, all from KMIT, Hyderabad, Dr. Charles Savage of Munich, Germany, Ms. Rachna Sehgal of New Delhi, Dr. P. Radhakrishna of NIT, Warangal, Mr. Madhu Reddy, Hyderabad, Mr. Rajesh Thomas, New Delhi, Mr. S. Balakrishna, Pondicherry, for their support and assistance in various stages and phases involved in the development of the manuscript of this book.

The authors thank the managements of the following institutions for supporting the authors:

1. KMIT, Hyderabad
2. KL University, Guntur
3. VIT, Chennai
4. Advance Analytics Institute, University of Technology, Sydney, (475), Sydney, Australia.

About This Book

Big Data Analytics is an Internet-scale commercial high-performance parallel computing paradigm for data analytics.

This book is a comprehensive textbook on all the multifarious dimensions and perspectives of Big Data Analytics: the platforms, systems, algorithms and applications, including case studies

This book presents data-derived technologies, systems and algorithmics in the areas of machine learning, as applied to Big Data Analytics.

As case studies, this book covers briefly the analytical techniques useful for processing data-driven workflows in various industries such as health care, travel and transportation, manufacturing, energy, utilities, telecom, banking and insurance, in addition to the IT sector itself.

The Big Data-driven computational systems described in this book have carved out, as discussed in various chapters, the applications of Big Data Analytics in various industry application areas such as IoT, social networks, banking and financial services, insurance, capital markets, bioinformatics, advertising and recommender systems. Future research directions are also indicated.

This book will be useful to both undergraduate and graduate courses in computer science in the area of Big Data Analytics.

Contents

1 Big Data Analytics . . . 1
C. S. R. Prabhu
1.1 Introduction . . . 1
1.2 What Is Big Data? . . . 2
1.3 Disruptive Change and Paradigm Shift in the Business Meaning of Big Data . . . 3
1.4 Hadoop . . . 4
1.5 Silos . . . 4
1.5.1 Big Bang of Big Data . . . 5
1.5.2 Possibilities . . . 5
1.5.3 Future . . . 6
1.5.4 Parallel Processing for Problem Solving . . . 6
1.5.5 Why Hadoop? . . . 7
1.5.6 Hadoop and HDFS . . . 7
1.5.7 Hadoop Versions 1.0 and 2.0 . . . 8
1.5.8 Hadoop 2.0 . . . 9
1.6 HDFS Overview . . . 10
1.6.1 MapReduce Framework . . . 11
1.6.2 Job Tracker and Task Tracker . . . 11
1.6.3 YARN . . . 12
1.7 Hadoop Ecosystem . . . 12
1.7.1 Cloud-Based Hadoop Solutions . . . 14
1.7.2 Spark and Data Stream Processing . . . 14
1.8 Decision Making and Data Analysis in the Context of Big Data Environment . . . 15
1.8.1 Present-Day Data Analytics Techniques . . . 15
1.9 Machine Learning Algorithms . . . 17
1.10 Evolutionary Computing (EC) . . . 21

1.11 Conclusion 21
1.12 Review Questions 21
References and Bibliography 22

2 Intelligent Systems 25
Aneesh Sreevallabh Chivukula
2.1 Introduction 25
2.1.1 Open-Source Data Science 26
2.1.2 Machine Intelligence and Computational Intelligence 29
2.1.3 Data Engineering and Data Sciences 34
2.2 Big Data Computing 37
2.2.1 Distributed Systems and Database Systems 37
2.2.2 Data Stream Systems and Stream Mining 40
2.2.3 Ubiquitous Computing Infrastructures 43
2.3 Conclusion 45
2.4 Review Questions 45
References 46

3 Analytics Models for Data Science 47
L. M. Jenila Livingston
3.1 Introduction 47
3.2 Data Models 47
3.2.1 Data Products 48
3.2.2 Data Munging 48
3.2.3 Descriptive Analytics 49
3.2.4 Predictive Analytics 50
3.2.5 Data Science 51
3.2.6 Network Science 54
3.3 Computing Models 54
3.3.1 Data Structures for Big Data 55
3.3.2 Feature Engineering for Structured Data 73
3.3.3 Computational Algorithm 78
3.3.4 Programming Models 78
3.3.5 Parallel Programming 79
3.3.6 Functional Programming 80
3.3.7 Distributed Programming 80
3.4 Conclusion 81
3.5 Review Questions 81
References 81

4 **Big Data Tools—Hadoop Ecosystem, Spark and NoSQL Databases** . . . 83
C. S. R. Prabhu
4.1 Introduction . . . 83
4.1.1 Hadoop Ecosystem . . . 83
4.1.2 HDFS Commands . . . 84
4.2 MapReduce . . . 93
4.3 Pig . . . 105
4.4 Flume . . . 133
4.5 Sqoop . . . 136
4.6 Mahout, The Machine Learning Platform from Apache . . . 142
4.7 GANGLIA, The Monitoring Tool . . . 142
4.8 Kafka, The Stream Processing Platform . . . 143
4.9 Spark . . . 144
4.10 NoSQL Databases . . . 151
4.11 Conclusion . . . 165
References . . . 165

5 **Predictive Modeling for Unstructured Data** . . . 167
Aditya Mogadala
5.1 Introduction . . . 167
5.2 Applications of Predictive Modeling . . . 169
5.2.1 Natural Language Processing . . . 169
5.2.2 Computer Vision . . . 174
5.2.3 Information Retrieval . . . 177
5.2.4 Speech Recognition . . . 178
5.3 Feature Engineering . . . 179
5.3.1 Feature Extraction and Weighing . . . 179
5.3.2 Feature Selection . . . 187
5.4 Pattern Mining for Predictive Modeling . . . 187
5.4.1 Probabilistic Graphical Models . . . 187
5.4.2 Deep Learning . . . 188
5.4.3 Convolutional Neural Networks (CNN) . . . 189
5.4.4 Recurrent Neural Networks (RNNs) . . . 190
5.4.5 Deep Boltzmann Machines (DBM) . . . 191
5.4.6 Autoencoders . . . 192
5.5 Conclusion . . . 192
5.6 Review Questions . . . 193
References . . . 193

6 Machine Learning Algorithms for Big Data 195
Aditya Mogadala
6.1 Introduction 195
6.2 Generative Versus Discriminative Algorithms 196
6.3 Supervised Learning for Big Data 198
6.3.1 Decision Trees 199
6.3.2 Logistic Regression 199
6.3.3 Regression and Forecasting 200
6.3.4 Supervised Neural Networks 200
6.3.5 Support Vector Machines 201
6.4 Unsupervised Learning for Big Data 202
6.4.1 Spectral Clustering 202
6.4.2 Principal Component Analysis (PCA) 203
6.4.3 Latent Dirichlet Allocation (LDA) 204
6.4.4 Matrix Factorization 205
6.4.5 Manifold Learning 206
6.5 Semi-supervised Learning for Big Data 207
6.5.1 Co-training 208
6.5.2 Label Propagation 208
6.5.3 Multiview Learning 209
6.6 Reinforcement Learning Basics for Big Data 209
6.6.1 Markov Decision Process 210
6.6.2 Planning 210
6.6.3 Reinforcement Learning in Practice 210
6.7 Online Learning for Big Data 210
6.8 Conclusion 213
6.9 Review Questions 213
References 213

7 Social Semantic Web Mining and Big Data Analytics 217
C. S. R. Prabhu
7.1 Introduction 217
7.2 What Is Semantic Web? 217
7.3 Knowledge Representation Techniques and Platforms in Semantic Web 218
7.4 Web Ontology Language (OWL) 219
7.5 Object Knowledge Model (OKM) 219
7.6 Architecture of Semantic Web and the Semantic Web Road Map 220
7.7 Social Semantic Web Mining 221
7.8 Conceptual Networks and Folksonomies or Folk Taxonomies of Concepts/Subconcepts 224
7.9 SNA and ABM 225

7.10 e-Social Science 226
7.11 Opinion Mining and Sentiment Analysis 228
7.12 Semantic Wikis 229
7.13 Research Issues and Challenges for Future 229
7.14 Review Questions 230
References 231

8 Internet of Things (IOT) and Big Data Analytics 233
C. S. R. Prabhu
8.1 Introduction 233
8.2 Smart Cities and IOT 234
8.3 Stages of IOT and Stakeholders 235
8.3.1 Stages of IOT 235
8.3.2 Stakeholders 235
8.3.3 Practical Downscaling 235
8.4 Analytics 236
8.4.1 Analytics from the Edge to Cloud 236
8.4.2 Security and Privacy Issues and Challenges in Internet of Things (IOT) 236
8.5 Access 237
8.6 Cost Reduction 238
8.7 Opportunities and Business Model 238
8.8 Content and Semantics 238
8.9 Data-Based Business Models Coming Out of IOT 239
8.10 Future of IOT 239
8.10.1 Technology Drivers 239
8.10.2 Future Possibilities 239
8.10.3 Challenges and Concerns 240
8.11 Big Data Analytics and IOT 241
8.11.1 Infrastructure for Integration of Big Data with IOT 241
8.12 Fog Computing 241
8.12.1 Fog Data Analytics 242
8.12.2 Fog Security and Privacy 244
8.13 Research Trends 245
8.14 Conclusion 246
8.15 Review Questions 246
References 246

9 Big Data Analytics for Financial Services and Banking 249
C. S. R. Prabhu
9.1 Introduction 249
9.2 Customer Insights and Marketing Analysis 250

9.3 Sentiment Analysis for Consolidating Customer Feedback 251
9.4 Predictive Analytics for Capitalizing on Customer Insights 252
9.5 Model Building 252
9.6 Fraud Detection and Risk Management 252
9.7 Integration of Big Data Analytics into Operations 253
9.8 How Banks Can Benefit from Big Data Analytics? 253
9.9 Best Practices of Data Analytics in Banking for Crises Redressal and Management 253
9.10 Bottlenecks 254
9.11 Conclusion 255
9.12 Review Questions 255
References 256

10 Big Data Analytics Techniques in Capital Market Use Cases 257
C. S. R. Prabhu
10.1 Introduction 257
10.2 Capital Market Use Cases of Big Data Technologies 258
10.2.1 Algorithmic Trading 258
10.2.2 Investors' Faster Access to Securities 259
10.3 Prediction Algorithms 259
10.3.1 Stock Market Prediction 259
10.3.2 Efficient Market Hypothesis (EMH) 260
10.3.3 Random Walk Theory (RWT) 260
10.3.4 Trading Philosophies 260
10.3.5 Simulation Techniques 261
10.4 Research Experiments to Determine Threshold Time for Determining Predictability 261
10.5 Experimental Analysis Using Bag of Words and Support Vector Machine (SVM) Application to News Articles 262
10.6 Textual Representation and Analysis of News Articles 262
10.7 Named Entities 263
10.8 Object Knowledge Model (OKM) 263
10.9 Application of Machine Learning Algorithms 263
10.10 Sources of Data 264
10.11 Summary and Future Work 264
10.12 Conclusion 265
10.13 Review Questions 265
References 265

11 Big Data Analytics for Insurance 267
C. S. R. Prabhu
11.1 Introduction 267

11.2 The Insurance Business Scenario ... 268
11.3 Big Data Deployment in Insurance ... 268
11.4 Insurance Use Cases ... 268
11.5 Customer Needs Analysis ... 269
11.6 Other Applications ... 270
11.7 Conclusion ... 270
11.8 Review Questions ... 270
References ... 270

12 Big Data Analytics in Advertising ... 271
C. S. R. Prabhu
12.1 Introduction ... 271
12.2 What Role Can Big Data Analytics Play in Advertising? ... 272
12.3 BOTs ... 272
12.4 Predictive Analytics in Advertising ... 272
12.5 Big Data for Big Ideas ... 273
12.6 Innovation in Big Data—Netflix ... 273
12.7 Future Outlook ... 273
12.8 Conclusion ... 273
12.9 Review Questions ... 274
References ... 274

13 Big Data Analytics in Bio-informatics ... 275
C. S. R. Prabhu
13.1 Introduction ... 275
13.2 Characteristics of Problems in Bio-informatics ... 276
13.3 Cloud Computing in Bio-informatics ... 276
13.4 Types of Data in Bio-informatics ... 276
13.5 Big Data Analytics and Bio-informatics ... 279
13.6 Open Problems in Big Data Analytics in Bio-informatics ... 279
13.7 Big Data Tools for Bio-informatics ... 282
13.8 Analysis on the Readiness of Machine Learning Techniques for Bio-informatics Application ... 282
13.9 Conclusion ... 283
13.10 Questions and Answers ... 283
References ... 284

14 Big Data Analytics and Recommender Systems ... 287
Rohit Ghosh
14.1 Introduction ... 287
14.2 Background ... 287
14.3 Overview ... 289
14.3.1 Basic Approaches ... 290
14.3.2 Content-Based Recommender Systems ... 291

14.3.3 Unsupervised Approaches . . . 291
14.3.4 Supervised Approaches . . . 291
14.3.5 Collaborative Filtering . . . 292
14.4 Evaluation of Recommenders . . . 294
14.5 Issues . . . 296
14.6 Conclusion . . . 297
14.7 Review Questions . . . 297
References . . . 297

15 Security in Big Data . . . 301
C. S. R. Prabhu
15.1 Introduction . . . 301
15.2 Ills of Social Networking—Identity Theft . . . 302
15.3 Organizational Big Data Security . . . 302
15.4 Security in Hadoop . . . 303
15.5 Issues and Challenges in Big Data Security . . . 303
15.6 Encryption for Security . . . 304
15.7 Secure MapReduce and Log Management . . . 304
15.8 Access Control, Differential Privacy and Third-Party Authentication . . . 304
15.9 Real-Time Access Control . . . 305
15.10 Security Best Practices for Non-relational or NoSQL Databases . . . 305
15.11 Challenges, Issues and New Approaches Endpoint Input, Validation and Filtering . . . 305
15.12 Research Overview and New Approaches for Security Issues in Big Data . . . 306
15.13 Conclusion . . . 307
15.14 Review Questions . . . 307
References . . . 308

16 Privacy and Big Data Analytics . . . 311
C. S. R. Prabhu
16.1 Introduction . . . 311
16.2 Privacy Protection . . . 311
16.3 Enterprise Big Data Privacy Policy and COBIT 5 . . . 312
16.4 Assurance and Governance . . . 313
16.5 Conclusion . . . 315
16.6 Review Questions . . . 315
References . . . 315

17 Emerging Research Trends and New Horizons 317
Aneesh Sreevallabh Chivukula
17.1 Introduction 317
17.2 Data Mining 317
17.3 Data Streams, Dynamic Network Analysis and Adversarial Learning 318
17.4 Algorithms for Big Data 318
17.5 Dynamic Data Streams 318
17.6 Dynamic Network Analysis 319
17.7 Outlier Detection in Time-Evolving Networks 319
17.8 Research Challenges 320
17.9 Literature Review of Research in Dynamic Networks 320
17.10 Dynamic Network Analysis 320
17.11 Sampling 321
17.12 Validation Metrics 322
17.13 Change Detection 323
17.14 Labeled Graphs 324
17.15 Event Mining 324
17.16 Evolutionary Clustering 325
17.17 Block Modeling 326
17.18 Surveys on Dynamic Networks 326
17.19 Adversarial Learning—Secure Machine Learning 328
17.20 Conclusion and Future Emerging Direction 329
17.21 Review Questions 329
References 330

Case Studies 333

Appendices 355

About the Authors

Dr. C. S. R. Prabhu has held prestigious positions with Government of India and various institutions. He retired as Director General of the National Informatics Centre (NIC), Ministry of Electronics and Information Technology, Government of India, New Delhi, and has worked with Tata Consultancy Services (TCS), CMC, TES and TELCO (now Tata Motors). He was also faculty for the Programs of the APO (Asian Productivity Organization). He has taught and researched at the University of Central Florida, Orlando, USA, and also had a brief stint as a Consultant to NASA. He was Chairman of the Computer Society of India (CSI), Hyderabad Chapter. He is presently working as an Advisor (Honorary) at KL University, Vijayawada, Andhra Pradesh, and as a Director of Research and Innovation at Keshav Memorial Institute of Technology (KMIT), Hyderabad.

He received his Master's degree in Electrical Engineering with specialization in Computer Science from the Indian Institute of Technology, Bombay. He has guided many Master's and doctoral students in research areas such as Big Data.

Dr. Aneesh Sreevallabh Chivukula is currently a Research Scholar at the Advanced Analytics Institute, University of Technology Sydney (UTS), Australia. Previously, he chiefly worked in computational data science-driven product development at Indian startup companies and research labs. He received his M.S. degree from the International Institute of Information Technology (IIIT), Hyderabad. His research interests include machine learning, data mining, pattern recognition, big data analytics and cloud computing.

Dr. Aditya Mogadala is a postdoc in the Language Science and Technology at Saarland University. His research concentrates on the general area of Deep/Representation learning applied for integration of external real-world/ common-sense knowledge (e.g., vision and knowledge graphs) into natural language sequence generation models. Before Postdoc, he was a PhD student and

Research Associate at the Karlsruhe Institute of Technology, Germany. He holds B.Tech and M.S. degree from the IIIT, Hyderabad, and has worked as a Software Engineer at IBM India Software Labs.

Mr. Rohit Ghosh currently works at Qure, Mumbai. He previously served as a Data Scientist for ListUp, and for Data Science Labs. Holding a B.Tech. from the IIT Mumbai, his work involves R&D areas in computer vision, deep learning, reinforcement learning (mostly related to trading strategies) and cryptocurrencies.

Dr. L. M. Jenila Livingston is an Associate Professor with the CSE Dept at VIT, Chennai. Her teaching foci and research interests include artificial intelligence, soft computing, and analytics.

Chapter 1
Big Data Analytics

1.1 Introduction

The latest disruptive trends and developments in digital age comprise social networking, mobility, analytics and cloud, popularly known as SMAC. The year 2016 saw Big Data Technologies being leveraged to power business intelligence applications. What holds in store for 2020 and beyond?

Big Data for governance and for competitive advantage is going to get the big push in 2020 and beyond. The tug of war between governance and data value will be there to balance in 2020 and beyond. Enterprises will put to use the enormous data or Big Data they already have about their customers, employees, partners and other stakeholders by deploying it for both regulatory use cases and non-regulatory use cases of value to business management and business development. Regulatory use cases require governance, data quality and lineage so that a regulatory body can analyze and track the data to its source all through its various transformations. On the other hand, the non-regulatory use of data can be like 360° customer monitoring or offering customer services where high cardinality, real time and mix of structured, semi-structured and unstructured data will produce more effective results.

It is expected that in 2020 businesses will shift to a data-driven approach. All businesses today require analytical and operational capabilities to address customers, process claims, use interfaces to IOT devices such as sensors in real time, at a personalized level, for each individual customer. For example, an e-commerce site can provide individual recommendations after checking prices in real time. Similarly, health monitoring for providing medical advice through telemedicine can be made operational using IOT devices for monitoring all individual vital health parameters. Health insurance companies can process valid claims and stop paying fraudulent claims by combining analytics techniques with their operational systems. Media companies can deliver personalized content through set-top boxes. The list of such use cases is endless. For achieving the delivery of such use cases, an agile platform is essentially required which can provide both analytical results and also operational efficiency so as to make the office operations more relevant and accurate, backed

C. S. R. Prabhu et al., *Big Data Analytics: Systems, Algorithms, Applications*,
https://doi.org/10.1007/978-981-15-0094-7_1

by analytical reasoning. In fact, in 2020 and beyond the business organizations will go beyond just asking questions to taking great strides to achieve both initial and long-term business values.

Agility, both in data and in software, will become the differentiator in business in 2020 and beyond. Instead of just maintaining large data lakes, repositories, databases or data warehouses, enterprises will leverage on data agility or the ability to understand data in contexts and take intelligent decisions on business actions based on data analytics and forecasting.

The agile processing models will enable the same instance of data to support batch analytics, interactive analytics, global messaging, database models and all other manifestations of data, all in full synchronization. More agile data analytics models will be required to be deployed when a single instance of data can support a broader set of tools. The end outcome will be agile development and application platform that supports a very broad spectrum of processing and analytical models.

Block chain is the big thrust area in 2020 in financial services, as it provides a disruptive way to store and process transactions. Block chain runs on a global network of distributive computer systems which any one can view and examine. Transactions are stored in blocks such that each block refers to previous block, all of them being time-stamped and stored in a form unchangeable by hackers, as the world has a complete view of all transactions in a block chain. Block chain will speed up financial transactions significantly, at the same time providing security and transparency to individual customers. For enterprises, block chain will result in savings and efficiency. Block chain can be implemented in Big Data environment.

In 2020, microservices will be offered in a big way, leveraging on Big Data Analytics and machine learning by utilizing huge amount of historical data to better understand the context of the newly arriving streaming data. Smart devices from IOT will collaborate and analyze each other, using machine learning algorithms to adjudicate peer-to-peer decisions in real time.

There will also be a shift from post-event and real-time analytics to pre-event and action (based on real-time data from immediate past).

Ubiquity of connected data applications will be the order of the day. In 2020, modern data applications will be highly portable, containerized and connected quickly replacing vertically integrated monolithic software technologies.

Productization of data will be the order of the day in 2020 and beyond. Data will be a product, a commodity, to buy or to sell, resulting in new business models for monetization of data.

1.2 What Is Big Data?

Supercomputing at Internet scale is popularly known as Big Data. Technologies such as distributed computing, parallel processing, cluster computing and distributed file system have been integrated to take the new avatar of Big Data and data science. Commercial supercomputing, now known as Big Data, originated at companies such as Google, Facebook, Yahoo and others, operates at Internet scale that needed to process the ever-increasing numbers of users and their data which was of very large volume, with large variety, high veracity and changing with high velocity which had a great value. The traditional techniques of handling data and processing it were found to be completely deficient to rise up to the occasion. Therefore, new approaches and a new paradigm were required. Using the old technologies, the new framework of Big Data Architecture was evolved by the very same companies who needed it. Thence came the birth of Internet-scale commercial supercomputing paradigm or Big Data.

1.3 Disruptive Change and Paradigm Shift in the Business Meaning of Big Data

This paradigm shift brought disruptive changes to organizations and vendors across the globe and also large social networks so as to encompass the whole planet, in all walks of life, in light of Internet of things (IOT) contributing in a big way to Big Data. Big Data is not the trendy new fashion of computing, but it is sure to transform the way computing is performed and it is so disruptive that its impact will sustain for many generations to come.

Big Data is the commercial equivalent of HPC or supercomputing (for scientific computing) with a difference: Scientific supercomputing or HPC is computation intensive with scientific calculations as the main focus of computing, whereas Big Data is only processing very large data for mostly finding out the patterns of behavior in data which were previously unknown.

Today, Internet-scale commercial companies such as Amazon, eBay and Filpkart use commercial supercomputing to solve their Internet-scale business problems, even though commercial supercomputing can be harnessed for many more tasks than simple commercial transactions as fraud detection, analyzing bounced checks or tracking Facebook friends! While the scientific supercomputing activity came downward and commercial supercomputing activity went upward, they both are reaching a state of equilibrium. Big data will play an important role in 'decarbonizing' the global economy and will also help work toward Sustainable Development Goals.

Industry 4.0, Agriculture or Farming 4.0, Services 4.0, Finance 4.0 and beyond are the expected outcomes of the application IOT and Big Data Analytics techniques together to the existing versions of the same sectors of industry, agriculture or farming, services, finance, by weaving together of many sectors of the economy to the one new order of the World 4.0. Beyond this, the World 5.0 is aimed to be achieved by the governments of China and Japan by deploying IOT and Big Data in a big way, a situation which may become 'big brothers,' becoming too powerful in tracking everything aiming to control everything! That is where we need to find a scenario of Humans 8.0 who have human values or Dharma, so as to be independent and yet have a sustainable way of life. We shall now see how the Big Data technologies based on Hadoop and Spark can handle practically the massive amounts of data that is pouring in modern times.

1.4 Hadoop

Hadoop was the first commercial supercomputing software platform that works at scale and also is affordable at scale. Hadoop is based on exploiting parallelism and was originally developed in Yahoo to solve specific problems. Soon it was realized to have large-scale applicability to problems faced across the Internet-scale companies such as Facebook or Google. Originally, Yahoo utilized Hadoop for tracking all user navigation clicks in web search process for harnessing it for advertisers. This meant millions of clickstream data to be processed on tens of thousands of servers across the globe on an Internet-scale database that was economical enough to build and operate. No existing solutions were found capable to handle this problem. Hence, Yahoo built, from scratch, the entire ecosystem for effectively handling this requirement. Thus was born Hadoops [1]. Like Linux, Hadoop was also in open source. Just as Linux spans over clusters of servers, clusters of HPC servers or Clouds, so also Hadoop has created the Big Data Ecosystem of new products, vendors, new startups and disruptive possibilities. Even though in open-source domain originally, today even Microsoft Operating System supports Hadoop.

1.5 Silos

Traditionally, IT organizations partition expertise and responsibilities which constrains collaboration between and among groups so created. This may result in small errors in supercomputing scale which may result in huge losses of time and money. A 1% error, say for 300 terabytes, is 3 million megabytes. Fixing such bugs will be an extremely expensive exercise.

In scientific supercomputing area, small teams managed well the entire environment. Therefore, it is concluded that a small team with a working knowledge of the entire platform works the best. Silos become impediments in all circumstances, both in scientific and in commercial supercomputing environments. Internet-scale computing can and will work only when it is taken as a single platform (not silos of different functions). A small team with complete working knowledge of the entire platform is essential. However, historically since the 1980s, the customers and user community were forced to look at computing as silos with different vendors for hardware, operating system, database and development platform. This leads to a silo-based computing. In Big Data and Hadoop, this is replaced with a single platform or a single system image and single ecosystem of the entire commercial supercomputing activities.

Supercomputers are Single Platforms

Originally, mainframes were single platforms. Subsequently, silos of products from a variety of vendors came in. Now again in Big Data, we are arriving at a single platform approach.

1.5.1 Big Bang of Big Data

Big Data will bring about the following changes:

1. Silo mentality and silo approach will be closed and will give rise to platform approach.
2. All the pre-Hadoop products will be phased out gradually since they will be ridiculously slow, small and expensive, compared to the Hadoop class of platforms.
3. Traditional platform vendors will therefore give way to Hadoop class of frameworks, either by upgrading or bringing out new platforms so as to meet the requirements.

1.5.2 Possibilities

The possibilities Big Data opens up are endless. Answers to questions hitherto never asked can be and will be answerable in the Big Data environment.

In the context of Internet of things (IOT), the data that can flow will be really big, in real time. In addition to the transactional data, the big time, big variety of data includes text, sensor data, audio and video data also. It expects processing and response in real time, which can be really delivered in Big Data Analytics. This means, while the data is still being collected, it can be analyzed in real time and plans or decisions can be made accordingly. This can enable the significant edge over competitors in terms of knowing in advance the trends, opportunities or impending dangers of problems much earlier than the competitors. Usage scenarios and use cases can be as follows.

Farmers get sensor data from smart farms to take decisions on crop management; automotive manufactures get real-time sensor data from cars sold and also monitor car health continuously through real-time data received from car-based sensor network. Global outbreaks of infectious diseases can be monitored in real time, so as to take preemptive steps to arrest their spread.

Previously, data was captured from different sources and accumulated in a supercomputer for being processed slowly, not in real time. The Big Data Ecosystem enables real-time processing of data in Hadoop clusters. Organizations are facing so massive volumes of data that if they do not know how to manage it, they will be overwhelmed by it. Whether the wall of data rises as a fog or as a tsunami, it can be collected with a common pool of data reservoir in Hadoop cluster, in real time, and processed in real time. This will be the superset of all individual sources of data in all organizations. Organizations can integrate their traditional internal data infrastructure as databases or data warehouses with a new Big Data infrastructure with multiple new channels of data. This integration is essential, along with the appropriate governance structure for the same.

1.5.3 Future

Big Data will change the course of history—the disruptive technology is thrusting computer science into a new vista away from the good old Von Neumann sequential computer model into the new Hadoop cluster model of parallel computing with real huge data being processed in real time.

1.5.4 Parallel Processing for Problem Solving

Conventionally, when large data is required to be processed adequately fast to meet the requirements of the application, parallel processing was identified to be the correct approach.

Parallel processing was achieved by multiple CPUs sharing the same storage in a network. Thus, we had the approaches of storage area network (SAN) or network access storage (NAS). Alternatively, 'shared nothing' architectures with each of the parallel CPUs having its own storage with stored data are also possible.

Due to rapid technology development, the processor speed shot up from 44 mips (million instructions per second) at 40 MHz in 1990 to 147,600 MIPS at 3.3 GHZ and beyond after 2010. RAM capacities went up from 640 KB in 1990 to 32 GB (8 such modules) and beyond after 2010. Storage disk capacities went up from 1 GB in 1990 to 1 TB and beyond after 2010 [2].

But, importantly, the disk latency speeds had not grown much beyond their 1990 ratings of about 80 MB/s.

While PC computing power grew 200,000% and storage disk capacity 50,000%, read/seek latency of the disk storage has not grown anywhere near that. Therefore, if we require to read 1 TB at 80 Mb/s, one disk takes 3.4 h, 10 disks take 20 min, 100 disks take 2 min, and 1000 disks take 12 s. This means that parallel reading of data from disks and processing them parallelly are the only answers.

Parallel data processing is really the answer. This was addressed earlier in grid computing where a large number of CPUs and disks are connected in a network for parallel processing purpose. The same was achieved in cluster computing with all CPUs being connected through a high-speed interconnection network (ICN).

While parallel processing, as a concept, may be simple, it becomes extremely challenging and difficult to write and implement parallel applications. Serious problems of data distribution for parallel computing followed by integration or summation of the results so generated also become very important. Since each node or CPU of the parallel CPU network computes only one small piece or part of the data, it becomes essential to keep track of the initial fragmentation of the data to be able to make sense during the integration of the data after the completion of computations. This means we will spend a lot of time and effort in management and housekeeping of the data much more than for computing itself.

Hardware failures in network need to be handled by switching over to standby machines. Disk failures also need to be considered. To process large data in parallel, we need to handle partial hardware failures without causing a total processing failure. If a CPU fails, we need to shift the job to a backup CPU.

1.5.5 *Why Hadoop?*

When data is stored in multiple locations, the synchronization of the changed data due to any update becomes a problem. If the same data is replicated (not for backup recovery but for processing purposes), then each replication location requires to be concerned with the backup of the data and the recovery of the data—this leads to greater complexity. In theory, if we can, we should keep only one single version of the data (as it happens in RDBMS). But in Hadoop environment, large volumes of data are stored in parallel and do not have an update capability.

What is the Answer?

Appropriate software that can handle all these issues effectively is the answer. That functionality is made available in Hadoop Distributed File System (HDFS).

1.5.6 *Hadoop and HDFS*

Hadoop and HDFS were initiated in Apache (under Notch project) developed at Yahoo by Doug Cutting for being able to process Internet-scale data. Since high-powered systems were expensive, commodity work stations were deployed. Large volumes of data were distributed across all these systems and processed in parallel. Failures of CPU and disk were common. Therefore, replication was done. In case of failure, the replicated backup node or disk will be utilized. Hadoop is a batch processing environment. No random access or update is possible. Throughput is given more importance.

Hadoop is an open-source project of Apache Foundation, and it is basically a framework written in Java [3]. Hadoop uses Google's MapReduce programming model and Google File System for data storage, as its basic foundations. Today, Hadoop is a core computing infrastructure for Yahoo, Facebook, LinkedIn, Twitter, etc.

Hadoop handles massive amounts of structured, semi-structured and unstructured data, using inexpensive commodity servers.

Hadoop is a 'shared nothing' parallel processing architecture.

Hadoop replicates its data across multiple computers (in a cluster), so that if one such computer server (node) goes down, the data it contained can still be processed by retrieving it from its replica stored in another server (node).

Hadoop is for high throughput, rather than low latency—therefore, Hadoop performs only batch operations, handling enormous quantity of data—response time in real time is not being considered.

Hadoop is not online transaction processing (OLTP) and also not online analytical processing (OLAP), but it complements both OLTP and OLAP. Hadoop is not the equivalent or replacement of a DBMS or RDBMS (other supporting environments over Hadoop as extensions such as Hive and other tools provide the database (SQL or similar) functionality over the data stored in Hadoop, as we shall see later in this chapter). Hadoop is good only when the work is parallelized [4]. It is not good to use Hadoop if the work cannot be parallelized (parallel data processing in large data environments). Hadoop is not good for processing small files. Hadoop is good for processing huge data files and datasets, in parallel.

What are the advantages of Hadoop and what is its storage structure?

(a) **Native Format Storage**: Hadoop's data storage framework called Hadoop Distributed File System (HDFS) can store data in its raw, native format. There is no structure that is imposed while keeping in data or storing data. HDFS is a schema-less storage structure. It is only later, when data needs to be processed, that a structure is imposed on the raw data.
(b) **Scalability**: Hadoop can store and distribute very large datasets (involving thousands of terabytes (or petabytes) of data).
(c) **Cost-Effectiveness**: The cost per terabyte of storage of data is the lowest in Hadoop.
(d) **Fault Tolerance and Failure Resistance**: Hadoop ensures replicated storage of data on duplicate server nodes in the cluster which ensures nonstop availability of data for processing, even upon the occurrence of a failure.
(e) **Flexibility**: Hadoop can work with all kinds of data: structured, semi-structured and unstructured data. It can help derive meaningful business insights from unstructured data, such as email conversations, social media data and postings and clickstream data.
(f) **Application**: Meaningful purposes such as log analysis, data mining, recommendation systems and market campaign analysis are all possible with Hadoop infrastructure.
(g) **High Speed and Fast Processing**: Hadoop processing is extremely fast, compared to the conventional systems, owing to 'move code to data' paradigm.

1.5.7 Hadoop Versions 1.0 and 2.0

Hadoop 1.0 and Hadoop 2.0 are the two versions. In Hadoop 1.0, there are two parts: (a) data storage framework which is the Hadoop Distributed File System (HDFS) which is schema-less storage mechanism; it simply stores the data files, and it stores in any format, whatsoever; the idea is to store data in its most original form possible; this enables the organization to be flexible and agile, without constraint on how to implement; and (b) data processing framework. This provides the functional programming model known as MapReduce. It has two functions: Map and Reduce functions to process data. The Mappers take in a set of key–value pairs and generate intermediate data (which is another set of key–value pairs). The Reduce function then acts on the input to process and produce the output data. The two functions, Map and Reduce, seemingly work in isolation from one another, so as to enable the processing to be highly distributed in a highly parallel, fault-tolerant and reliable manner.

1.5.7.1 Limitations of Hadoop 1.0

1. The requirement of proficiency in MapReduce programming along with proficiency in Java.
2. Only batch processing is supported, which can be useful only for typical batch applications such as log analysis and large-scale data mining and not useful for other applications.
3. Hadoop 1.0 is largely computationally coupled with MapReduce. Thus, DBMS has no option but to either deploy MapReduce programming in processing data or pull out data from Hadoop 1.0 and then process in DBMS. Both of these options are not attractive.

Therefore, Hadoop 2.0 attempted to overcome these constraints.

1.5.8 Hadoop 2.0

In Hadoop 2.0, the HDFS continues to be the data storage framework. However, a new and separate resource management framework called Yet Another Resource Negotiator or YARN has been added. Any application which is capable of dividing itself into parallel tasks is supported by YARN. YARN coordinates the allocation of subtasks of the submitted application, thereby enhancing the scalability, flexibility and efficiency of the application. It performs by deploying 'Application Master' in place of the old 'Job Tracker,' running application on resources governed by a new Node Manager (in place of old 'Task Tracker'). Application Master is able to run any application and not just MapReduce.

Therefore, MapReduce programming is not essential. Further, real-time processing is also supported in addition to the old batch processing. In addition to MapReduce model of programming, other data processing functions such as data standardization and master data management also can now be performed naturally in HDFS.

1.6 HDFS Overview

If large volumes of data are going to be processed very fast, then we essentially require: (i) Parallelism: Data needs to be divided into parts and processed in parts simultaneously or parallelly in different nodes. (ii) Fault tolerance through data replication: Data needs to be replicated in three or more simultaneously present storage devices, so that even if some of these storage devices fail at the same time, the others will be available (the number of replication as three or more are decided by the replication factor given by the administrator or developer concerned). (iii) Fault tolerance through node (server) replication: In case of failure of the processing nodes, the alternate node takes over the processing function. We process the data on the node where the data resides, thereby limiting transferring of the data between all the nodes (programs to process the data are also accordingly replicated in different nodes).

Hadoop utilizes Hadoop Distributed File System (HDFS) and executes the programs on each of the nodes in parallel [5]. These programs are MapReduce jobs that split the data into chunks which are processed by the 'Map' task in parallel. The 'framework' sorts the output of the 'Map' task and directs all the output records with the same key values to the same nodes. This directed output hence then becomes the input into the 'Reduce' task (summing up or integration) which also gets processed in parallel.

- HDFS operates on the top of an existing file system (of the underlying OS in the node) in such a way that HDFS blocks consist of multiple file system blocks (thus, the two file systems simultaneously exist).
- No updates are permitted.
- No random access is permitted (streaming reads alone are permitted).
- No caching is permitted.
- Each file is broken into blocks and stored in three or more nodes in HDFS to achieve reliability through redundancy by replication.
- Master node (also known as name node) carries a catalogue or directory of all the slave nodes (or data nodes).
- Slave nodes (or data nodes) contain the data.
- Limited file security.

Data read by the local OS file system gets cached (as it may be called up for reading again any time, as HDFS cannot perform the caching of the data).

HDFS performs only batch processing using sequential reads. There is no random reading capability, nor there is any capability to update the data in place.

The master node includes name node, Job Tracker and secondary name node for backup.

The slave node consists of data nodes and Task Tracker. Data nodes are replicated for fault tolerance.

HDFS uses simple file permissions (similar to Linux) for read/write purposes.

1.6.1 *MapReduce Framework*

HDFS described above works on MapReduce framework.

What is MapReduce? It is a simple methodology to process large-sized data by distributing across a large number of servers or nodes. The master node will first partition the input into smaller subproblems which are then distributed to the slave nodes which process the portion of the problem which they receive. (In principle, this decomposition process can continue to many levels as required). This step is known as Map step.

In the Reduce step, a master node takes the answers from all the subproblems and combines them in such a way as to get the output that solves the given application problem.

Clearly, such parallel processing requires that there are no dependencies in the data. For example, if daily temperature data in different locations in different months is required to be processed to find out the maximum temperature among all of them, the data for each location for each month can be processed parallelly and finally the maximum temperature for all the given locations can be combined together to find out the global maximum temperature. The first phase of sending different locations of data to different nodes is called Map Phase, and the final step of integrating all the results received from different nodes into the final answer is called Reduce Phase.

MapReduce framework also takes care of other tasks such as scheduling, monitoring and re-executing failed tasks HDFS and MapReduce framework run in the same set of nodes. Configuration allows effective scheduling of tasks on the nodes where data is present (data locality). This results in very high throughput. Two daemons (master) Job Tracker and (slow) Task Tracker for cluster nodes are deployed as follows.

1.6.2 *Job Tracker and Task Tracker*

Job Tracker performs

(1) Management of cluster and
(2) Application management.

In managing the cluster, it keeps free and busy notes and assigns the tasks accordingly.

In application management, it receives the application problem from the client (by the user) and replicates the same into all the nodes. It will split the input data into blocks which will be sent to the Task Trackers in data nodes (Fig. 1.1).

The Task Tracker is responsible for executing the individual tasks assigned by the Job Tracker. A single Task Tracker exists per slave node and spawns multiple MapReduce tasks in parallel. Task Tracker sends continuous heartbeat messages to Job Tracker. If heartbeat message is not received indicating failure of node, then the task will be assigned to another node by Job Tracker.

(a) MapReduce Framework:

Phases	Daemons
Map Reduce	Job Tracker Task Tracker

(b) Client

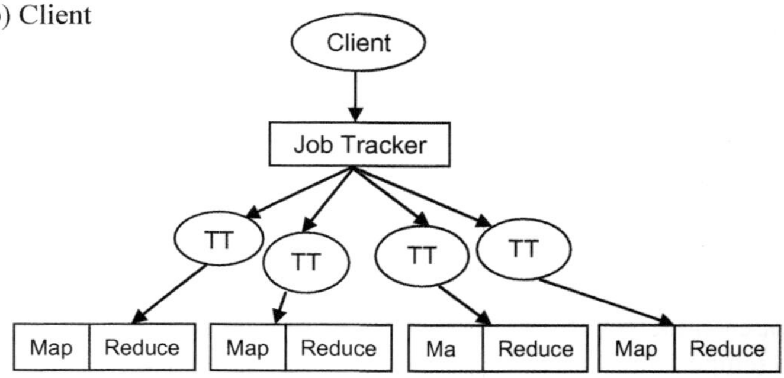

Fig. 1.1 MapReduce framework and Job Tracker

1.6.3 YARN

YARN, which is the latest version of MapReduce or MapReduce 2, has two tasks (1) resource manager and (2) application manager.

Resource manager is fixed and static. It performs node management for free and busy nodes for allocating the resources for Map and Reduce phases.

For every application, there is a separate **application manager** dynamically generated (on any data node). Application manager communicates with the **resource manager,** and depending on the availability of data nodes (or node managers in them) it will assign the Map Phase and Reduce Phase to them.

1.7 Hadoop Ecosystem

1. Hadoop Distributed File System (**HDFS**) simply stores data files as close to the original format as possible.
2. **HBase** is a Hadoop database management system and compares well with RDBMS. It supports structured data storage for large tables.
3. **Hive** enables analysis of large data with a language similar to SQL, thus enabling SQL type of processing of data in a Hadoop cluster.
4. **Pig** is an easy-to-understand data flow language, helpful in analyzing Hadoop-based data. Pig scripts are automatically converted to MapReduce jobs by the

Pig Interpreter, thus enabling SQL-type processing of Hadoop data [6]. By using Pig, we overcome the need of MapReduce-level programming.

5. **ZooKeeper** is a coordinator service for distributed applications.
6. **Oozie** is a workflow schedule system to manage Apache Hadoop Jobs.
7. **Mahout** is a scalable machine learning and data mining library.
8. **Chukwa** is a data collection system for managing large distributed systems.
9. **Sqoop** is used to transfer bulk data between Hadoop and as structured data management systems such as relational databases.
10. **Ambari** is a web-based tool for provisioning, managing and monitoring Apache Hadoop clusters.
11. **Ganglia** is the monitoring tool.
12. **Kafka** is the stream processing platform.

We will be covering all the above later in Chap. 4 (Fig. 1.2).

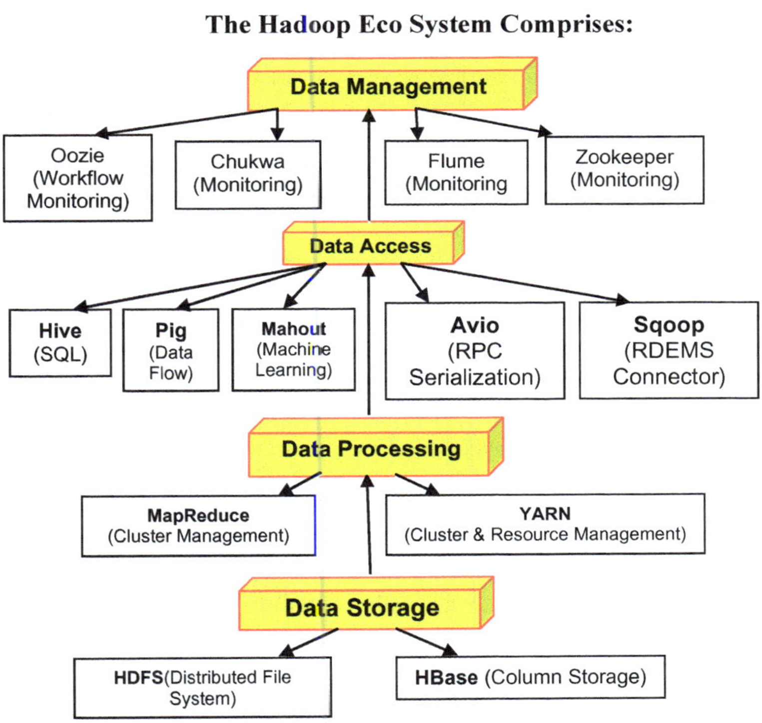

Fig. 1.2 Hadoop ecosystem elements—various stages of data processing

1.7.1 *Cloud-Based Hadoop Solutions*

a. Amazon Web Services (AWS) offers Big Data services on cloud for very low cost.
b. Google BigQuery or Google Cloud Storage connector for Hadoop empowers performing MapReduce jobs on data in Google Cloud Storage.

1.7.2 *Spark and Data Stream Processing*

Batch processing of ready-to-use historical data was one of the first use cases for big data processing using Hadoop environment. However, such batch processing systems have high latency, from a few minutes to few hours to process a batch. Thus, there is a long wait before we see the results of such large volume batch processing applications.

Sometimes, data is required to be processed as it is being collected. For example, to detect fraud in an e-commerce system, we need real-time and instantaneous processing speeds. Similarly, network intrusion or security breach detection must be in real time. Such situations are identified as applications of data stream processing domain. Such data stream processing applications require handling of high velocity of data in real time or near real time.

A data stream processing application executing in a single machine will not be able to handle high-velocity data. A distributed stream processing framework on a Hadoop cluster can effectively address this issue.

Spark Streaming Processing

Spark streaming is a distributed data stream processing framework. It enables easy development of distributed applications for processing live data streams in near real time. Spark has a simple programming model in Spark Core. On top of Spark Core, Spark streaming is a library. It provides scalable, fault-tolerant and high-throughput distributed stream processing platform by inheriting all features of Spark Core. Spark libraries include: Spark SQL, Spark MLlib, Spark ML and GraphX. We can analyze a data stream using Spark SQL. We can also apply machine learning algorithms on a data stream using Spark ML. We can apply graph processing algorithms on a data stream using GraphX.

Architecture

A data stream is divided into microbatches of very small fixed time intervals. Data in each microbatch is stored as a RDD which is processed by Spark Core. Any RDD operations can be applied to an RDD and can be created by Spark streaming. The final results of RDD operations are streamed out in batches.

Sources of Data Streams

Sources of data streams can be any basic source as TCP sockets and actors, files or advanced streams such as KIKA, Flume, MQTT, ZeroMQ and Twitter. For advanced sources, we need to acquire libraries from external sources for deployment, while for basic sources we can have library support within spark itself.

API

Even though Spark library is written in Scala, APIs are provided for multiple languages such as Java and Python, in addition to native Scala itself.

We cover Spark in greater detail in Chap. 4.

1.8 Decision Making and Data Analysis in the Context of Big Data Environment

Big Data is characterized by large volume, high speed or velocity, and large accuracy or veracity and variety. Current trend is that of data flowing in from a variety of unstructured sources such as sensors, mobile phones and emails, in addition to the conventional enterprise data in structured forms such as databases and data warehouses. There exists a need for correct decision making while taking an integrated, unified and comprehensive view of the data coming from all these different and divergent sources. Even the regular data analysis techniques such as data mining algorithms are required to be redefined, extended, modified or adapted for Big Data scenarios. In comparison with the conventional statistical analysis techniques, the Big Data Analytics differ in the sense of scale and comprehensiveness of the data availability. In traditional statistical analysis approach, the data processed was only a sample. This was so due to the fact that data was scarce and comprehensive data was not available. But in today's context, the situation is exactly opposite. There is a 'data deluge.' Data, both structured or semi-structured and unstructured, flows in nonstop, either as structured data through various information systems and databases and data warehouses or as unstructured data in social networks, emails, sensors in Internet of things (IOT). All this data needs to be processed and sensible conclusions to be drawn from that deluge of data. Data analytics techniques which have been around for processing data need to be extended or adapted for the current Big Data scenario.

1.8.1 Present-Day Data Analytics Techniques

Knowledge Discovery in Database (KDD) or data mining techniques are aimed at automatically discovering previously unknown, interesting and significant patterns in given data and thereby build predictive models. Data mining process enables us to find out gems of information by analyzing data using various techniques such as decision trees, clustering and classification, association rule mining and also advanced techniques such as neural networks, support vector machines and genetic algorithms. The respective algorithms for this purpose include apriori and dynamic item set counting. While the algorithms or the processes involved in building useful models can be well understood, the implementation of the same calls for large efforts. Luckily, open-source tool kits and libraries such as Weka are available based on Java Community Processes JSR73 and JSR247 which provide a standard API for data mining. This API is known as Java Data Mining (JDM).

The data mining processes and algorithms have strong theoretical foundations, drawing from many fields such as mathematics, statistics and machine learning. Machine learning is a branch of artificial intelligence (AI) which deals with developing algorithms that machines can use to automatically learn the patterns in data. Thus, the goal and functionality of data mining and machine learning coincide. Sometimes, advanced data mining techniques with the machine being able to learn are also called machine learning techniques. Data mining differs from the conventional data analysis. Conventional data analysis aims only at fitting the given data to already existing models. In contrast, data mining finds out new or previously unknown patterns in the given data. Online analytical processing (OLAP) aims at analyzing the data for summaries and trends, as a part of data warehousing technologies and its applications. In contrast, data mining aims at discovering previously unknown, non-trivial and significant patterns and models present within the data. Therefore, data mining provides a new insight into the data being analyzed. However, all these related topics of data warehousing, OLAP and data mining are broadly identified as business intelligence (BI).

Present-day data analytics techniques are well identified as: (1) clustering, (2) classification, (3) regression, etc. The core concepts involved in data mining are as follows:

(1) Attributes: numerical, ordinal and nominal,
(2) Supervised and unsupervised learning and
(3) The practical process of data mining.

Attributes

A learning algorithm learns from the patterns in the input data and then can perform prediction. Input data is in the form of examples where each example consists of a set of attributes or dimensions. Each attribute should be independent of all other attributes, and therefore, its value cannot be computed on the basis of the values of other attributes. Attributes can be numeric, as real numbers which are continuous numbers (as age) or discrete values (as number of people). Closeness of two numbers is given by their difference.

Attributes can be ordinal values which are discrete, but in an order within them (as small, medium, large) with no specific or exact measurement involved.

Attributes can be nominal values which are discrete values with no particular order such as categorical values (color of eyes as black, brown or blue) with each category being independent and with no distance between any two categories being identifiable.

Algorithms that discover relationship between different attributes in a dataset are known as 'association rule algorithms.' Algorithms which are capable of predicting the value of an attribute based on the value of another attribute, based on its importance in clustering, are called 'attribute importance algorithms.'

Classification of learning is supervised learning and unsupervised learning.

In supervised learning, we have and use training datasets with a set of instances as example, for which the predicted value in known. Each such example consists of a set of input attributes and one predicted attribute. The objective of the algorithms is to build a mathematical model that can predict the output attribute value given a set of input attribute values.

Simple predictive models can be **decision trees** or **Bayesian belief networks** or **rule induction**.

Advanced predictive models can be **neural networks** and **regression**. **Support vector machine (SVM)** also provides advanced mathematical models for prediction.

The accuracy of prediction on new data is based on the prediction accuracy of the training data. When the target or predicted attribute is a categorical value, then the prediction model is known as classifier and the problem being tackled is called classification. On the other hand, when the attribute is a continuous variable it is called regression and the problem being tackled is also known as regression.

In the **unsupervised learning,** there is no predicted value to be learned. The algorithm for **unsupervised learning** simply analyzes the given input data and distributes them into clusters. A two-dimensional cluster can be identified such that the elements in each cluster are more closely related each other than to the elements in another cluster. Algorithms such as K-means clustering, hierarchical clustering and density-based clustering are various clustering algorithms.

1.9 Machine Learning Algorithms

We shall now examine the machine learning algorithms as a very brief overview. Later in Chap. 6, a detailed treatment will be provided.

Decision trees are simplest of learning algorithms which can perform classification, dealing with only nominal attributes. As an example, let us examine a very simple problem for a classification of gender-wise customers to be identified for purchasing a lottery ticket based on their income and age as follows.

Figure 1.3 shows an example of a decision tree to decide whether a customer will purchase a lottery ticket or not. Nodes are testing points, and branches are outcomes of the testing points. Usually, '>' (greater than) sign is placed to the right-hand side and '<' (less than) sign is placed to the left-hand side as indicated below:

It is easy to convert a decision tree into a rule as follows:

If gender is male and income is $\geq$100,000, then 'Yes', else 'No'.

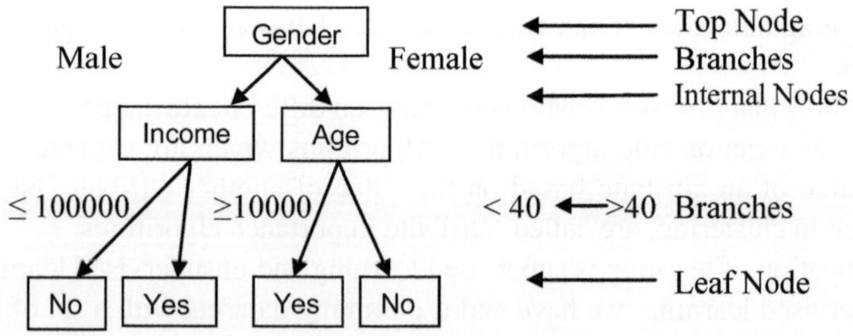

Fig. 1.3 Decision tree example

If gender is female and age is ≤40, then 'Yes', else 'No'.

(a) **Clustering algorithms** are aiming at creating a cluster of data elements which are most closely related, more than the elements in another cluster. *K*-means is the most popular clustering algorithms, where *K*-randomly selected clusters are seeded where *K* is the number of predefined clusters. Each example element is then associated with *K*th cluster whose center is the closest to that element. At the end of the iteration, the means of *K*-clusters are recomputed by looking at all the points associated with the cluster. This process is repeated until the elements do not move any more between the clusters.

(b) **Hierarchical clustering algorithm** has data points, and each data point starts as its own cluster. Next two points that are most similar are combined together into the parent's node. This process is continued and repeated until we have no points left to continue.

(c) **Density-based clustering algorithms** find out high-density areas that are separated from the low-density areas. The number of high-density clusters alone is counted as the total number of clusters, ignoring low-density areas as just noise.

(d) **Regression** If we have two points in two-dimensional space to be joined together as a line, it is represented by the linear equation $y = ax + b$. Same approach can be extended for higher-dimensional function that best fit multiple points in multidimensional space. Regression-based algorithms represent data in a matrix form and transform the matrix to compute the required parameters. They require numerical attributes to create predictive models. The squared error between the predicted and the actual values is minimized for all cases in the training datasets. Regression can be linear or polynomial in multiple dimensions. Logistic regression is the statistical technique utilized in the context of prediction.

(e) **Neural Networks** Neural networks can function as classifiers and also as predictive models. Multilayered perceptron (MLP) and radial basis function (RBF) are the most common neural networks. A multilayered perceptron (MLP) consists of many layers beginning an input layer as shown in Fig. 1.4. The number of inputs is the same as the number of attributes. The input values may be varying between −1 and 1 depending on nature of transformation function used by the node. Links in the network correspond to a weight by which the output from the

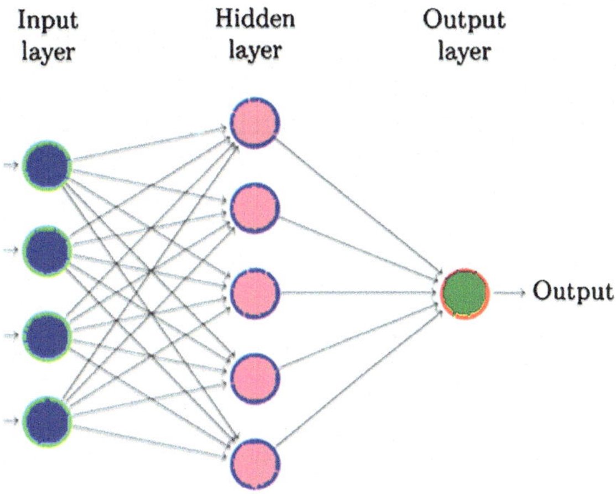

Fig. 1.4 Multi layered perceptron (three layers)

node is multiplied. The second layer is known as hidden layer in a three-layered network.

The input to a given node is the sum of the outputs from two nodes multiplied by the weight associated with the link. The third layer is output layer, and it predicts the attribute of interest. Building a predictive model for MLP comprises of estimating the weights associated with each of the links. The weights can be decided by a learning procedure known as back-propagation. Since there is no guarantee that global minimum will be found by this procedure, the learning process may be enhanced to run in conjunction with some optimization methodologies such as genetic algorithms.

In contrast, in radial basic function or RBF, the data is clustered into *K*-clusters using *K*-means clustering algorithm. Each cluster then corresponds to a node in the network, the output from which is dependent on the nearness or proximity of input to the center of the node the output from this layer is finally transformed into the final output using Weights. Learning the weights associated with such links is a problem of linear regression.

Modular Neural Networks

Modular constructive neural networks are more adaptive and more generalized in comparison with the conventional approaches.

In applications such as moving robots, it becomes necessary to define and execute a trajectory to a predefined goal while avoiding obstacles in an unknown environment. This may require solutions to handle certain crucial issues as overflow of sensorial information with conflicting objectives. Modularity may help in circumventing such problems. A modular approach, instead of a monolithic approach, is helpful since it combines all the information available and navigates the robot through an unknown environment toward a specific goal position.

(f) Support vector machine (SVM) and **relevance vector machine (RVM)** algorithms for learning are becoming popular. They come under kernel methods, and they combine principles from statistics, optimization and learning in a sound mathematical framework capable of solving large and complex problems in the areas of intelligent control and industry, information management, information security, finance, business, bioinformatics and medicine. If we consider a two-dimensional space with a large number of points, there exist a large number of lines that can be used to divide the points into two segments; let us call these lines as separating lines. Now we define 'margin' as the distance between a separating line and a parallel line that passes through the closest point to the lines. SVM selects the line that has the maximum margin associated with it. Points that this second parallel line passed through are known as **support vector points**. This model can be extended to be generalized for multiple dimensions, and its performance was observed to be very good, with output variables either as continuous or as discrete containing two values.

(g) Bayesian algorithms are algorithms based on probability theory. Naïve Bayesian classifier is one such simple but good classifier. In this algorithm, it is assumed that input attributes are independent among themselves and the prediction is made based on estimating the probability for the training data. Bayesian belief networks (BBNs) are directed acyclic graphs where a link denotes this conditional distributed function between the parent node and the child node.

Applications of SVMs

SVMs have a large variety of applications as follows:

(a) **Intelligent Control and Industrial Applications**: Intelligent signal processing aiming at quality monitoring, fault detection and control in an industrial process is performed by SVMs as part of quality monitoring tools is analyzing complex data patterns. Results showed that SVMs performed better than neural networks in these contexts. SVMs can handle feature spaces of high dimension. In regression problems, also SVMs have been deployed for control of a robot in a single plane.

(b) **Information Management**: In areas such as text classification for search process, SVMs have been deployed for statistical pattern analysis and inductive learning based on a large number of pattern instances. In the context of digital data classification and learning and ensemble kernel-based learning, the techniques using SVMs are also deployable.

(c) **Information Security**: In steganography, the JPEG images are utilized to Camouflage the secret data for sending secret messages. SVMs have been deployed to address this problem by constructing models by extracting correctly a set of features based on image characteristics for reliable validations of a set of JPEG images.

1.10 Evolutionary Computing (EC)

Evolutionary computing (EC) and evolutionary algorithm (EA) are denoting a kind of optimization methodology which is only inspired by (and not exactly the same as) nature-based phenomena such as biological evolution and also behavior of living organisms. These techniques which attempt to mimic or simulate evolution in nature for applying to real-world problems include genetic algorithm (GA), evolutionary programming (EP), evolutionary strategies (ES), genetic programming (GP), learning classifier system (LCS), differential evolution (DE), estimation of distributed algorithm (BDA), swarm intelligence (SI) algorithms like ant colony optimization (ACO) and practical swarm optimization (PSO). All these different algorithms have similar framework in algorithmic characteristics and also in their implementation details as three fundamental essential operations and two optional operations:

The first step is initialization followed by
the second step 'fitness evaluation and selection' followed by
the third step of 'population reproduction and variation.'

The new population is evaluated again, and iterations continued until a termination criterion is satisfied. In addition, some of these algorithms may have local search (LS) procedure and such algorithms are called mimetic algorithms (MAs). Further, techniques from machine learning have also been applied to EC to enhance their functionality and also vice versa; i.e., EC algorithmic technique is applied to machine learning (ML) algorithms. Thus, machine learning (ML) has become an effective and powerful area with the applications in wide-ranging fields of life [7].

1.11 Conclusion

In this chapter, we have presented the background and motivation for data science and Big Data Analytics. We have also presented a very brief introductory overview of all the major algorithms prevalent today for data mining and machine learning including neural networks, SVMs and evolutionary computing algorithms. The detailed treatment of the same can be found elsewhere. Luckily, we have ready-to-use open-source data mining and machine learning platforms such as Weka even though many other proprietary tools are also available (see http://www.donoz.org//computers/software/databases/Data_mining/Tool_Vendors/).

1.12 Review Questions

1. What are the current trends in IT? Explain in detail.
2. Where and how Big Data Analytics stands in current trends in IT?

3. How Big Data Analytics has business value? What are the possible sectors of IT in which Big Data Analytics can be deployed?
4. Explain a file processing model and development.
5. What is block chain technology and how it can be used?
6. Define the paradigm of Big Data and the role of Big Data Analytics in current business scenarios.
7. What is Hadoop? What is its importance? What are its benefits?
8. Which are the various scenarios and when Hadoop can be deployed?
9. Explain HDFS architecture and functionality.
10. What is Hadoop 2.0? What are its benefits over Hadoop 1.0?
11. Explain MapReduce framework and paradigm.
12. What is YARN? Why it is required?
13. Explain the various modules in Hadoop ecosystem.
14. Explain Spark and its architecture. What are its benefits?
15. What is machine learning? Compare data mining with machine learning?
16. Describe decision trees.
17. Describe clustering and its algorithm.
18. Describe regression and its application.
19. Explain neural networks and their categories.
20. Explain relevance vector machines.
21. Explain support vector machines.
22. What is evolutionary computing and what are its algorithms? Explain with examples.

References and Bibliography

1. C.B.B.D Manyika, *Big Data: The Next Frontier for Innovation, Competition and Productivity* (McKinsey Global Institute, 2011)
2. IBM, *Big Data and Netezza Channel Development* (2012)
3. http://hadoop.apache.org [online]
4. D.R John Gantz, *The Digital Universe in 2020: Big Data, Bigger Digital Shadows, and Biggest Growth in the Far in the East* (IDC, 2013)
5. http://ercoppa.github.io/HadoopInternals [online]
6. S. Acharya, S. Chellappan, *Big Data and Analytics* (2015)
7. A. Ghoting, *SystemML: Declarative Machine Learning on Map Reduce* (IBW Watson research center, 2011)
8. K.D. Strang, Z. Sun, Big data paradigm: what is the status of privacy and security? Ann. Data Sci. Heidelb. **4**(1), 1–17 (2017)
9. L. Cui, F.R. Yu, Q. Yan, When big data meets software-defined networking: SDN for big data and big data for SDN. IEEE Netw. **30**(1), 58 (New York, Jan–Feb 2016)
10. R.M. Alguliyev, R.T. Gasimova, R.N. Abbasli, The obstacles in big data process. Int. J. Modern Edu. Comput. Sci. **9**(3), (Hong Kong, Mar 2017)
11. F. Pourkamali-Anaraki, S. Becker, Preconditioned Data Scarification for Big Data with Applications to PCA and K-Means. IEEE Trans. Inf. Theory **63**(5), 2954–2974 (New York, 2017)
12. K. Vishwa, *Trends 2016: Big Data, IOT take the plunge* (Voice & Data, New Delhi, Apr 5, 2016

13. J. Kremer, K. Stensbo-Smidt, F. Gieseke, K.S. Pedersen, C. Igel, Big Universe, big data: machine learning and image analysis for Astronomy. IEEE Intell. Syst. **32**(2),16–22 (Los Alamitos, 2017)
14. R. Varshney, *Why Enterprises will En-route India for Big Data Analytics* (Express Computer, Mumbai, Jul 15, 2016)
15. G. Guruswamy, *How to Avoid the Common Big Data Follies in 2016* (Express Computer, Mumbai, Apr 22, 2016)
16. Big Data, Big Science: Students Share 'Big Data' Research at Poster Session US Fed News Service, Including US State News; Washington, D.C. (Washington, D.C, 01 May 2017)
17. Electronics and Telecommunications Research Institute; Researchers Submit Patent Application, *Big Data Distribution Brokerage System Using Data Verification and Method Thereof, for Approval (USPTO 20170140351) Information Technology* (Newsweekly, Atlanta, Jun 6, 2017), p. 910
18. P. Lade, R. Ghosh, S. Srinivasan, Manufacturing analytics and industrial internet of things. IEEE Intell. Syst. **32**(3), 74–79 (Los Alamitos, 2017)
19. *Research and Markets; Securing Big Data Infrastructure: An Evolving Market Ecosystem-Research and Markets Information Technology* (Newsweekly, Atlanta, Feb 23, 2016), p. 453
20. N.A. Shozi, J. Mtsweni, Big data privacy and security: a systematic analysis of current and future challenges, in *International Conference on Cyber Warfare and Security*; Reading: 296-XI. (Academic Conferences International Limited, Reading, 2016)
21. *Big Data in Leading Industry Verticals: Retail, Insurance, Healthcare, Government, and Manufacturing 2015–2020—Research and Markets* (Business Wire, New York, 27 Jan 2016)
22. *Securing Big Data Infrastructure: An Evolving Market Ecosystem* (PR Newswire, New York, 08 Feb 2016)
23. *Big data report 2016—Global Strategic Business Report 2014–2022: The Need to Turn Big Data' Into Big Advantage Drives Focus on Big Data Technologies & Services NASDAQ OMX's News Release Distribution Channel* (New York, 19 Dec 2016)
24. K. Yang, H. Qi, H. Li, K. Zheng, S. Zhou, et al., An efficient and fine-grained big data access control scheme with privacy-preserving policy. IEEE Internet Th. J. **4**(2), 563–571 (Piscataway, 2017)
25. C.P. Chullipparambil, *Big Data Analytics Using Hadoop Tools* (San Diego State University, ProQuest Dissertations Publishing, 2016). 10106013
26. M. Pascalev, Privacy exchanges: restoring consent in privacy self-management. Eth. Inf. Technol. **19**(1), 39–48 (Dordrecht, 2017)
27. W. Feng, E.A. Mack, R. Maciewjewski, Analyzing entrepreneurial social networks with big data wang. Ann. Am. Assoc. Geogr. **107**(1), 130–150 (Washington, 2017)

Chapter 2
Intelligent Systems

2.1 Introduction

In Chap. 1, we presented a total overview of Big Data Analytics. In this chapter, we delve deeper into Machine Learning and Intelligent Systems.

By definition, an algorithm is a sequence of steps in a computer program that transforms given input into desired output. Machine learning is the study of artificially intelligent algorithms that improve their performance at some task with experience. With the availability of big data, machine learning is becoming an integral part of various computer systems. In such systems, the data analyst has access to sample data and would like to construct a hypothesis on the data. Typically, a hypothesis is chosen from a set of candidate patterns assumed in the data. A pattern is taken to be the algorithmic output obtained from transforming the raw input. Thus, machine learning paradigms try to build general patterns from known data to make predictions on unknown data.

According to a standard definition [1], patterns must (i) be easily understood (ii) validate a hypothesis (iii) be useful in a particular domain of application (iv) be novel or unknown. In the context of data engineering, the various patterns found by machine learning algorithms are engineered from data mining tasks. While machine learning is concerned with the design of intelligent algorithms, data mining is concerned with the analysis and process of applying intelligent algorithms. The common pattern's output by data mining is summarization, correlation, association, clustering, classification, regression and optimization. Thus, computer systems using machine learning are built around data mining patterns. The context in which data is applied determines the data engineering products and data analytics tasks built around these patterns.

Based on the various informatics expected to be output, the data analytics tasks maybe also further categorized into descriptive, predictive, prescriptive, embedded and causal analytics tasks. The output informatics is at least as useful as the semantics of input given to the analytics tasks. Whereas Knowledge Discovery in Databases is a standard process for deriving various semantics from big data, complex event processing is a standard process for validating the various informatics from big data.

C. S. R. Prabhu et al., *Big Data Analytics: Systems, Algorithms, Applications*,
https://doi.org/10.1007/978-981-15-0094-7_2

The declarative programming languages used for data analytics have the following programming constructs to help the data analyst derive semantics and validate informatics in big data.

- **Queries**: Depending on the data and metadata organization in the data model, various queries allow the data scientist to explore (i) transactional datasets via selection, projection, joining of relational table (ii) analytic datasets via drill down (list more data warehousing queries) of data warehouses.
- **Algorithms**: Depending on the data science patterns in the concept model, various algorithms allow the data scientist to analyze (i) descriptive modeling via patterns such as correlation, association, clustering (ii) predictive modeling via patterns such as classification, regression, forecasting.
- **Techniques**: Depending on computational complexity of the algorithms and information loss in the concepts, various machine learning techniques are designed to profile the performance, model and algorithm used in the data science patterns. Common machine learning techniques include unsupervised learning, supervised learning and reinforcement learning. Emerging machine learning techniques include semi-supervised learning, metric learning and transfer learning. Beyond the classical algorithm design techniques such as divide and conquer, greedy and dynamic programming, machine learning techniques also need to consider algorithms that are either recursive or iterative; either serial or parallel; either exact or approximate.
- **Pipelines**: Data pipelines are the architectural construct commonly used to combine the informatics obtained from queries, algorithms and techniques suitable for a particular data science system or application. A data pipeline is a sequence of steps to incrementally refine the raw data into meaningful output. The pipelines (i) are built to have preprocessing, modeling, post-processing steps (ii) are reused across various modeling algorithms (iii) handle large data from multiple data sources that is varying quickly. Lambda architecture is one of the commonly used architectural construct for designing data pipelines.

2.1.1 Open-Source Data Science

Data science programming is largely driven by descriptive programming languages such as SQL, R, Scala and object-oriented programming languages such as Java, Python and Julia. Depending on the choice of programming language, the following data science programming stacks are available:

- **JavaScript**: JavaScript libraries are suitable for data visualization and exploratory data analysis. D3.js, Protovis, jQuery, OpenRefine, Tableau and knockout are some of the popular JavaScript libraries.
- **R**: R has an extensive suite of libraries for all the data analytics tasks. It is the de facto standard platform for statistical modeling. However, R is most useful only

for datasets that can fit in the main memory of a computer. Thus, R is suitable for rapid prototyping than large-scale development of the data science patterns. Functionally, R is comparable to open-source analytics tools such as Pentaho, RapidMiner.

- **Python**: Python is an object-oriented programming language. Python is an alternative to R when the dataset is too large for the main memory. Most of the models built in Python can reuse CPU and disk for analyzing big data. However, the python packages are specialized to a particular analytics task or model. NumPy, SciPy, scikits, , matplotlib, SageMath, PyTables, IPython are some of the popular Python libraries.
- **Java**: Java is an object-oriented programming language. Java is useful for performing Big Data Analytics on commodity hardware. The Hadoop software stack released by Apache Foundation is written in Java. Hadoop stack has components for data storage, data querying, parallel processing and machine learning. Research stacks such as Weka, LIBSVM Tools and ELKI are also written in Java.
- **Scala**: Scala is a functional programming language. The programming constructs in Scala are especially useful for machine learning on big data. The Apache Spark stack is written in Scala. Spark stack has components for real-time computing and machine learning.
- **C**: In terms of sheer performance, C remains the best programming language for data analytics. However, mixing the low-level memory management of C with the abstractions of data science patterns is a difficult task. Many analytics packages produced by research communities are written in either C or C++. Vowpal Wabbit, Lnknet, megam, VFML and BLAS are some of the popular C libraries.
- **Julia**: Julia is said to have the usability of Python and performance of C. Julia was built for technical computing and statistical computing. Julia's packages for analytics include those made for parallel execution, linear algebra and signal processing.

To scale machine learning with multithread programming, we can reuse one or more of main memory, CPU cores and hard disk. Following is the open-source software designing scalable machine learning algorithms for data science with parallel processing over a hardware cluster. The design is a mixture of data parallel and task parallel approaches to parallel processing. The resulting libraries like Mahout and MLLib are widely used in the data science community.

- **Iterative and Real-Time Applications**: Hadoop is the most common iterative computing framework. Hadoop-like software is mainly built for embarrassingly parallel problems with iterative computations and in-memory data structures for caching data across iterations. The software is not suitable if sufficient statistics cannot be defined within each data or task split. As a design pattern, MapReduce has origin in functional programming languages. MapReduce is built for embarrassingly parallel computations. Machine learning algorithms that can be expressed in as statistical queries over summations are suitable for MapReduce programming model. Furthermore, the algorithms for real-time processing can be categorized by the number of MapReduce executions needed per iteration. Examples for iterative algorithms are ensemble algorithms, optimization algorithms, time–frequency analysis methods and graph algorithms over social networks, neural networks and tensor networks.

- **Graph Parallel Processing Paradigms**: Extensions to iterative computing frameworks using MapReduce, add loop aware task scheduling, loop invariant caching and in-memory directed acyclic graphs like the Resilient Distributed Datasets hat may be cached in memory and reused across iterations. By comparison, R programming environment is designed for single threaded, single node execution over a shared array architecture suitable for iterative computing. R has a large collection of serial machine learning algorithms. R extensions integrate R with Hadoop to support distributed execution over Hadoop clusters. Beyond key-value set-based computing in MapReduce frameworks, we also have iterative computing frameworks and programming models building on graph parallel systems based on think-like-a-vertex programming models such as the Bulk Synchronous Parallel (BSP) model found in Apache Hama, Apache Giraph, Pregel, GraphLab and GoFFish. In contrast to MR programming model, BSP programming model is suitable for machine learning algorithms such as conjugate gradient and support vector machines. A real-time data processing environment such as Kafka-Storm can help integrate real-time data processing with machine learning. BSP is suitable for implementing deterministic algorithms on graph parallel systems. However, user must architect the movement of data by minimizing the dependencies affecting parallelism and consistency across jobs in the graph. Such dependencies in a graph may be categorized as asynchronous, iterative, sequential and dynamic; whereas, data is associated with edges within a machine and vertices across machines. In contrast to BSP model, Spark has RDDs that allow parallel operations on data partitioned across machines in a cluster. RDDs can then be stored and retrieved from the Hadoop Distributed File System.

In applying various algorithms in open-source software to a particular data science problem, the analyst building data products has to focus on the algorithm properties with respect to two steps, namely feature engineering and model fitting. Features encode information from raw data that can be consumed by machine learning algorithms. Algorithms that build features from data understood by domain knowledge and exploratory data analysis are called feature construction algorithms. Algorithms that build features from data understood by scientific knowledge and model evaluation are called feature extraction algorithms. Both feature construction and feature extraction algorithms are commonly used in Big Data Analytics. When many features are available for analysis, a compact mapping of the features to data is derived by feature selection or variable selection algorithms. For a successful solution, the data analyst must obtain maximum amount of information about the data science problem by engineering a large number of independent relevant features from data. Feature engineering can be integrated from multiple statistical analysis in a loosely coupled manner with minimum coordination between the various analysts. Depending on statistical analysis being performed, feature engineering is an endless cycle of iterative code changes and tests. The templates and factors affecting variability and dependencies of data, respectively, may be categorized into desired–undesired, known–unknown, controllable–uncontrollable and observable–unobservable factors involved in feature design and validation. For big data systems focused on scalability, feature engineering is focused on throughput than latency. Distant supervision and crowd sourcing are two popular techniques that have been used to define features on web scale. Several different feature engineering approaches can be compared in terms of the feature interdependence, feature subset selection, feature search method, feature evaluation method, parametric feature stability and predictive performance. Better variable selection, noise reduction and class separation may be obtained by adding interdependent features into an objective function.

Univariate and multivariate algorithms for feature engineering are implemented using filter, wrapper and embedded programming paradigms. Feature engineering from filters ranks features or feature subsets independent of a predictor or classifier. Feature engineering from wrappers uses a classifier to assess features or feature subsets. To search for feature or feature subsets with the aid of a machine learning process, embedded feature engineering combines the search criteria evaluating the goodness of a feature or feature subset with the search criteria generating a feature or feature subset. Common ways to evaluate goodness of feature are statistical significance tests, cross validation and sensitivity analysis. Common ways to generate features are exhaustive search, heuristic or stochastic search, feature or feature subset ranking and relevance, forward selection and backward elimination. Heuristic preprocessing, feature normalization, iterative sampling, parson windows, risk minimization, minimum description length, maximum likelihood estimation, generalized linear models, convolutional filters, clustering, decision trees, rule induction, neural networks, kernel methods, ensemble methods, matrix decomposition methods, regression methods and Bayesian learning are some of the most commonly used feature engineering models. Applications of feature engineering include data visualization, case-based reasoning, data segmentation and market basket analysis.

2.1.2 Machine Intelligence and Computational Intelligence

Machine intelligence is the study of applying machine learning, artificial intelligence to Big Data Analytics. While Big Data Analytics is a broader term encompassing data, storage and computation, machine intelligence is specialized to intelligent programs that can be built for big data. Machine intelligence technologies are being used for a variety of problem types like classification and clustering for natural language processing, modeling support vector machines and neural networks. Innovations in machine intelligence span technologies, industries, enterprises and societies. Machine intelligence technologies under development include natural language processing, deep learning, predictive modeling, signal processing, computer vision, speech recognition, robotics and augmented reality. Machine intelligence is being applied to industries such as agriculture, education, finance, legal, manufacturing, medical, media, automotive and retail. Machine intelligence is developing enterprises in sales, security, recruitment, marketing and fraud detection.

Computational intelligence refers to the specific type of algorithms that are useful for machine intelligence. Computational intelligence algorithms are derived from ideas in computational sciences and engineering. Computational sciences involve the study of computer systems built from the knowledge that is a mixture of programming, mathematics and domains of application. Whereas the focus of computer sciences is theory and mathematics, the focus of computational sciences is application and engineering. The data mining tasks in computational sciences depend on the domain of application and computational complexity of the application.

Commonly used computational intelligence techniques construct intelligent computing architectures. These architectures commonly involve heuristics built over complex networks, fuzzy logic, probabilistic inference and evolutionary computing. Furthermore, the various computing architectures are integrated in a structured manner to form hybrid architectures. Hybrid architecture models the best result of each data mining task by assuming that a combination of tasks is better than any single task. In the literature, hybrid architectures are also referred to by the umbrella term soft computing architectures. Here, soft computing refers to the frameworks of knowledge representation and decision making that explore imprecise strategies that become computationally tractable on real-word problems. The guiding principle of soft computing and computational intelligence is to provide human-like expertise at a low solution cost. Human-like expertise is provided by validating the computational intelligence against domain knowledge. In modern science, most of the application domains reason with time variable data that is uncertain, complex and noisy. A two-step problem-solving strategy is employed to deal with such data. In the first step, a stand-alone intelligent system is built to rapidly experiment with possible developments into a prototype system. On successful completion of the first step, a second step builds a hybrid architecture that is more complex yet stable. A more complex hybrid model could use an evolutionary algorithm to train a neural network which in turn acts as preprocessing system to a fuzzy system that produces the final output. Modeling performance in any one data mining task of the hybrid architecture has a corresponding incremental or detrimental effect on the final performance of the entire solution.

Many machine learning problems reduce to optimization or mathematical programming problems in operations research. In optimization, the problem is to find the best set of parameters of the model that solve the problem over a solution search space. The optimization algorithm is designed and implemented with respect to standard objective function formulations. In the area of optimization, objective functions typically model the information loss in the problem. Hence, the objective functions are also called loss functions or cost functions. Typically, the parameters found by solving for the objective function are also normalized by regularization functions and relaxation functions that model prior information about the parameters with respect to records in the data. The parametric normalization is supposed to prevent the dominance of one or few parameters over the rest of the parameters in the modeling process. Such a parametric dominance is referred to as the bias-variance trade-off in the machine learning literature. Furthermore, constrained optimization imposes constraints on searching for the solutions of the objective functions.

There are many types of standard objective functions used in constrained optimization. Typical formulations include objectives and constraints that are one or more of linear, quadratic, nonlinear, convex, goal, geometric, integer, fractional, fuzzy, semi-definite, semi-infinite and stochastic [2]. Convex programming optimizes objective functions that are solved for the same solution by local search and global search. In real-world problems, convex programming is preferred as a solution technique because it is computationally tractable on a wide variety of problems. Novel methods of machine learning algorithms that use convex optimization include kernel-based ranking, graph-based clustering and structured learning. If the problem cannot be directly mapped onto a convex objective function, the machine learning algorithms attempt to decompose the problem into subproblems that can be mapped onto a convex objective function. Most of the convex objectives can be solved easily by using standard optimization algorithms. In any case, error is introduced into the solutions found by these machine learning algorithms. The error is either due to a finite amount of data that is available for algorithm or because the optimal solution underlying error distribution is unknown in advance. Another source of error is the approximations made by the search technique used in the machine learning algorithm. Since search techniques cannot find all possible solutions in finite resources of the computer, they reduce the problem space of search by finding only important solutions. The importance of a solution is then formulated as a convex optimization problem by correct choice of objective functions. In real-world problems, machine learning algorithms require optimization algorithms that have properties of generalization, scalability, performance, robustness, approximation, theoretically known complexity and simple but accurate implementation [2].

A class of computational intelligence algorithms that are commonly used by both researchers and engineers for global optimization in machine learning are the evolutionary computing algorithms. Another class of common computational intelligence algorithms is multiobjective or multicriterion optimization algorithms. In such algorithms, evolution is defined as a two-step process of random variation and selection in time. The process of evolution is modeled through evolutionary algorithms. Evolutionary algorithms are useful method for optimization when direct analytical discovery is not possible. Multiobjective optimization considers the more complicated scenario of many competing objectives [3]. A classical optimization algorithm applied to the single objective optimization problems can find only a single solution in one simulation run. By contrast, evolutionary algorithms deal with a population of solutions in one simulation run. In evolutionary algorithms that are applied to multiobjective optimization, we first define conditions for a solution to become an inferior or dominated solution and then we define the conditions for a set of solutions to become the Pareto-optimal set. In the context of classical optimization, weighted sum approach, perturbation method, goal programming, Tchybeshev method, min—max method are popular approaches to multiobjective optimization. These algorithms not only find a single optimum solution in one simulation but also have difficulties with nonlinear non-convex search spaces.

Evolutionary computing simulates the mechanisms of evolutionary learning and complex adaptation found in natural processes. To be computationally tractable, evolutionary learning systems ought to efficiently search the space of solutions and exploit domain knowledge to produce results that are intelligible. The techniques used for searching a solution include genetic mutation and recombination, symbolic reasoning and structured communication through language, morphology and physiology determining organism's behavior in nervous systems. To model complex adaptation, objects in the environment are grouped into set of properties or concepts. Concepts are combined to form propositions, and propositions are combined to form reasons and expressions. Syntactic, semantic and preference properties of logical expressions can then reduce the candidate solutions that are searched with respect to the training examples. Such data properties are represented in computational intelligence algorithms by defining the architecture of neural networks and rules in symbolic representations [4].

As discussed in [5], evolutionary computing algorithms can be categorized into genetic algorithms and programming, evolutionary strategies and programming, differential evolution and estimation of distribution, classifier systems and swarm intelligence. Each of these algorithms has initialization steps, operational steps and search steps that are iteratively evaluated until a termination criterion is satisfied. Machine learning techniques that have been used in evolutionary computing include case-based reasoning and reinforcement learning, clustering analysis and competitive learning, matrix decomposition methods and regression, artificial neural networks, support vector machines and probabilistic graphical models. As discussed in [5], these techniques are used in various evolutionary steps of the evolutionary computing algorithm, namely population initialization, fitness evaluation and selection, population reproduction and variation, algorithm adaptation and local search. They could use algorithm design techniques such as sequential approach, greedy approach, local search, linear programming, and dynamic programming and randomized algorithms.

Exact optimization algorithms are guaranteed to find an optimal solution. To maintain convergence guarantees, exact algorithms must search the entire solution space. Although many optimization techniques efficiently eliminate large number of solutions at each iteration, lot of real-world problems cannot be tackled by exact optimization algorithms. Because of the impracticality of exact search, heuristic search algorithms have been developed for global optimization in machine learning. Algorithms for global optimization that are based on heuristics fall under the category of hyper-heuristics and meta-heuristics. These optimization algorithms are yet another set of approaches for solving computational search problems. They search for a solution by designing and tuning heuristic methods. Hyper-heuristics indirectly finds solution for the optimization problem by working on a search space of heuristics. By comparison, meta-heuristics directly works on a solution space that is same as search space on which the objective function is defined. Difficulties in using heuristics for search arise out of the parameter or algorithm selections involved in the solution.

Hyper-heuristics attempts to find correct method of heuristics that could solve a given problem. Hyper-heuristic approaches can be categorized into low-level heuristics, high-level heuristics, greedy heuristics and meta-heuristics. Genetic programming is the most widely used hyper-heuristic technique. To build low-level heuristics, one can use mixed integer programming paradigms like branch-and-bound, branch-and-cut local search heuristics, graph coloring heuristics like largest-weighted degree and saturation degree. By contrast, high-level heuristics has an evaluation mechanism that allocates high priority to most relevant features in the data. High-level heuristics defined on a search space of heuristics builds on optimization algorithms like iterative local search, steepest descent and Tabu Search.

By contrast, meta-heuristics process guides and modifies the operations of problem-specific heuristics while avoiding the disadvantages of iterative improvement whose local search cannot escape local optima. The goal of meta-heuristic search is to find near global solutions to the optimization problem. It is implemented by efficient restarting of the local search and introducing a bias into the local search. As discussed in [6], the bias can be of various types such as descent bias (based on objective function), memory bias (based on previously made decisions) and experience bias (based on prior performance). While meta-heuristics ranges from local search to machine learning processes, the underlying heuristic can be a local search. Moreover, meta-heuristics may use domain knowledge to control the search procedure. Meta-heuristics like genetic algorithms and ant algorithms models nature. Algorithms like Iterated Local Search and Tabu Search work with single solutions in the local search procedure. Some meta-heuristics algorithms such as Guided Local Search change the search criteria by utilizing information collected during the search. Some meta-heuristics builds algorithms with memory of the past searching. Memory-less algorithms assume Markov condition on the underlying probability distribution to determine next step in the search algorithm. Iterative improvement, Simulated Annealing, Tabu Search, Guided Local Search, Variable Neighborhood Search and Iterated Local Search are some of the most common meta-heuristics [6].

Neural networks offer a structured technique for algebraically combining the input features into progressively more abstract features that eventually lead to the expected output variables. The different ways to arrive at abstract features are determined by the various possible neural network architectures. Thus, the abstract features in neural network allow machine learning with generalization capabilities. Neural networks store knowledge in the weights connecting neurons. The weights are determined by training a learning algorithm on given data. By contrast, fuzzy inference systems offer a framework for approximate reasoning. Fuzzy algorithms modeled data as a set of rules and interpolate the reasoning output as a response to new inputs. Understanding fuzzy rules in terms of domain knowledge is simple. However, only complex training algorithms can learn the fuzzy rules. Probabilistic reasoning in Bayesian belief networks updates previous estimates of the outcome by conditioning with new evidence. Evolutionary computing is an optimization method that finds candidate solutions by iterative, generative and adaptive processes built on the known samples. By contrast, classical optimization techniques like gradient descent, conjugate gradient and quasi-Newton techniques use gradients and Hessians of the objective function to search for optimum parameters. Depending on nonlinearity of the objective function, these techniques have been adapted into algorithms training neural networks such as feed forward, back propagation, Hebbian and recurrent learning.

2.1.3 Data Engineering and Data Sciences

Knowledge discovery in databases (KDD) and pattern recognition (PR) are the core problem-solving techniques used for applying data engineering techniques in data science problems. Before understanding KDD, let us review problem-solving techniques. In general, the output of problem-solving methodologies is knowledge. Knowledge can satisfy one or more themes such as empirical, rational and pragmatic knowledge. Individuals can build coherence between different pieces of such knowledge. Or consensus can be the ultimate criteria for judging such knowledge. Different methodologies for acquiring knowledge include ontological, epistemological, axiological and rhetorical methodologies.

In data engineering problems, various notions of knowledge discovery crystallize into the three steps of problem definition, problem analysis and solution implementation. To do problem definition, we need to be able to specify the end data science problem as a composition of well-defined statistical problems. To do problem analysis, we need to understand the present and new system of algorithms implementing the statistical problems. To do solution implementation, we need to generate conceptual solutions that consider the subject matter knowledge related to implementation/deployment detail. Additionally, the implementation/deployment detail can be validated against decision-making processes to determine triggers for redesign or reconfiguration or decommissioning.

The three steps of problem definition, problem analysis and solution implementation can also benefit from the various standards in systems life cycles. The triggers for problem definition can result from situation analysis and objective formulation. The triggers for problem analysis can result from solution search and evaluation. The triggers for solution implementation can result from solution selection and decision making. Systems life cycle is a systematic and structured procedure of formulating objectives and finding solutions given a situation analysis. Evaluating decisions and avoiding frequent errors are also part of problem solving with a system's life cycle. An iterative situation analysis is to be summarized as a problem definition derived from one or more of mining tasks, system architectures and future goals. Ideas obtained from a strength–weakness–opportunity–threat, cost-benefit, root cause analysis from system's thinking are also useful in problem definition. Each method of systems analysis can emphasize one or more of institution, discourse, solution and convention in the problem-solving processes.

The problem-solving techniques applied to the scenario of data engineering are called as KDD. According to a standard definition by [7], KDD process consists of the following nine steps. KDD that deals with data mining and machine intelligence in the computational space can be fruitfully combined with human–computer interaction (HCI) in cognitive space that deals with questions of human perception, cognition, intelligence and decision making when humans interact with machines. Consequently, variants of steps in KDD and HCI have been incorporated in industry standards like complex event processing (CEP), real-time business intelligence (RTBI), cross industry standard process for data mining (CRISP-DM), Predictive Model Markup Language (PMML) and Sample, Explore, Modify, Model and Assess (SEMMA).

- **Learning from the Application Domain**: includes understanding previous knowledge, goals of application and a certain amount of domain expertise;
- **Creating a Target Dataset**: includes selecting a dataset or focusing on a subset of variables or data samples on which discovery shall be performed;
- **Data Cleansing (and Preprocessing)**: includes removing noise or outliers, strategies for handling missing data, etc.;
- **Data Reduction and Projection**: includes finding useful features to represent the data, dimensionality reduction, etc.;
- **Choosing the Function of Data Mining**: includes deciding the purpose and principle of the model for mining algorithms (e.g., summarization, classification, regression and clustering);
- **Choosing the Data Mining Algorithm**: includes selecting method(s) to be used for searching for patterns in the data, such as deciding which models and parameters may be appropriate and matching a particular data mining method with the criteria of KDD process;
- **Data Mining**: includes searching for patterns of interest in a representational form or a set of such representations, including classification rules or trees, regression, clustering, sequence modeling, dependency and line analysis;
- **Interpretation**: includes interpreting the discovered patterns and possibly returning to any of the previous steps, as well as possible visualization of the extracted patterns, removing redundant or irrelevant patterns and translating useful patterns into terms understandable by user;
- **Using Discovered Knowledge**: includes incorporating this knowledge into the performance of the system, taking actions based on the knowledge or documenting it and reporting it to interested parties, as well as checking for and resolving, potential conflicts with previously believed knowledge.

In practice, problem solving for data science with pattern recognition is defined in two steps. The first step is the study of problem context for problem formulation, data collection and feature extraction. The second step is the study of mathematical solution for modeling, evaluation and decision making. Whereas features (or key data dimensions) are output from first step, models (or key complexity measures) are output from the second step. The features engineered from the first step give the analyst an idea of the data properties that capture the data complexity. Typically, the data properties are obtained by data sampling, data transformation and feature selection. The models engineered from the second step give the analyst an idea of the data properties that impact the analytics solution. Accuracy of the solution depends on goodness of fit between model and problem. By removing irrelevant dimensions and compressing discriminatory information, feature selection may change computational complexity of the model. Thus, the data properties in the model are quantitatively described through ideas measuring computational complexity such as degree of linear separability, length of class boundary, shapes of class manifolds, uncertainty in feature space and uncertainty in complexity space. Research in either computational machine learning or statistical machine learning uses the data properties to do a systematic evaluation of algorithms against problems.

In data mining project development, often the methodology translating business objectives into data mining objectives is not immediately available. The common pitfalls of data mining implementation are addressed by questions on data mining cost estimation. Main factors affecting the cost estimation are data sources, development platform and expertise of the development team. Technical elements include data quality, data ownership, data generation processes and type of data mining problems. As discussed by [8], the drivers for cost estimation in data mining project development can be listed as following. Until all these drivers are sufficiently addressed, KDD and PR are conducted in a non-sequential non-ending circle of iterations where backtracking to previous phases is usually necessary.

- Data Drivers
- Model Drivers
- Platform Drivers
- Tools and Techniques Drivers
- Project Drivers
- People Drivers.

At the beginning of the data mining project, we need to be able to design hypothesis and experiments that evaluate and validate the impact of implementation and deployment in detail. In this context, agile and lean software development methodologies, like SCRUM and parallel thinking, need to be flexible enough for managing data product development, management paradigms and programming paradigms in terms of features, personnel and cost. From a modeling perspective, the development process centers around a system model with executable specifications of design in continuous tests and verification of implementation. Feasibility and compatibility of design goals are to be analyzed by multidomain simulation. Each organization has varying levels of formality in the modeling processes driven by people culture, legal regulation, best practices and latest trends.

Six Sigma is a popular method to monitor the quality and cost of organizational processes and activities. The focus of Six Sigma is on increasing throughput while reducing bottlenecks in scalability. Six Sigma is implemented by defining the skill sets necessary to successfully coordinate people's activities around organizational objectives. The formal approach to Six Sigma is defined by the acronym, DMAIC, which stands for Define, Measure, Analyze, Improve and Control. DMAIC cycle is monitored by statistical process control (SPC) tools. Problem definition and root cause analysis are the most difficult parts of Six Sigma method. The ideas taken from

KDD and PR can help us with these parts of Six Sigma. Control charts, flow charts, scatter diagrams, concept maps and infographics are also popular data visualization techniques for Six Sigma.

2.2 Big Data Computing

2.2.1 Distributed Systems and Database Systems

Machine learning has been traditionally associated with algorithm design, analysis and synthesis. However, with the advent of big data, machine learning is becoming integral part of many computer systems. In this section, we cover two such computer systems, namely database systems and distributed systems.

Database systems have the ability to store a wide variety of data. Traditionally, data was stored in flat files on the operating system. With the advent of data that is structured according to relations between sets, relational database systems have been widely used. The original relational databases are not designed for storing the voluminous big data in a cost-effective manner. To deal with the problems of big data, relational database has spawned new generations of systems such as in-memory databases and analytic databases. With the availability of even more data generated by humans and machines, there is a bigger need for databases that can deal with the streaming, unstructured, spatiotemporal nature of the data sourced from documents, emails, photos, videos, web logs, click streams, sensors and connected devices. Therefore, from a standpoint of database systems, big data has been defined in terms of volume, variety, velocity and veracity. To be intelligible, the big data must be mined using machine learning algorithms. These algorithms require lots of processing power to effectively mine the data while considering various trade-offs of speed, scale and accuracy. The processing power is becoming available through distributed systems of software built for coordinated management of a network of commodity hardware clusters. That is why there is a close relationship between database systems and distributed systems used for supporting Big Data Analytics in the cloud.

The traditional data processing model of distributed systems is composed of two tiers. A first 'storage cluster' tier is used to clean and aggregate the data. A second 'compute cluster' tier is used to copy and process the data. However, this data processing model does not work for storing and processing big data that is operating at scale. One solution is to merge the two tiers into one tier where data is both stored and processed. The computing over this one tier is called cloud computing. To deal with the problems of speed, scale and accuracy, cloud computing defines a loosely connected coordination of various hardware clusters that allow well-defined protocols for utilizing the hardware toward both storage and processing.

Modern distributed systems like Hadoop are built for big data processing. These systems do not impose prior conditions on structure of the unstructured data, scale horizontally so that more processing power is achieved by adding more hardware nodes to the cloud, integrate distributed storage and computing. Moreover, these systems are cost effective by allowing the end user to perform reasonably complex computations on off-the-shelf commodity hardware. This in turn allows organizations to capture and store data at a reasonable cost for longer durations. However, the real benefit of the big data captured over such distributed systems is in the big data analysis frameworks provided by them. For example, the big data framework of Apache Hadoop is MapReduce, and the big data frame work of Apache Spark is Resilient Distributed Datasets. These frameworks support analytics in programming languages like R, Java, Python and Scala.

Hadoop is an open-source implementation of proprietary distributed computing frameworks. It consists of Hadoop Distributed File System (HDFS) and Hadoop MapReduce. Like Spring and Struts, MapReduce is a Java programming framework that originated in functional programming constructs. MapReduce takes care of distributing the computations in a program. This allows the programmer to concentrate on programming logic rather than organizing multiple computers into a cluster. MapReduce also takes care of hardware and network failures to avoid loss in data and computation. To use MapReduce, the programmer has to break the computational problem into Map and Reduce functions that are processed in parallel on a shared nothing architecture. Many useful storage and computer components have been built on top of Hadoop. Some of these components that are widely used include Apache HBase, Apache Hive, Apache Pig, Apache Mahout, Apache Giraph, Apache Yarn, Apache Tez, Apache Presto, Apache Drill and Apache Crunch [9].

HDFS allows horizontal scalability with fault tolerance on commodity hardware. MapReduce does distributed computing in a fault-tolerant manner. HBase is a key-value database that scales to millions of columns and billions of rows. Hive is a data warehouse that uses a SQL-like query language. Through stored procedures, Pig allows analysis of large volumes of data to create new derivative datasets that solve sophisticated large-scale problems without having to code at the level of MapReduce. Mahout is a machine learning library that has common algorithms for analytics in a distributed environment. Drill and Crunch also allow analytics at scale. Graph is graph-driven distributed programming framework implementing the Bulk Synchronous Parallel Model on computations. ETL tools to pull data into Hadoop include Flume, Chukwa, Sqoop and Kafka. Thus, we can see that Hadoop ecosystem provides for a rich functionality as far as big data analysis is concerned. Moreover, custom MapReduce programs can be written to extend the capabilities of Hadoop software stack. We shall survey the above modules in detail in Chap. 4. For storing key–value datasets, some of the alternatives for Hadoop include columnar NoSQL databases like MongoDB, Cassandra, Vertica and Hypertable. Grid computing frameworks like Web Services Resource Framework (WSRF), Grid Gain and JavaSpaces are some of the more complex open-source alternatives to MapReduce. Just like Linux distributions, Hadoop software stack has been packaged into Hadoop distributions for easy management of software. Some of the common Hadoop distributions are provided by Apache Foundation, Cloudera, Horton Works and MapR.

Spark defines the big data processing framework in terms of Scala transformations and actions on data stored into Resilient Distributed Datasets (RDDs). Internally, RDDs are stored as directed acyclic graphs (DAGs). DAGs are supposed to allow a better representation of data and compute dependencies in a distributed program. Also, Scala is easier to program than Java. Thus, for Big Data Analytics, the functional programming constructs of Spark RDD are an alternative to Hadoop MapReduce. Data caching is also possible in Spark. The in-memory processing of Spark also has better benchmarks than Hadoop. Spark has components for accessing data from either local file system or cluster file system. Spark framework can be accessed and programmed in R, Python, Java and Scala. By allowing functional programming over distributed processing, Spark is quickly becoming the de facto standard for machine learning algorithms applied to big data integrated in a distributed system. By contrast, Hadoop is already the standard for database systems storing big data. Spark can also interoperate with Hadoop components for storage and processing like HDFS and YARN. Like Hadoop software stack, Spark software stack has components like Spark Streaming, Shark, MLLib, GraphX, SparkR for data storage, data processing and data analytics [10]. The novelty in Spark is a software stack that has compute models for both data streams and analytics in shared memory architecture. Spark stream processing also integrates well with both batch processing and interactive queries. As compared to record-at-a-time stream processing models, stream computation in Spark is based on a series of very small deterministic batch jobs. State between batches of data is stored in memory as a fault-tolerant RDD dataset. The data parallel MapReduce framework is not suitable for algorithms that require results of Map functions as input to other Map functions in same procedure. MapReduce framework is not suitable for computations between datasets that are not independent and impose a specific chronology on data dependencies. Moreover, MapReduce is not designed to provide iterative execution of Map and Reduce steps. Thus, MapReduce is not suitable for iterative processing of machine learning algorithms like expectation–maximization algorithms and belief propagation algorithms. The disadvantages of MapReduce have led to richer big data frameworks where more control is left to the framework user at the cost of more programming complexity. Spark filter, join, aggregate functions for distributed analytics allow better parameter tuning and model fitting in machine learning algorithms that are executed in a distributed fashion. Typically, these algorithms are CPU intensive and need to process data streams. Some of the popular machine learning algorithms in Spark include logistic regression, decision trees, support vector machines, graphical models, collaborative filtering, topic modeling, structured prediction and graph clustering. GraphLab and DryadLINQ like alternatives to Spark are software components that provide scalable machine learning with graph parallel processing. All such graph parallel systems reduce the communication cost of distributed processing by utilizing asynchronous iterative computations. They exploit graph structure in datasets to reduce inefficiencies in data movement and duplication to achieve orders-of-magnitude performance gains over more general data parallel systems. To deal with trade-offs in speed, scale and accuracy, graph-based distributed systems combine advances in machine learning with low-level system design constructs like asynchronous distributed graph computation, prioritized scheduling and efficient data structures. Moreover, graph parallel systems are computationally suitable for graphs that arise in social networks (e.g., human or animal), transaction networks (e.g., Internet, banking), molecular biological networks (e.g., protein–protein interactions) and semantic networks (e.g., syntax and compositional semantics).

2.2.2 Data Stream Systems and Stream Mining

With the growth of unstructured data in forms such as spatial, temporal, multimedia and hypertext, there is a need for intelligent computer systems that develop upon the data mining techniques used in relational databases, analytic databases and data warehouses. This data mining task is one of mining complex types of data, including sequential pattern mining, subgraph pattern mining, and data stream mining. The sources of such complex data are sensor streams, time series, decoded sequences found, for example, in web applications, network monitoring and security, bioinformatics, telecommunications and data management, manufacturing, power grids. Streaming data is uncertain, massive, temporal, high-dimensional and fast changing. As discussed in [11], streaming data arrives in multiple, continuous, rapid, time-varying, unpredictable and unbounded data streams of tuples and events. Traditional OLTP, OLAP and data mining techniques require multiple scans for indexing and searching the stream data. Traditional database management systems (DBMS) do not support continuous, approximate and adaptive query processing over data streams. This leads us to the study of analytic techniques that can deal with heterogeneous, incremental and dynamic data requiring more precise and accurate analysis over search spaces issued from complex random processes. These analytic techniques include data mining models for vector auto regression, wavelet decomposition, similarity search, dependency modeling and finite mixture modeling.

Following solutions have been proposed for the problem of data stream mining:

- Redesign the analytics platform to provide fast and scalable computation infrastructure. An ideal computation system would balance the usage of memory and disk to provide fast and scalable computations. Data stream management systems (DSMS) have been proposed that process and discard/archive elements arriving in a data stream. Query processing over data streams may require an unlimited amount of memory as data grows without bounds. Therefore, approximate query processing is done by relaxing the system requirements on memory usage. Ad hoc query processing would then require the history of approximate query processing. Data model and query semantics in data streams must allow streaming operations that do not need the entire input. Since results must be produced on part of the data, the streaming operations must depend on the order in which records arrive. The order can be a chronological order in time or any other sorted order of records. To ensure scalability, the query evaluation model must implement a shared memory system executing many continuous queries. Another challenge with DSMS is concept drift in data streams where certain elements in the data stream are no longer consistent with the current concepts built from query processing and algorithmic analysis.

- Compression and summarization of streaming data is an effective way of tackling the problem of scalability. However, the challenge is to select the suitable synopsis data structure and algorithm that gives the required summarization and compression. The synopsis data structures are substantially smaller than stream data and can be used to return approximate answers to queries. Some of the common synopsis data structures are obtained from random sampling, sliding windows, counting techniques, hashing techniques, sketching techniques, hierarchical clusters, sequence similarity search, wavelet decomposition, matrix decompositions and correlation graphs. Some of the compression algorithms are obtained from decision trees, ensemble analysis, probabilistic graphical models, latent variable models and deep learning.
- Transforming the serial algorithms for analytics into distributed algorithms. A challenge is to break the programming logic in the serial algorithm into a form that is suitable for processing under one or more of embarrassingly parallel, data parallel, task parallel and graph parallel computing frameworks. Another challenge is to transform exact, inefficient algorithms into approximate, efficient algorithms while providing reasonable accuracy.
- Typically a data stream varies in time. So, one common solution to the problem of stream mining is time series analysis. A time series database consists of sequence of values or events obtained over repeated measurements of time. The values are measured at equal time intervals like hours, days and weeks. Time series data is typically analyzed for trends, bursts and outliers found over time where response time is near real time. Trend analysis finds components in time series that characterize the time series. It is generally performed by modeling time series and forecasting time series. Common trends include seasonality, periodicity and irregular variations found in the short-term and long-term time series. Auto regression analysis and similarity search analysis are commonly used models for time series analysis.
- A transactional database consisting of ordered elements or events is called a sequence database. In transactional databases, the data steam mining problem converts to sequence pattern mining problem. The problem of sequential pattern mining is that of finding frequent subsequences with the techniques in association rules mining. The candidate rules can be generated by a variety of searching, pruning and optimization criteria. The candidate rules are output as association rules if they satisfy certain constraints on the search criteria. Typical constraints are expressed over aggregations of attributes, attribute values or candidate patterns that are either partially or fully periodic. Sequential pattern mining is also useful to find interesting information in multidimensional and multilevel databases. Examples of such applications include targeted marketing and customer retention, production quality control, web access control, bio-sequence discovery and phylogenetics.

- With the advent of Internet of Things, distributed stream processing systems (DSPS) are expected to become increasingly important. DSPS is a new class of data stream systems that support large-scale real-time data analytics. Stream processing systems require data processing before storing; whereas, batch processing systems require processing after storing. Thus, distributed data processing over data streams is different from MapReduce like batch processing on large datasets. Stream processing engines create a logical network of streaming tuples, events in a directed acyclic graph (DAG) of processing elements (PE). Tuples of the data stream flow through the DAG. A PE in the DAG can choose to emit none, one or more tuples after it has seen the input tuples. Depending on available resources, each PE executes independently and communicates with other PEs through messaging. The communication between PEs is based on a push or pull messaging mechanism. PEs running in parallel are not synchronized in general. Load balancing and scheduling among processing tasks are done by either a central server or by a fully distributed peer-to-peer system. The various DSPS differ on latency and availability requirements. The data partitioning in DSPS affects the scale at which the DSPS can process data in parallel. Some of the PEs in DSPS allow non-deterministic operations. Some of the open-source DSPS include Apache Storm, Apache S4 and Apache Samza. The system architecture of the open-source DSPS provides for a variety of data flow mechanisms, query models and programming models. In comparison to DSPS, complex event processing (CEP) engines are used for analyzing enterprise data. Whereas DSPS input events are from a single event stream, the events input to a CEP engine belong to an event cloud. Thus, DSPS do more streamlined and efficient processing as compared to CEP engines. CEP engines also consume more memory because events have to be remembered to discover the causal relationships and ordering among events.

In summary, data stream mining techniques are classified into data-based techniques and task-based techniques [12, 13]. Prominent data-based techniques are sampling, aggregation and synopsis data structures. Prominent task-based techniques are approximation algorithms for clustering, classification, frequent pattern mining and time series analysis. The continuous data in data stream mining is dealt with the aid of database systems and distributed systems. Because of the dynamic nature of data streams that change over time, conflicting accuracy and scalability requirements must be satisfied by the stream mining algorithms. Such algorithm design requirements include stream preprocessing, model over fitting and real-time evaluation. Computational and statistical learning theory is the potential basis for designing loss functions that satisfy the conflicting design requirements. Moreover, interactive versions of the stream mining algorithms are highly specific to the domains of application. For the interactive algorithms, the algorithm analysis objectives should be achieved after considering the application requirements.

2.2.3 Ubiquitous Computing Infrastructures

In this section, we turn our attention to big data management and analysis platforms, products and systems for the next generation of technological advances in science, medicine and business. Although big data is defined in terms of the 4Vs—volume, variety, velocity and veracity—a definition that takes into account the algorithmic computations required by most big data frameworks leads us to the study of big data that is embedded in Ubiquitous Computing Infrastructures. Ubiquitous Computing is an amalgamation of a wide range of research areas like distributed computing, mobile computing, human–computer interaction, wireless networking and artificial intelligence. Ubiquitous computing Infrastructure refers to the ongoing shift from centralized mainframes and stand-alone microcomputers to decentralized and pervasive computing. Both data science and ubiquitous computing attempt to enhance human intelligence by computational intelligence. Like data science, ubiquitous computing needs open specifications for software infrastructures that provide higher level computing abstractions for creating, assessing and accessing data models, language constructs and system architectures, respectively. The big data frameworks in ubiquitous computing require novel programming models and paradigms that mitigate the data complexity resulting from constrained resources and large, dynamic network topologies. Information sharing opportunity on the Internet is expected to become embedded into daily life to such an extent that Internet will become as common as electricity. Along with World Wide Web (WWW), emerging applications of ubiquitous computing on the Internet have led to ideas such as Internet of hings (IoT) [14, 15]. The idea of IoT is to extend the reach of Internet into the physical world to map the physical and social world with artificial intelligence and Big Data Analytics. To do this, sensor networks and communication networks are embedded into the physical environment. In such sensor networks, mobile phones are most common gateways to data analysis. The scale and dynamics of information space being searched is the main analytics problem posed by IoT. Security and privacy will also be a concern as IoT and big data applications become pervasive in the various sectors of the economy. To understand IoT, we must understand devices and device connectivity embedded into urban, rural and natural settings. The meaning produced by IoT data is critically dependent on the analytics performed on data collections and compute capacities. Therefore, the evolution of IoT is tied up with the evolution of computational and storage capacities. Some of the early adopters of IoT and big data include science, enterprises, health care, education, energy, finance, telecom, retail, manufacturing and automotive sectors of the economy. The IoT devices used in these sectors include sensors and processors, personal computers and satellites, industrial equipment and machine tools, commercial vehicles and telecom towers, consumer electronics and utility appliances. The IoT applications include waste management, traffic safety, refrigerated transport, intelligent parking, demand response for utilities, monitoring energy and data warehousing.

Heterogeneity, scale, complexity, privacy and usability are problems with creating value from these big data pipelines. In data collection, we encounter unstructured data lacking semantic content required for search problems. In data sharing, linking data obtained from multiple sources presents data integration problems. Data organization, retrieval, analysis and modeling also present different kinds of problems with big data. In an enterprise context, data analysis presents scalability problems and data modeling presents visualization problems. Following are some of the common application areas where Big data technology is being used.

- **Science**: In sciences like astronomy and biology to discover hidden patterns in scientific data that has been stored in public data repositories.
- **Education**: Approaches to academic effectiveness of educational activities are being designed by considering the student's interests and performance in both basic and advanced level courses.
- **Health**: Information technology has been used to reduce the cost of healthcare management. Big data allows personalized health care in preventive and prescriptive medicine.
- **Society**: Big Data Analytics on geographical information systems allows better urban planning, environmental modeling, energy usage, transportation and security.
- **Enterprise**: Innovations driven by Big Data Analytics can enhance and transform business processes, policies and architectures. These innovations allow new ways and means of defining price, product, market and customer in an enterprise.
- **Machines**: All over the world, economic sectors like Energy, Automotive, Aerospace, Manufacturing, Telecom, Retail, Transportation and Industrial Automation use several complex physical machines in their infrastructure. The data collected from these machines is consumed either by machines or processes or both. Analyzing the context in which the machine data is produced and consumed allows for decision making about efficiency and revenue driven by the mechanisms of data collection, processing, examination and interpretation.

An emerging technology landscape fits into the larger landscape comprising policies, technologies and markets. Companies that create IoT are specialized to optimizing the impact and managing the scale for one or more of hardware, software, device and service. To illustrate the potential advantages and disadvantages of Big Data Analytics in academia and industry, we further discuss the areas of semantic web [16, 17] and smart grids [18–20], respectively.

Semantic web mixes unstructured and structured data for consumption by humans and machines, respectively. The semantic web is defined as a computing infrastructure for supplying the Internet with formalized knowledge in addition to its actual information content. In the semantic web, formalized knowledge discovery is supported by machines; whereas, knowledge management techniques are used by humans. In general, semantic web requires work into Big Data Analytics applied to language and infrastructure. This involves layered and modular representation of human and machine languages. Developing computational models for trust, proof and reward on the web is also needed. Electronic commerce and knowledge grid systems are common applications of the semantic web.

Power grids are concerned with generation, distribution and consumption of electricity. With the increasing use of electric vehicles and price increases in oil production, smart power grids that use Big Data Analytics are expected to become a widespread reality. In power plants that utilize smart power grids, the electricity is generated from renewable sources of energy like wind, solar and tidal than coal and natural gas. Since renewable generation intermittent and distributed, the electricity production is determined by local environmental conditions. Thus, across both transmission and distribution networks in smart power grids, there is a need to match demand for electricity against available supply. Automatic pattern matching between demand and supply allows a transparent yet controlled exchange of both information and electricity between power producers and power consumers. Thus, the intelligent system in a smart power grid is a distributed control and communication system. It enables real-time market transactions. It presents many challenges in terms of systems engineering, artificial intelligence and communication networks. Big data technologies in smart grids will require algorithms that can solve problems with competing objectives having significant levels of uncertainty.

2.3 Conclusion

In this chapter, we have presented an overview of data science, machine learning and intelligent systems. We have presented concepts of machine intelligence, computational intelligence, data engineering, data science and finally big data computing. Looking at future, we have presented data streams, stream data mining and finally Ubiquitous Computing Infrastructures in the upcoming era of IOT and Fog Computing. We shall delve deeper into analytics models for data streams in the next chapter.

2.4 Review Questions

1. Explain what compares data science and open-source data science.
2. Explain the main benefits of machine intelligence and computational intelligence.
3. Explain data experiencing in contact to data science.
4. Explain big data computing and its features.
5. What are data streams? How to mix data streams?
6. What are the developments in Ubiquitous Computing Infrastructures?

References

1. U.M. Fayyad, G. Piatetsky-Shapiro, P. Smyth, *From Data Mining to Knowledge Discovery: An Overview* (AAAI, 1996)
2. K.P. Bennett, E. Parrado-Hernández, The interplay of optimization and machine learning research. J. Mach. Learn. Res. (2006)
3. K. Deb, *Evolutionary Algorithms for Multi-Criterion Optimization in Engineering Design* (1999)
4. P.G.K. Reiser, Computational models of evolutionary learning, in *Apprentissage: des principes naturels aux methodes artificielles* (1998)
5. J. Zhang, Z.-H. Zhan, Y. Lin, N. Chen, Y.-J. Gong, J.-h. Zhong, H.S.H. Chung, Y. Li, Y.-h. Shi, Evolutionary computation meets machine learning: a survey. *Computational Intelligence Magazine* (IEEE, 2011)
6. C. Blum, A. Roli, Meta-heuristics in combinatorial optimization: overview and conceptual comparison. ACM Comput. Surv. (2003)
7. U.M. Fayyad, G. Piatetsky-Shapiro, P. Smyth, The KDD process for extracting useful knowledge from volumes of data. Commun. ACM (1996)
8. P. Gonzlez-Aranda, E. Menasalvas, S. Milln, C. Ruiz, J. Segovia, Towards a methodology for data mining project development: the importance of abstraction, in *Data Mining: Foundations and Practice, Studies in Computational Intelligence* (2008)
9. J. Lin, C. Dyer, *Data-Intensive Text Processing with MapReduce* (Morgan and Claypool Publishers, 2010)
10. V. Agneeswaran, *Big Data Analytics Beyond Hadoop: Real-Time Applications with Storm, Spark, and More Hadoop Alternatives* (Pearson FT Press, 2014)
11. B. Babcock, S. Babu, M. Datar, R. Motwani, J. Widom, Models and issues in data stream systems, in *PODS '02* (2002)
12. M.M. Gaber, A. Zaslavsky, S. Krishnaswamy, Mining data streams: a review. SIGMOD Rec. (2005)
13. L. Golab, M.T. Özsu, Issues in data stream management. SIGMOD Rec. (2003)
14. P. Misra, Y. Simmhan, J. Warrior, Towards a practical architecture for India centric internet of things. CoRR (2014)
15. N. Kaka, A. Madgavkar, J. Manyika, J. Bughin, P. Parameswaran, India's Tech opportunity: transforming work, empowering people. *McKinsey Global Institute Report* (2014)
16. H. Zhuge, The knowledge grid and its methodology, in *First International Conference on Semantics, Knowledge and Grid* (2005)
17. Euzenat, J., Research challenges and perspectives of the Semantic Web. *Intelligent Systems* (IEEE, 2002)
18. S.C. Chan, K.M. Tsui, H.C. Wu, Y. Hou, Y.-C. Wu, F.F. Wu, Load/price forecasting and managing demand response for smart grids: methodologies and challenges. *Signal Processing Magazine* (IEEE, 2012)
19. H. Farhangi, The path of the smart grid. *Power and Energy Magazine* (IEEE, 2010)
20. S. Ramchurn, D. Sarvapali, P. Vytelingum, A. Rogers, N.R. Jennings, Putting the 'smarts' into the smart grid a grand challenge for artificial intelligence. Commun. ACM (2012)

Chapter 3
Analytics Models for Data Science

3.1 Introduction

The ultimate goal of data science is to turn raw data into data products. Data analytics is the science of examining the raw data with the purpose of making correct decisions by drawing meaningful conclusions. The key differences of traditional analytics versus Big Data Analytics are shown in Table 3.1.

3.2 Data Models

Data model is a process of arriving at the diagram for deep understanding, organizing and storing the data for service, access and use. The process of representing the data

Table 3.1 Traditional analytics versus Big Data Analytics

Concept	Traditional analytics	Big Data Analytics
Focus on	• Descriptive analytics • Diagnosis analytics	• Predictive analytics • Data science • Innovate with machine learning
Datasets	• Limited datasets • Less types of data • Cleansed data • Structured data	• Large-scale/unlimited datasets • More types of data • Raw data • Semi-structured/unstructured data
Data models	• Simple data models	• Complex data models
Data architecture	• Centralized database architecture in which complex and large problems are solved in a single system	• Distributive database architecture in which complex and large problems are solved by dividing into many chunks
Data schema	• Fixed/static schema for data storage	• Dynamic schema for data storage

C. S. R. Prabhu et al., *Big Data Analytics: Systems, Algorithms, Applications*,
https://doi.org/10.1007/978-981-15-0094-7_3

in a pictorial format helps the business and the technology experts to understand the data and get to know how to use the data. This section deals with data science and four computing models of data analytics.

3.2.1 Data Products

Data products define which type of profiles would be desirable to deliver the resultant data products that are collected during data gathering process. Data gathering normally involves collecting unstructured data from various sources. As an example, this process could involve acquiring raw data or reviews from a web site by writing a crawler. Three types of data products are as follows:

- Data used to predict,
- Data used to recommend,
- Data used to benchmark.

3.2.2 Data Munging

Data munging, sometimes referred to as data wrangling, is the process of converting and mapping data from one format (raw data) to another format (desired format) with the purpose of making it more suitable, easier and valuable for analytics. Once data retrial is done from any source, for example, from the web, it needs to be stored in an easy to use format. Suppose a data source provides reviews in terms of rating in stars (1–5 stars); this can be mapped with the response variable of the form $\mathbf{x} \in \{\mathbf{1}, \mathbf{2}, \mathbf{3}, \mathbf{4}, \mathbf{5}\}$. Another data source provides reviews using thumb rating system, thumbs-up and thumbs-down; this can be inferred with a response variable of the form $\mathbf{x} \in \{\mathbf{positive}, \mathbf{negative}\}$ (Fig. 3.1). In order to make a combined decision, first data source response (five-point star rating) representation has to be converted to the second form (two-point logical rating), by considering one and two stars as

Fig. 3.1 Thumb and star rating system

negative and three, four and five stars as positive. This process often requires more time allocation to be delivered with good quality.

3.2.3 Descriptive Analytics

Descriptive analytics is the process of summarizing historical data and identifying patterns and trends. It allows for detailed investigation to response questions such as 'what has happened?' and 'what is currently happening?' This type of analytics uses historical and real-time data for insights on how to approach the future. The purpose of descriptive analytics is to observe the causes/reasons behind the past success or failure.

The descriptive analysis of data provides the following:

- Information about the certainty/uncertainty of the data,
- Indications of unexpected patterns,
- Estimates and summaries and organize them in graphs, charts and tables and
- Considerable observations for doing formal analysis.

Once the data is grouped, different statistical measures are used for analyzing data and drawing conclusions. The data was analyzed descriptively in terms of

1. Measures of probability,
2. Measures of central tendency,
3. Measures of variability,
4. Measures of divergence from normality,
5. Graphical representation.

A measure of central tendency indicates the central value of distribution which comprises **mean, median and mode**. However, the central value alone is not adequate to completely describe the distribution. We require a measure of the spread/scatter of actual data besides the measures of centrality. The **standard deviation** is more accurate to find the measure of dispersion. The degree of dispersion is measured by the measures of variability, and that may vary from one distribution to another. Descriptive analysis is essential by the way it helps to determine the normality of the distribution. A measure of divergence from normality comprises **standard deviation, skewness and kurtosis**. A measure of probability includes standard error of mean and fiduciary limits used to set up limits for a given degree of confidence. Statistical techniques can be applied for inferential analysis, to draw inferences/ make predictions from the data. Graphic methods are used for translating numerical facts into more realistic and understandable form.

3.2.4 Predictive Analytics

Predictive analytics is defined to have data modeling for making confident predictions about the future or any unknown events using business forecasting and simulation which depends upon the observed past occurrences. These address the questions of 'what will happen?' and 'why will it happen?'

Predictive model uses statistical methods to analyze current and historical facts for making predictions [1]. Predictive analytics is used in actuarial science, capacity planning, e-commerce, financial services, insurance, Internet security, marketing, medical and health care, pharmaceuticals, retail sales, transportation, telecommunications, supply chain and other fields. Applications of predictive analytics in business intelligence comprise customer segmentation, risk assessment, churn prevention, sales forecasting, market analysis, financial modeling, etc.

Predictive modeling process is divided into three phases: plan, build and implement. Planning includes scoping and preparing. Predictive modeling process for a wide range of businesses is depicted well in Fig. 3.2.

To build a predictive model, one must set clear objectives, cleanse and organize the data, perform data treatment including missing values and outlier fixing, make a descriptive analysis of the data with statistical distributions and create datasets used for the model building. This may take around 40% of the overall time. The subphases

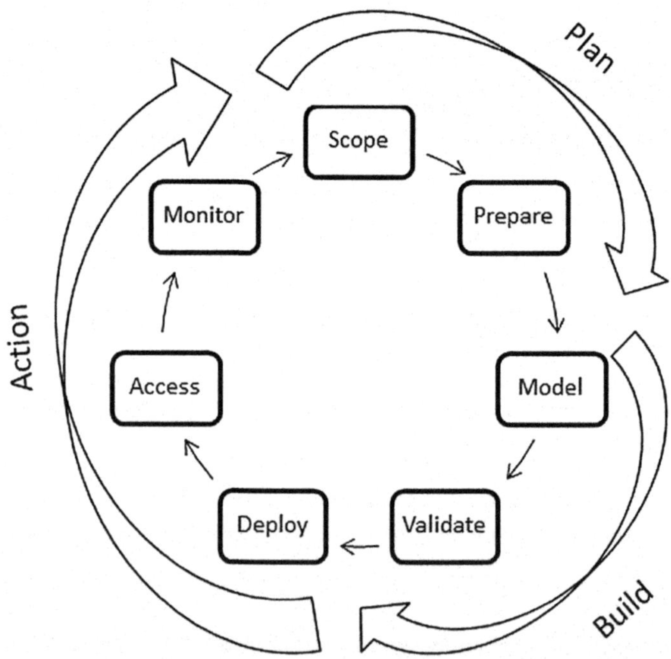

Fig. 3.2 Predictive modeling process

of build the model are model and validate. Here, you will write model code, build the model, calculate scores and validate the data. This is the part that can be left with the data scientists or technical analysts. This might take around 20% of the overall time. The subphases of action are deploy, assess and monitor. It involves deploy and apply the model, generate ranking scores for customers or products (most popular product), use the model outcomes for business purpose, estimate model performance and monitor the model. This might take around 40% of the overall time.

Sentiment analysis is the most common model of predictive analytics. This model takes plain text as input and provides sentiment score as output which in turn determines whether the sentiment is neutral, positive or negative. The best example of predictive analytics is to compute the credit score that helps financial institutions like banking sectors to decide the probability of a customer paying credit bills on time. Other models for performing predictive analytics are

- Time series analysis,
- Econometric analysis,
- Decision trees,
- Naive Bayes classifier,
- Ensembles,
- Boosting,
- Support vector machines,
- Linear and logistic regression,
- Artificial neural network,
- Natural language processing,
- Machine learning.

3.2.5 Data Science

Data science is a field associated with data capturing, cleansing, preparation, alignment and analysis to extract information from the data to solve any kind of problem. It is a combination of statistics, mathematics and programming. One way of assessing how an organization currently interacts with data and determines where they fit to achieve increasingly accurate hindsight, insight and foresight and its different ecosystems in four fundamental ways [2] (i) descriptive analytics, (ii) diagnostic analytics, (iii) predictive analytics and (iv) prescriptive analytics by Gartner's analytics maturity model is illustrated in Fig. 3.3.

The four types of analytics [3] and their association along with the dimensions from rule-based to probability-based and the dimensions of time (past, present and future) are depicted in Fig. 3.4.

The focus of the top left and the bottom right quadrants, i.e., diagnostic and prescriptive analytics, is past and future, while the focus of the bottom left and top right quadrants, i.e., descriptive and predictive analytics, is past and future. The summary is provided in Table 3.2.

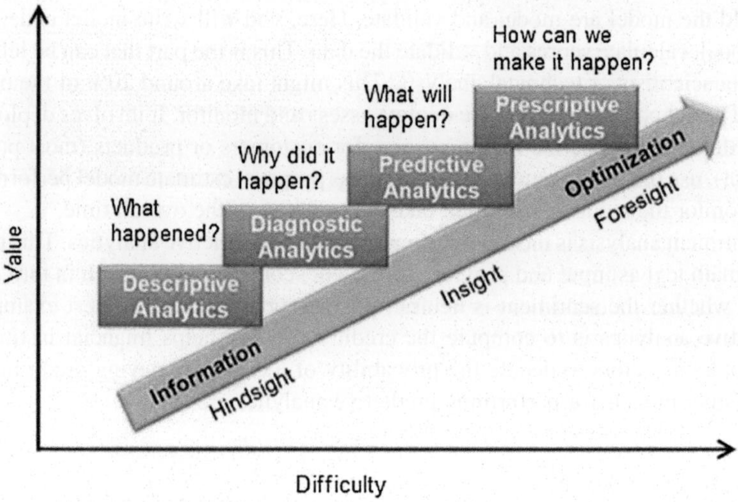

Fig. 3.3 Gartner's analytics maturity model

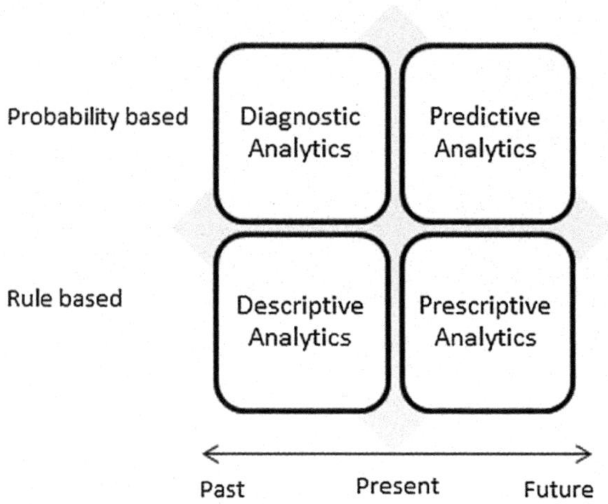

Fig. 3.4 Four types of analytics

Table 3.2 Key differences of four types of analytics

Descriptive analytics	Diagnostic analytics	Predictive analytics	Prescriptive analytics
Backward focus	Backward focus	Forward focus	Forward focus
Rule-based	Probability-based	Probability-based	Rule-based
Live data that is comprehensive, accurate and good visualization	Find out the root cause of the problem and remove all confusing information	Historical pattern to predict specific outcomes using algorithms	Applying advanced analytical techniques to make specific recommendations

Consider an example of the story of The Jungle Book. Baloo a bear hired a data analyst (Mowgli) to help find his food honey. Mowgli had access to a large database, which consisted of data about the jungle, road map, its creatures, mountains, trees, bushes, places of honey and events happening in the jungle. Mowgli presented Baloo with a detailed report summarizing where he found honey in the last six months, which helped Baloo decide where to go for hunting next; this process is called descriptive analytics.

Subsequently, Mowgli has estimated the probability of finding honey at certain places with a given time, using advanced machine learning techniques. This is predictive analytics. Next, Mowgli has identified and presented the areas that are not fit for hunting by finding their root cause why the places are not suitable. This is diagnostic analytics. Further, Mowgli has explored a set of possible actions and suggested/recommended actions for getting more honey. This is prescriptive analytics. Also, he identified shortest routes in the jungle for Baloo to minimize his efforts in finding food. This is called as optimization. We already had enough discussion on descriptive and predicative analytics; further let go through the explanatory details of diagnostic and prescriptive analytics.

Diagnostic Analytics

It is a form of analytics which examines data to answer the question 'why did it happen?' It is kind of root cause analysis that focuses on the processes and causes, key factors and unseen patterns.

Steps for diagnostic analytics:

(i) Identify the worthy problem for investigation.

(ii) Perform the Analysis: This step finds a statistically valid relationship between two datasets, where the upturn or downturn occurs. Diagnostic analytics can be done by the following techniques

- Multiregression,
- Self-organizing maps,
- Cluster and factor analysis,
- Bayesian clustering,
- k-nearest neighbors,
- Principal component analysis,
- Graph and affinity analysis.

(iii) Filter the Diagnoses: Analyst must identify the single or at most two influential factors from the set of possible causes.

Prescriptive Analytics

This model determines what actions to take in order to change undesirable trends. Prescriptive analytics is defined as deriving optimal planning decisions given the predicted future and addressing questions such as 'what shall we do?' and 'why shall we do it?' Prescriptive analytics is based on [4]

- Optimization that helps achieving the best outcomes,
- Stochastic optimization that helps understanding how to identify data uncertainties to make better decisions and accomplish the best outcome.

Prescriptive analytics is a combination of data, business rules and mathematical models. It uses optimization and simulation models such as sensitivity and scenario analysis, linear and nonlinear programming and Monte Carlo simulation.

3.2.6 *Network Science*

In networked systems, standardized graph-theoretic methods are used for analyzing data. These methods have the assumption that we precisely know the microscopic details such as how the nodes are interconnected with each other. Mapping network topologies can be costly, time consuming, inaccurate, and the resources they demand are often unaffordable, because the datasets comprised billions of nodes and hundreds of billions of links in large decentralized systems like Internet. This problem can be addressed by combining methods from statistical physics and random graph theory.

The following steps are used for analyzing large-scale networked systems [5]:

Step 1: To compute the aggregate statistics of interest. It includes the number of nodes, the density of links, the distribution of node degrees, diameter of a network, shortest path length between pair of nodes, correlations between neighboring nodes, clustering coefficient, connectedness, node centrality and node influence. In distributed systems, this can be computed and estimated efficiently by sensible sampling techniques.

Step 2: To determine the statistical entropy ensembles. This can be achieved by probability spaces by assigning probabilities to all possible network realizations that are consistent with the given aggregate statistics using analytical tools like computational statistics methods of metropolis sampling.

Step 3: Finally, derive the anticipated properties of a system based on aggregate statistics of its network topology.

3.3 Computing Models

Big data grows super-fast in four dimensions (4Vs: volume, variety, velocity and veracity) and needs advanced data structures, new models for extracting the details and novel algorithmic approach for computation. This section focuses on data structure for big data, feature engineering and computational algorithm.

3.3.1 Data Structures for Big Data

In big data, special data structures are required to handle huge dataset. Hash tables, train/atrain and tree-based structures like B trees and K-D trees are best suited for handling big data.

Hash table

Hash tables use hash function to compute the index and map keys to values. Probabilistic data structures play a vital role in approximate algorithm implementation in big data [6]. These data structures use hash functions to randomize the items and support set operations such as union and intersection and therefore can be easily parallelized. This section deals with four commonly used probabilistic data structures: membership query—Bloom filter, HyperLogLog, count–min sketch and MinHash [7].

Membership Query—Bloom filter

A Bloom filter proposed by Burton Howard Bloom in 1970 is a space-efficient probabilistic data structure that allows one to reduce the number of exact checks, which is used to test whether an element is 'a member' or 'not a member' of a set. Here, the query returns the probability with the result either 'may be in set' or 'definitely not in set.'

Bit vector is the base data structure for a Bloom filter. Each empty Bloom filter is a bit array of 'm' bits and is initially unset.

0	**0**	**0**	**0**	**0**	**0**	**0**	**0**	**0**	**0**	**0**	**0**	**0**
0	1	2	3	4	5	6	7	8	9	10	11	12

When an element is added to the filter, it is hashed by 'k' functions, h_1, h_2 … h_k mod by 'm', resulting in 'k' indices into the bit array, and the respective index is set to '1' [8]. Figure 3.5 shows how the array gets updated for the element with five distinct hashing functions h_1, h_2, h_3, h_4 and h_5.

To query the membership of an element, we hash the element again with the same hashing functions and check if each corresponding bit is set. If any one of them is zero, then conclude the element is not present.

Suppose you are generating an online account for a shopping Web site, and you are asked to enter a username during sign-up; as you entered, you will get an immediate response, 'Username already exists.' Bloom filter data structure can perform this task very quickly by searching from the millions of registered users.

Consider you want to add a username 'Smilie' into the dataset and five hash functions, h_1, h_2, h_3, h_4 and h_5 are applied on the string. First apply the hash function as follows

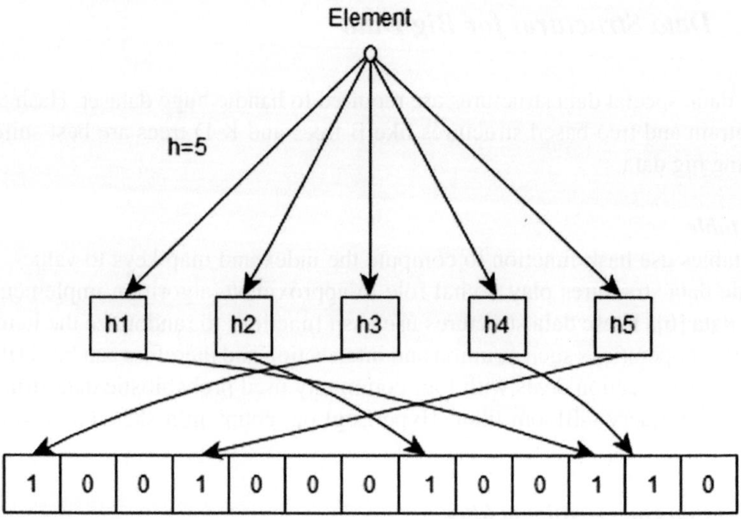

Fig. 3.5 Bloom filter hashing

h_1("Smilie")%13 = 10
h_2("Smilie")%13 = 4
h_3("Smilie")%13 = 0
h_4("Smilie")%13 = 11
h_5("Smilie")%13 = 6

Set the bits to 1 for the indices 10, 4, 0, 11 and 6 as given in Fig. 3.6.

Similarly, enter the next username 'Laughie' by applying the same hash functions.

h_1("Laughie")%13 = 3
h_2("Laughie")%13 = 5
h_3("Laughie")%13 = 8
h_4("Laughie")%13 = 10
h_5("Laughie")%13 = 12

Set the bits to 1 for the indices 3, 5, 8, 10 and 12 as given in Fig. 3.7.

Now check the availability of the username 'Smilie' is presented in filter or not. For performing this task, apply hashing using h_1, h_2, h_3, h_4 and h_5 functions on the string and check if all these indices are set to 1. If all the corresponding bits are set, then the string is 'probably present.' If any one of them indicates 0, then the string is 'definitely not present.'

Perhaps you will have a query, why this uncertainty of 'probably present', why not 'definitely present'? Let us consider another new username 'Bean.' Suppose we want to check whether 'Bean' is available or not. The result after applying the hash functions h_1, h_2, h_3, h_4 and h_5 is as follows

h_1("Bean")%13 = 6
h_2("Bean")%13 = 4

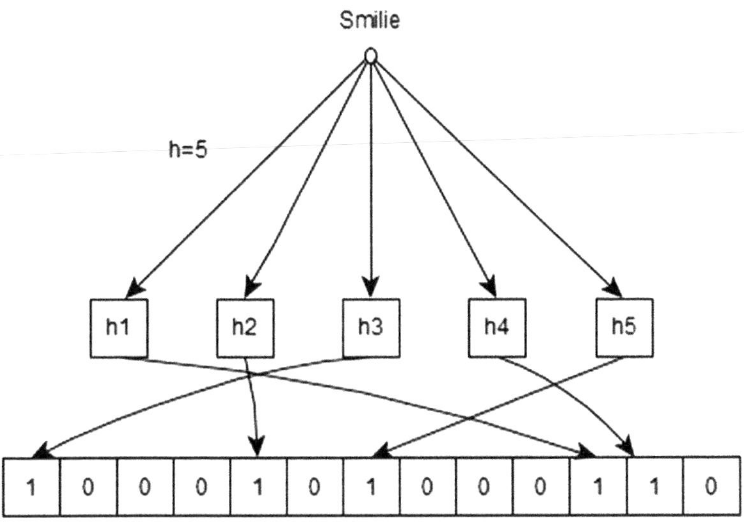

Fig. 3.6 Bloom filter after inserting a string 'Smilie'

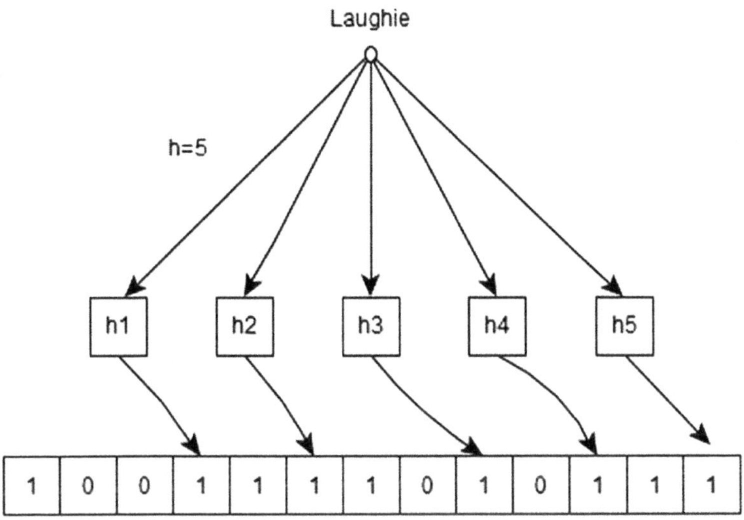

Fig. 3.7 Bloom filter after inserting a string 'Laughie'

h_3("Bean")%13 = 0
h_4("Bean")%13 = 11
h_5("Bean")%13 = 12

If we check the bit array after applying hash function to the string 'Bean,' bits at these indices are set to 1 but the string 'Bean' was never added earlier to the Bloom filter. As the indices are already set by some other elements, Bloom filter incorrectly claims that 'Bean' is present and thus will generate a false-positive result (Fig. 3.8).

We can diminish the probability of false-positive result by controlling the size of the Bloom filter.

- More size/space decrease false positives.
- More number of hash functions lesser false positives.

Consider a set element $A = \{a_1, a_2, \ldots, a_n\}$ of n elements. Bloom filter defines membership information with a bit vector 'V' of length 'm'. For this, 'k' hash functions, h_1, h_2, h_3 … h_k with h_i: $X\{1 \ldots m\}$, are used and the procedure is described below.

```
Procedure BloomFilter (a_i elements in set Arr, hash-functions h_j, integer m)
filter = Initialize m bits to 0
  foreach a_i in Arr:
    foreach hash-function h_j:
      filter[h_j(a_i)] = 1
    end foreach
  end foreach
return filter
```

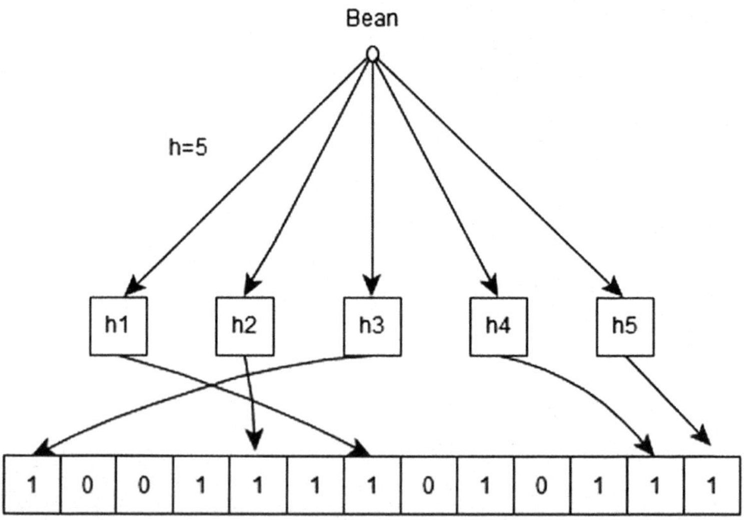

Fig. 3.8 Bloom filter for searching a string 'Bean'

Probability of false positivity 'P' can be calculated as:

$$P = \left(1 - \left[1 - \frac{1}{m}\right]^{kn}\right)^k$$

where

'm' is the size of bit array,
'k' is the number of hash functions and
'n' is the number of expected elements to be inserted.

Size of bit array 'm' can be calculated as [9]:

$$m = \frac{n \ln P}{(\ln 2)^2}$$

Optimum number of hash functions 'k' can be calculated as:

$$k = \frac{m}{n} \ln 2$$

where 'k' must be a positive integer.

Cardinality—HyperLogLog

HyperLogLog (HLL) is an extension of LogLog algorithm derived from Flajolet—Martin algorithm (1984). It is a probabilistic data structure used to estimate the cardinality of a dataset to solve the count-distinct problem, approximating the number of unique elements in a multiset. It required an amount of memory proportional to the cardinality for calculating the exact cardinality of a multiset, which is not practical for huge datasets. HLL algorithm uses significantly less memory at the cost of obtaining only an approximation of the cardinality [10]. As its name implies, HLL requires $O(\log_2\log_2 n)$ memory where n is the cardinality of the dataset.

HyperLogLog algorithm is used to estimate how many unique items are in a list. Suppose a web page has billions of users and we want to compute the number of unique visits to our web page. A naive approach would be to store each distinctive user id in a set, and then the size of the set would be considered by cardinality. When we are dealing with enormous volumes of datasets, counting cardinality by the said way will be ineffective because the dataset will occupy a lot of memory. But if we do not need the exact number of distinct visits, then we can use HLL as it was designed for estimating the count of billions of unique values.

Four main operations of HLL are:

1. Add a new element to the set.
2. Count for obtaining the cardinality of the set.
3. Merge for obtaining the union of two sets.
4. Cardinality of the intersection.

```
HyperLogLog [11]
  def add(cookie_id: String): Unit
  def cardinality( ):
    //|A|
  def merge(other: HyperLogLog):
    //|A ∪ B|
  def intersect(other: HyperLogLog):
    //|A ∩ B| = |A| + |B| − |A ∪ B|
```

Frequency—Count–Min sketch

The count–min sketch (CM sketch) is a probabilistic data structure which is proposed by G. Cormode and S. Muthukrishnan that assists as a frequency table of events/elements in a stream of data. CM sketch maps events to frequencies using hash functions, but unlike a hash table it uses only sublinear space to count frequencies due to collisions [12]. The actual sketch data structure is a matrix with w columns and d rows [13]. Each row is mapped with a hash function as in Fig. 3.9.

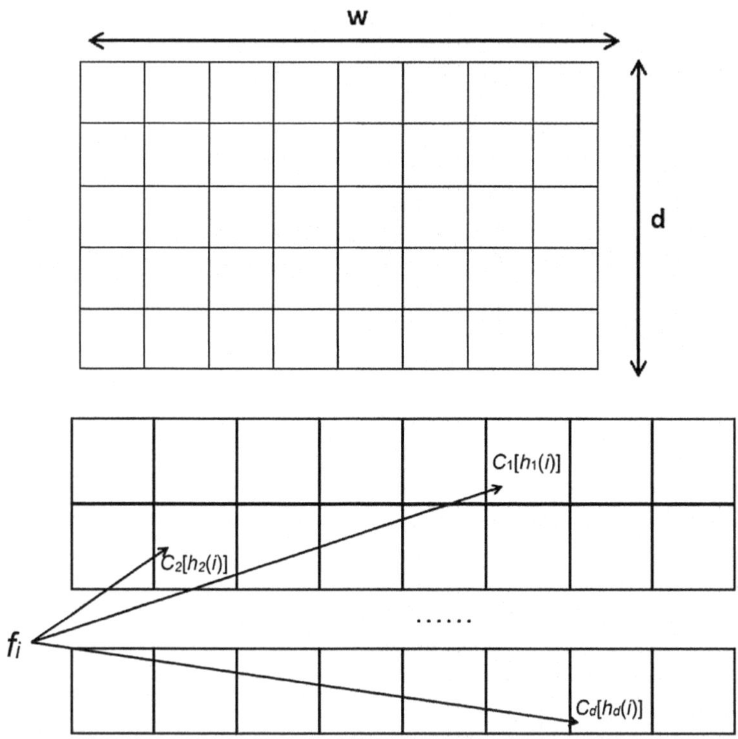

Fig. 3.9 Count–min data structure

Initially, set all the cell values to 0.

0	0	0	0	0	0	0	0
0	0	0	0	0	0	0	0
0	0	0	0	0	0	0	0
0	0	0	0	0	0	0	0
0	0	0	0	0	0	0	0

When an element arrives, it is hashed with the hash function for each row and the corresponding values in the cells (row d and column w) will be incremented by one.

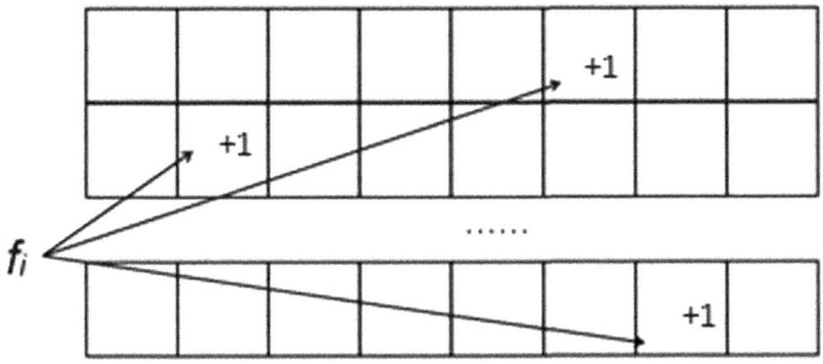

Let us assume elements arrive one after another and the hashes for the first element f_1 are: $h_1(f_1) = 6$, $h_2(f_1) = 2$, $h_3(f_1) = 5$, $h_4(f_1) = 1$ and $h_5(f_1) = 7$. The following table shows the matrix current state after incrementing the values.

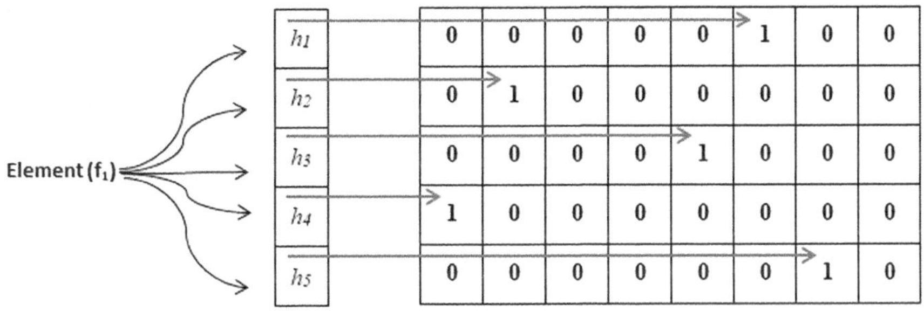

Let us continue to add the second element f_2, $h_1(f_2) = 5$, $h_2(f_2) = 1$, $h_3(f_2) = 5$, $h_4(f_2) = 8$ and $h_5(f_2) = 4$, and the table is altered as

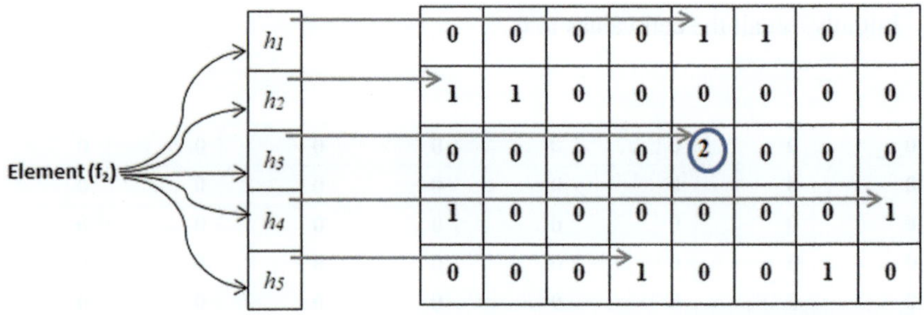

In our contrived example, almost element f_2 hashes map to distinct counters, with an exception being the collision of $h_3(f_1)$ and $h_3(f_2)$. Because of getting the same hash value, the fifth counter of h_3 now holds the value 2.

The CM sketch is used to solve the approximate Heavy Hitters (HH) problem [14]. The goal of HH problem is to find all elements that occur at least n/k times in the array. It has lots of applications.

1. Computing popular products,
2. Computing frequent search queries,
3. Identifying heavy TCP flows and
4. Identifying volatile stocks.

As hash functions are cheap to compute and produce, accessing, reading or writing the data structure is performed in constant time. The recent research in count–min log sketch which was proposed by G. Pitel and G. Fouquier (2015) essentially substitutes CM sketch linear registers with logarithmic ones to reduce the relative error and allow higher counts without necessary to increase the width of counter registers.

Similarity—MinHash

Similarity is a numerical measurement to check how two objects are alike. The essential steps for finding similarity are: (i) Shingling is a process of converting documents, emails, etc., to sets, (ii) min-hashing reflects the set similarity by converting large sets to short signature sets, and (iii) locality sensitive hashing focuses on pairs of signatures likely to be similar (Fig. 3.10).

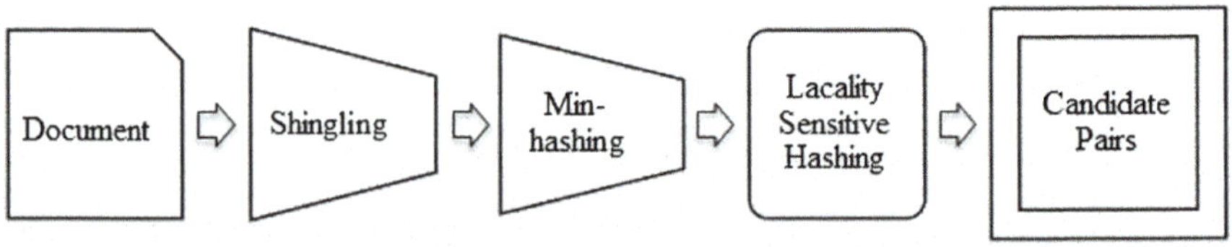

Fig. 3.10 Steps for finding similarity

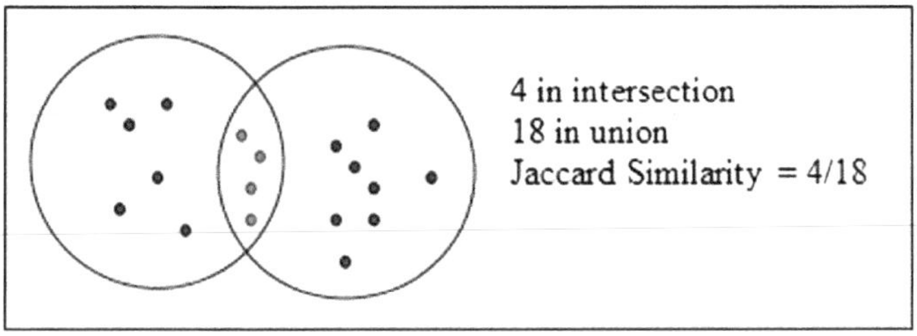

Fig. 3.11 Jaccard similarity

MinHash was invented by Andrei Broder (1997) and quickly estimates how similar two sets are. Initially, AltaVista search engine used MinHash scheme to detect and eliminate the duplicate web pages from the search results. It is also applied in association rule learning and clustering documents by similarity of their set of words. It is used to find pairs that are 'near duplicates' from a large number of text documents. The applications are to locate or approximate mirror web sites, plagiarism check, web spam detection, ordering of words, finding similar news articles from many news sites, etc.

MinHash provides a fast approximation to the Jaccard similarity [15]. The Jaccard similarity of two sets is the size of their intersection divided by size of their union (Fig. 3.11).

$$J(A, B) = \frac{|A \cap B|}{|A \cup B|}$$

If the sets are identical, then $J = 1$; if they do not share any member, then $J = 0$; if they are somewhere in between, then $0 \leq J \leq 1$.

MinHash uses hashing function to quickly estimate Jaccard similarities. Here, the hash function (h) maps the members of A and B to different integers. $h\text{min}(S)$ finds the member 'x' that results the lowest of a set S. It can be computed by passing every member of a set S through the hash function h. The concept is to condense the large sets of unique shingles into much smaller representations called 'signatures.' We can use these signatures alone to measure the similarity between documents. These signatures do not give the exact similarity, but the estimates they provide are close.

Consider the below input shingle matrix where column represents documents and rows represent shingles.

Documents

Shingles

1	1	0	1
2	0	1	1
3	1	0	0
4	1	0	1
5	0	1	0

Define a hash function h as by permuting the matrix rows randomly. Let perm1 = (12345), perm2 = (54321) and perm3 = (34512). The MinHash function $h\min(S)$ = the first row in the permuted order in which column C has '1'; i.e., find the index that the first '1' appears for the permuted order.

1	1	0	1
2	0	1	1
3	1	0	0
4	1	0	1
5	0	1	0

5	0	1	0
4	1	0	1
3	1	0	0
2	0	1	1
1	1	0	1

3	1	0	0
4	1	0	1
5	0	1	0
1	1	0	1
2	0	1	1

For the first permutation Perm1 (12345), first '1' appears in column 1, 2 and 1. For the second permutation Perm2 (54321), first '1' appears in column 2, 1 and 2. Similarly for the third permutation Perm3 (34512), first '1' appears in column 1, 3 and 2. The signature matrix [16] after applying hash function is as follows.

Perm1 = (12345)	1	2	1
perm2 = (54321)	2	1	2
perm3 = (34512)	1	3	2

The similarity of signature matrix can be calculated by the number of similar documents (s)/no. of documents (d). The similarity matrix is as follows.

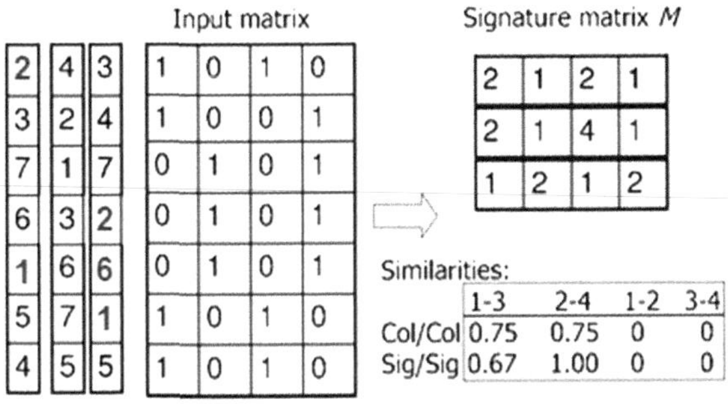

Fig. 3.12 Min-hashing

	1, 2	1, 3	2, 3
col/col $\left(\frac{\vert A\cap B\vert}{\vert A\cup B\vert}\right)$	0/5 = 0	2/4 = 0.5	1/4 = 0.25
sig/sig (s/d)	0/3 = 0	2/3 = 0.67	0/3 = 0

The other representation (the same sequence of permutation will be considered for calculation) of signature matrix for the given input matrix is as follows:

Perm1 = (12345)	1	2	1
perm2 = (54321)	4	5	4
perm3 = (34512)	3	5	4

Another way of representing signature matrix for the given input matrix is as follows. Consider 7 × 4 input matrix with three permutations after applying hash functions, perm1 = (3472615), perm2 = (4213675), perm3 = (2376154), and the corresponding 3 × 4 signature matrix is given in Fig. 3.12. Here, the signature matrix is formed by elements of the permutation based on first element to map a '1' value from the element sequence start at 1. First row of the signature matrix is formed from the first element 1 with row value (1 0 1 0). Second element 2 with row values (0 1 0 1) can replace the above with (1 2 1 2). The row gets completed if there is no more '0's. Second row of the signature matrix is formed from the first element (0 1 0 1) which will be further replaced as (2 1 0 1), and then third column '0' is updated with 4 since fourth element of the permutation is the first to map '1'.

With min-hashing, we can effectively solve the problem of *space complexity* by eliminating the sparseness and at the same time preserve the similarity.

Tree-based Data Structure

The purpose of a tree is to store naturally hierarchical information, such as a file system. B trees, M trees, R trees (R*, R+ and X tree), T trees, K-D trees, predicate trees, LSM trees and fractal tree are the different forms of trees to handle big data.

B trees are efficient data structure for storing big data and fast retrieval. It was proposed by Rudolf Bayer for maintaining large database. In a binary tree, each node has at most two children, and time complexity for performing any search operation is $O(\log_2 N)$. B tree is a variation of binary tree, which is a self-balancing tree, each node can have M children, where M is called fan-out or branching factor, and because of its large branching factor it is considered as one of the fastest data structures. It thus attains a time complexity of $O(\log_M N)$ for each search operation.

B tree is a one-dimensional index structure that does not work well and is not suitable for spatial data because search space is multidimensional. To resolve this issue, a dynamic data structure R tree was proposed by Antonin Guttman in 1982 for the spatial searching. Consider massive data that cannot fit in main memory. When the number of keys is high, the data is read in the form of blocks from the disk. So, disk access time is higher than main memory access time. The main motive of using B trees is to decrease the number of disk accesses by using a hierarchical index structure. Internal nodes may join and split whenever a node is inserted or deleted because range is fixed. This may require rebalancing of tree after insertion and deletion.

The order in B tree is defined as the maximum number of children for each node. A B tree of order 'n' (Fig. 3.13) has the following properties [17]:

1. A B tree is defined by minimum degree 'n' that depends upon the disk block size.
2. Every node has maximum 'n' and minimum '$n/2$' children.
3. A non-leaf node has 'n' children, and it contains '$n - 1$' keys.
4. All leaves are at the same level.
5. All keys of a node are sorted in increasing order. The children between two keys '$k1$' and '$k2$' contain the keys in the range from '$k1$' and '$k2$'.
6. Time complexity to search, insert and delete in a B tree is $O(\log_M N)$.

K-D Trees

A K-D tree or K-dimensional tree was invented by Jon Bentley in 1970 and is a binary search tree data structure for organizing some number of points in a 'K'-dimensional space. They are very useful for performing range search and nearest

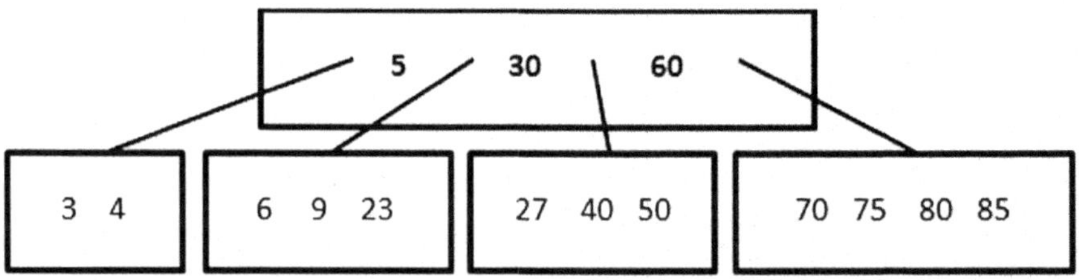

Fig. 3.13 A B tree of order 5

neighbor search. K-D trees have several applications, including classifying astronomical objects, computer animation, speedup neural networks, data mining and image retrieval.

The algorithms to insert and search are same as BST with an exception at the root we use the x-coordinate. If the point to be inserted has a smaller x-coordinate value than the root, go left; otherwise go right. At the next level, we use the y-coordinate, and then at the next level we use the x-coordinate, and so forth [18].

Let us consider the root has an x-aligned plane, then all its children would have y-aligned planes, all its grandchildren would have x-aligned planes, all its great-grandchildren would have y-aligned planes and the sequence alternatively continues like this. For example, insert the points (5, 7), (17, 15), (13, 16), (6, 12), (9, 1), (2, 8) and (10, 19) in an empty K-D tree, where $K = 2$. The process of insertion is as follows:

- Insert (5, 7): Initially, as the tree is empty, make (5, 7) as the root node and X-aligned.

- Insert (17, 15): During insertion firstly compare the new node point with the root node point. Since root node is X-aligned, the X-coordinate value will be used for comparison for determining the new node to be inserted as left subtree or right subtree. If X-coordinate value of the new point is less than X-coordinate value of the root node point, then insert the new node as a left subtree else insert it as a right subtree. Here, (17, 15) is greater than (5, 7), this will be inserted as the right subtree of (5, 7) and is Y-aligned.

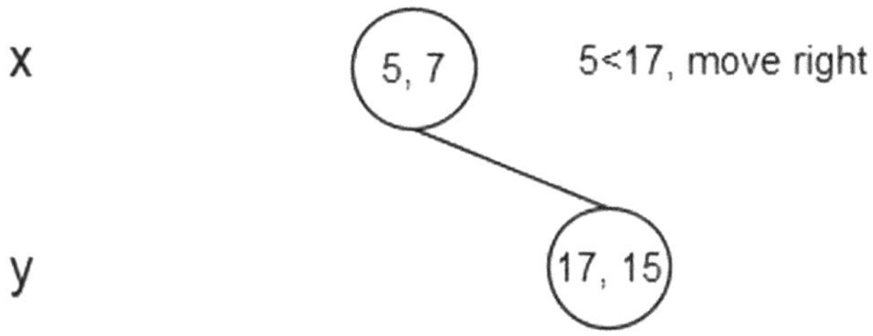

- Insert (13, 16): X-coordinate value of this point is greater than X-coordinate value of root node point. So, this will lie in the right subtree of (5, 7). Then, compare Y-coordinate value of this point with (17, 15). As Y-coordinate value is greater than (17, 15), insert it as a right subtree.

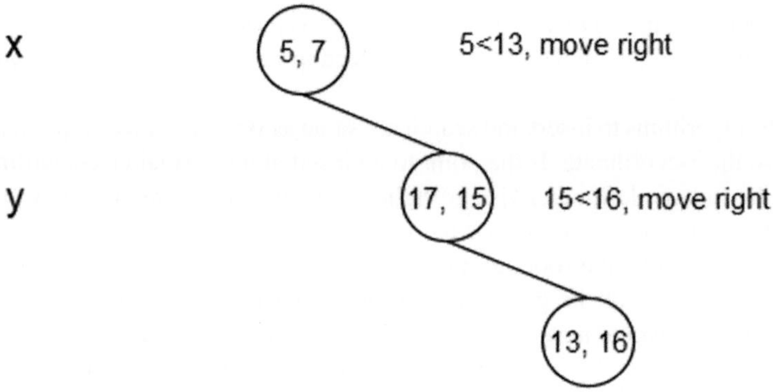

- Similarly insert (6, 12).

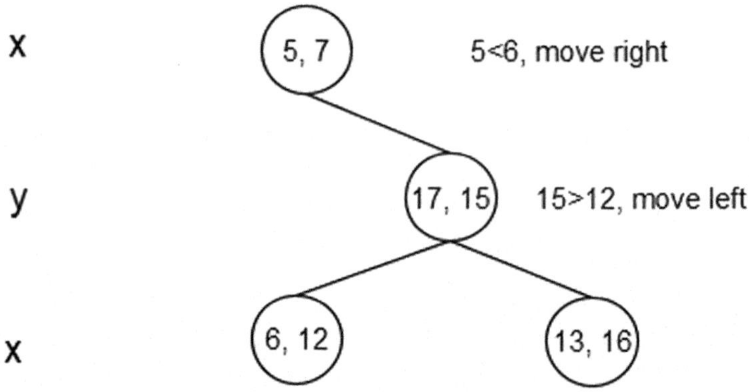

- Insert other points (9, 1), (2, 8) and (10, 19).

The status of 2-D tree after inserting elements (5, 7), (17, 15), (13, 16), (6, 12), (9, 1), (2, 8) and (10, 19) is given in Fig. 3.14 and the corresponding plotted graph is shown in (Fig. 3.15).

Algorithm for insertion

```
Insert (Keypoint key, KDTreeNode t, int level)
{
  typekey key[];
  if (t = = null)
    t = new KDTreeNode (key)
  else if (key = = t.data)
    Error // Duplicate, already exist
  else if (key[level] < t.data[level])
    t.left = insert (key, t.left, (level + 1) % D)
  else
```

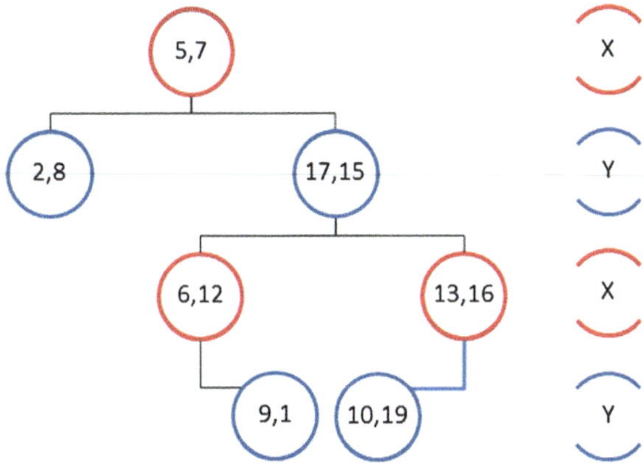

Fig. 3.14 2-D tree after insertion

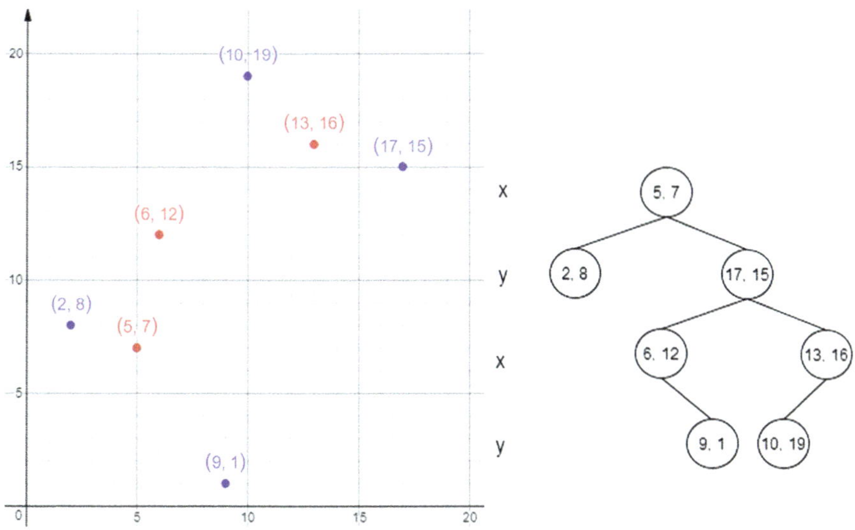

Fig. 3.15 K-D tree with a plotted graph

```
        t.right = insert (key, t.right, (level + 1) % D)
        //D-Dimension;
    return t
}
```

The process of deletion is as follows: If a target node (node to be deleted) is a leaf node, simply delete it. If a target node has a right child as not NULL, then find the minimum of current node's dimension (*X* or *Y*) in right subtree, replace the node with minimum point, and delete the target node. Else if a target node has left child as not NULL, then find minimum of current node's dimension in left subtree, replace

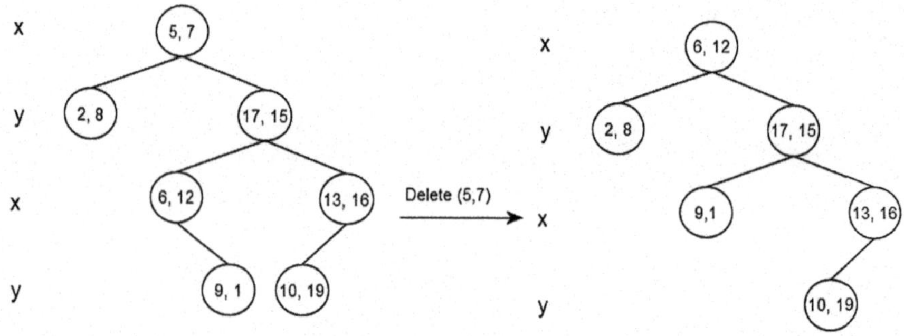

Fig. 3.16 K-D tree after deletion of *X*-coordinate point

the node with minimum point, delete the target node, and then make the new left subtree as right child of current node [19].

- Delete (5, 7): Since right child is not NULL and dimension of node is *x*, we find the node with minimum *x* value in right subtree. The node (6, 12) has the minimum *x* value, and we replace (5, 7) with (6, 12) and then delete (5, 7) (Fig. 3.16).
- Delete (17, 15): Since right child is not NULL and dimension of node is *y*, we find the node with minimum *y* value in right subtree. The node (13, 16) has a minimum *y* value, and we replace (17, 15) with (13, 16) and delete (17, 15) (Fig. 3.17).
- Delete (17, 15)—no right subtree: Since right child is NULL and dimension of node is *y*, we find the node with minimum *y* value in left subtree. The node (9, 1) has a minimum *y* value, and we replace (17, 15) with (9, 1) and delete (17, 15). Finally, we have to modify the tree by making new left subtree as right subtree of (9, 1) (Fig. 3.18).

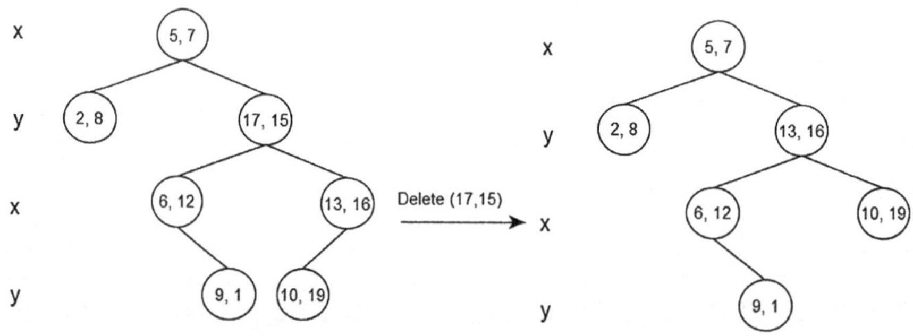

Fig. 3.17 K-D tree after deletion of *Y*-coordinate point

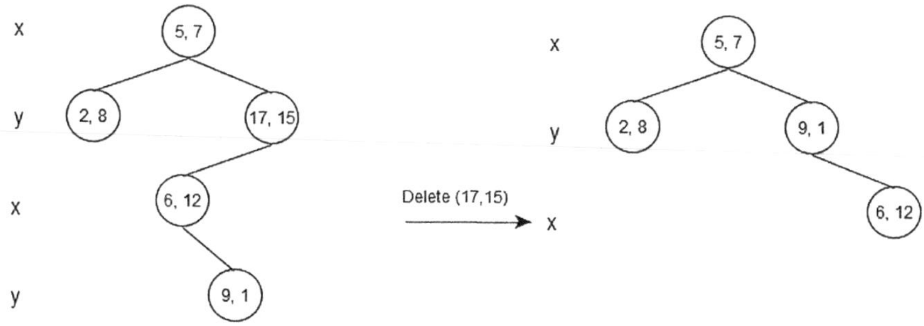

Fig. 3.18 K-D tree after deletion of *Y*-coordinate point with NULL right tree

Algorithm for finding a minimum

```
Keypoint findmin (KDTreeNode t, int d, int level)
{
        if (t = = NULL)  // empty tree
                return NULL
// t splits on a same dimension; search only in left subtree
        if (level = = d)
                if (t.left = = NULL)
                        return t.data
                else
                        return findmin (t.left, d, (level+1)%D)
// t splits on a different dimension; search both subtrees
        else
                return minimum (t.data,
                                findmin (t.left, d, (level+1)%D),
                                findmin (t.right, d, (level+1)%D))
}
```

Algorithm for deletion:

```
Keypoint delete (Keypoint key, KDTreeNode t, int level)
{
        if (t = = NULL)
                //element not found!
        next_level= (level+1)%D
        if( key = = t.data)
                // find min(level) from right subtree
                if (t.right != NULL)
                        t.data = findmin (t.right, level, next_level)
                        t.right = delete (t.data, t.right, next_level)
                // swap subtrees and use min(level) from new right:
                else if (t.left != NULL)
                        t.data = findmin (t.left, level, next_level)
                        t.right = delete (t.data, t.left, next_level)
                else
                        t = = null // leaf node, just remove it
                // search for the point
        else if( key[level] < t.data[level])
                t.left = delete (key, t.left, next_level)
        else
                t.right = delete (key, t.right, next_level)
        return t
}
```

K-D trees are useful for performing nearest neighbor (NN) search and range search. The NN search is used to locate the point in the tree that is closest to a given input point. We can get *k*-nearest neighbors, *k* approximate nearest neighbors, all neighbors within specified radius and all neighbors within a box. This search can be done efficiently by speedily eliminating large portions of the search space. A range search finds the points lie within the range of parameters.

Train and Atrian

Big data in most of the cases deals with heterogeneity types of data including structured, semi-structured and unstructured data. 'r-train' for handling homogeneous data structure and 'r-atrain' for handling heterogeneous data structure have been introduced exclusively for dealing large volume of complex data. Both the data structures 'r-train' ('train,' in short) and 'r-atrain' ('atrain,' in short), where *r* is a natural number, are new types of robust dynamic data structures which can store big data in an efficient and flexible way (Biswas [20]).

3.3.2 *Feature Engineering for Structured Data*

Feature engineering is a subset of the data processing component, where necessary features are analyzed and selected. It is vital and challenging to extract and select the right data for analysis from the huge dataset.

3.3.2.1 Feature Construction

Feature construction is a process that finds missing information about the associations between features and expanding the feature space by generating additional features that is useful for prediction and clustering. It involves automatic transformation of a given set of original input features to create a new set of powerful features by revealing the hidden patterns and that helps better achievement of improvements in accuracy and comprehensibility.

3.3.2.2 Feature Extraction

Feature extraction uses functional mapping to extract a set of new features from existing features. It is a process of transforming the original features into a lower-dimensional space. The ultimate goal of feature extraction process is to find a least set of new features through some transformation based on performance measures. Several algorithms exist for feature extraction. A feedforward neural network approach and principal component analysis (PCA) algorithms play a vital role in feature extraction by replacing original 'n' attributes by other set of 'm' new features.

3.3.2.3 Feature Selection

Feature selection is a data preprocessing step that selects a subset of features from the existing original features without a transformation for classification and data mining tasks. It is a process of choosing a subset of 'm' features from the original set of 'n' features, where $m \leq n$. The role of feature selection is to optimize the predictive accuracy and speed up the process of learning algorithm results by reducing the feature space.

Algorithms for feature selection:

1. Exhaustive and Complete Approaches

Branch and Bound (BB): This technique of selection guaranteed to find and give the optimal feature subset without checking all possible subsets. Branching is the construction process of tree, and bounding is the process of finding optimal feature set by traversing the constructed tree [21]. First start from the full set of original features, and then remove features using depth-first strategy. Three features reduced to two features as depicted in Fig. 3.19.

For each tree level, a limited number of subtrees are generated by deleting one feature from the set of features from the parent node (Fig. 3.20).

2. Heuristic Approaches

In order to select a subset of available features by removing unnecessary features to the categorization task novel heuristic algorithms such as sequential forward selection, sequential backward search and their hybrid algorithms are used.

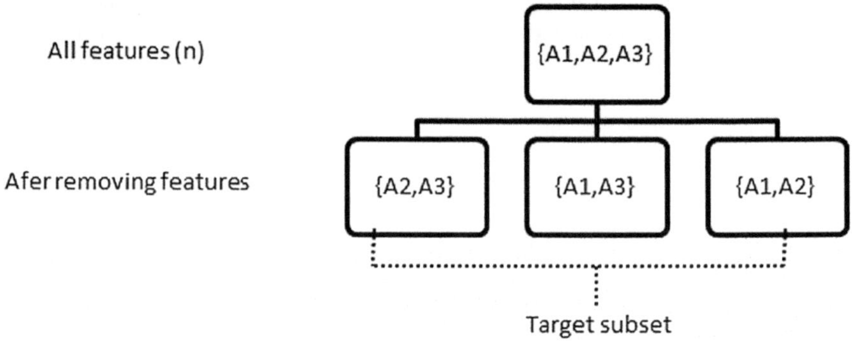

Fig. 3.19 Subtree generation using BB

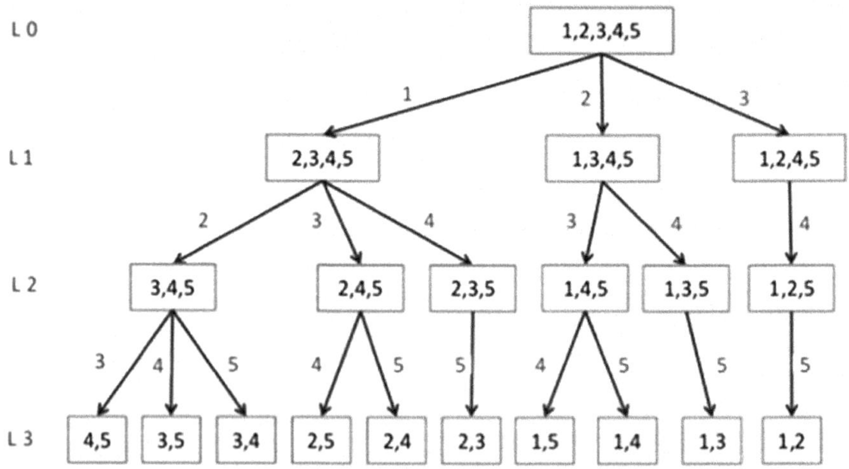

No. of levels = n-r = 5-2 = 3

No. of leaf nodes = $nC_r = 5C_3 = 10$

Fig. 3.20 Reduce five features into two features using BB

Sequential Forward Search (SFS): Sequential forward method started with an empty set and gradually increased by adding one best feature at a time. Starting from the empty set, it sequentially adds the new feature x^+ that maximizes $(Y_k + x^+)$ when combined with the existing features Y_k.

Steps for SFS:

1. Start with an empty set $Y_0 = \{\emptyset\}$.
2. Select the next best feature $x^+ = \arg x \notin Y_k \max J(Y_k + x)$.
3. Update $Y_{k+1} = Y_k + x^+; k = k + 1$.
4. Go to step 2.

Sequential Backward Search (SBS): SBS is initiated with a full set and gradually reduced by removing one worst feature at a time. Starting from the full set, it sequentially removes the unwanted feature x^- that least reduces the value of the objective function $(Y - x^-)$.

Steps for SBS:

1. Start with the full set $Y_0 = X$.
2. Remove the worst feature $x^- = \arg x \in Y_k \max J(Y_k - x)$.
3. Update $Y_{k+1} = Y_k - x^- = Y_k - x^-; k = k + 1$.
4. Go to step 2.

Bidirectional Search (BDS): BDS is a parallel implementation of SFS and SBS. SFS is executed from the empty set, whereas SBS is executed from the full set. In BDS, features already selected by SFS are not removed by SBS as well as features already removed by SBS are not selected by SFS to guarantee the SFS and SBS converge to the same solution.

3. Non-deterministic Approaches

In this stochastic approach, features are not sequentially added or removed from a subset. These allow search to follow feature subsets that are randomly generated. Genetics algorithms and simulated annealing are two often-mentioned methods. Other stochastic algorithms are Las Vegas Filter (LVF) and Las Vegas Wrapper (LVW). LVF is a random procedure to generate random subsets and evaluation procedure that checks that is each subset satisfies the chosen measure. One of the parameters here is an inconsistency rate.

4. Instance-based Approaches

ReliefF is a multivariate or instance-based method that chooses the features that are the most distinct among other classes. ReliefF ranks and selects top-scoring features for feature selection by calculating a feature score for each feature. ReliefF feature scoring is based on the identification of feature value differences between nearest neighbor instance pairs.

3.3.2.4 Feature Learning

Representation learning or feature learning is a set of techniques that automatically transform the input data to the representations needed for feature detection or classification [22]. This removes the cumbersome of manual feature engineering by allowing a machine to both learn and use the features to perform specific machine learning tasks.

It can be either supervised or unsupervised feature learning. Supervised learning features are learned with labeled input data and are the process of predicting an output variable (Y) from input variables (X) using suitable algorithm to learn the mapping function from the input to the output; the examples include supervised neural networks, multilayer perceptron and supervised dictionary learning. In unsupervised feature learning, features are learned with unlabeled input data and are used to find hidden structure in the data; the examples include dictionary learning, independent component analysis, autoencoders, matrix factorization and other clustering forms.

3.3.2.5 Ensemble Learning

Ensemble learning is a machine learning model where the same problem can be solved by training multiple learners [23]. Ensemble learning model tries to build a set of hypotheses and combine them to use. Ensemble learning algorithms are general methods that enhance the accuracy of predictive or classification models.

Ensemble learning techniques

Bagging: It gets its name because it combines bootstrapping and aggregation to form an ensemble model. Bagging implements similar learners on small sample populations by taking an average of all predictions. In generalized bagging, different learners can be used on a different population. This helps us to reduce the variance error. The process of bagging is depicted in Fig. 3.21.

Random forest model is a good example of bagging. Random forest models decide where to split based on a random selection of features rather than splitting the same features at each node throughout. This level of differentiation gives a better ensemble aggregation producing a more accurate predictor results. Final prediction by aggregating the results from different tress is depicted in Fig. 3.22.

Boosting: It is an iterative technique which adjusts the weight of an observation in each iteration based on the previous last classification. That is, it tries to increase/decrease the weight of the observation if it was wrongly or imperfectly classified. Boosting in general aims to decrease the bias error and builds strong predictive models.

Stacking: It is a combining model which combines output from different learners [24]. This decreases bias or variance error depending on merging the learners (Fig. 3.23).

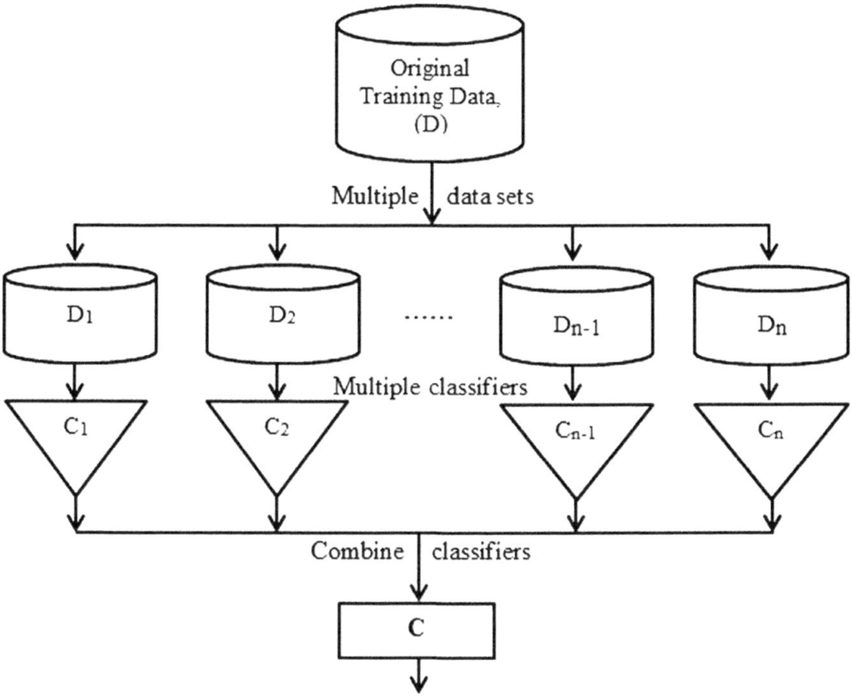

Fig. 3.21 Bagging

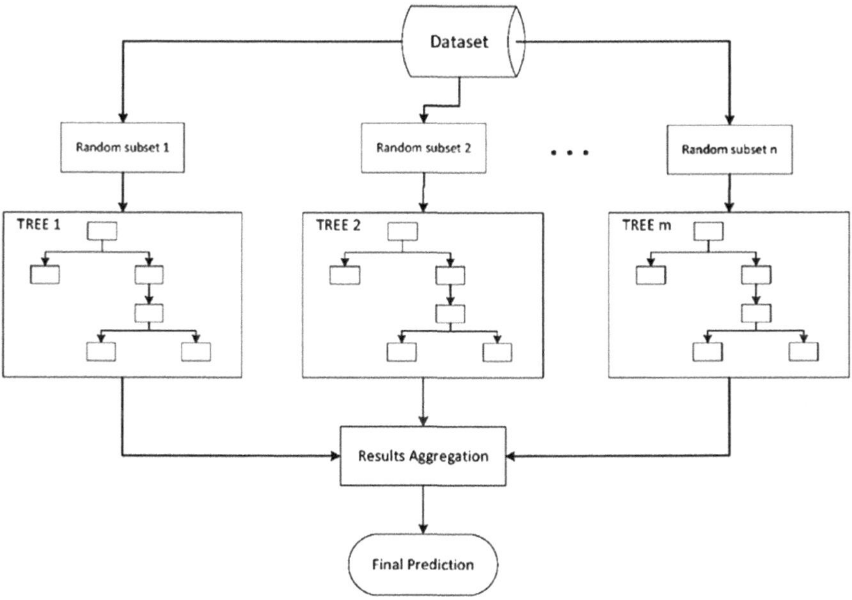

Fig. 3.22 Bagging—random forest [28]

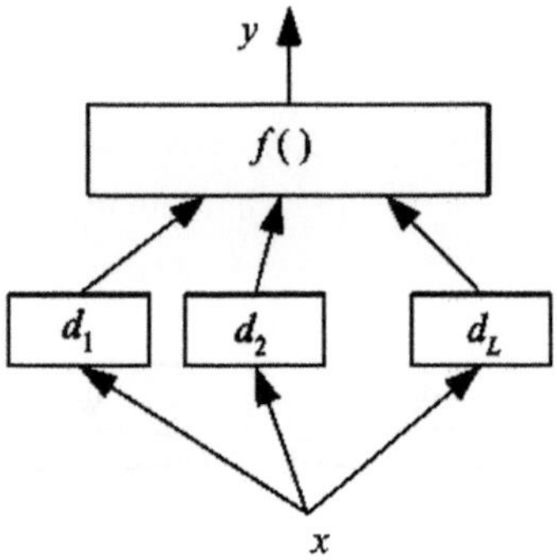

Fig. 3.23 Stacking

3.3.3 Computational Algorithm

Businesses are increasingly relying on the analysis of their massive amount of data to predict consumer response and recommend products to their customers. However, to analyze such enormous data a number of algorithms have been built for data scientists to create analytic platforms. Classification, regression and similarity matching are the fundamental principles on which many of the algorithms are used in applied data science to address big data issues.

There are a number of algorithms that exist; however, the most commonly used algorithms [25] are

- *K*-means clustering,
- Association rule mining,
- Linear regression,
- Logistic regression,
- C4.5,
- Support vector machine (SVM),
- Apriori,
- Expectation–maximization (EM),
- AdaBoost and
- Naïve Bayesian.

3.3.4 Programming Models

Programming model is an abstraction over existing machinery or infrastructure. It is a model of computation to write computer programs effectively and efficiently over distributed file systems using big data and easy to cope up with all the potential issues using a set of abstract runtime libraries and programming languages.

Necessary requirements for big data programming models include the following.

1. Support big data operations.
   - Split volumes of data.
   - Access data fast.
   - Distribute computations to nodes.
   - Combine when done.
2. Handle fault tolerance.
   - Replicate data partitions.
   - Recover files when needed.
3. Enable scale-out by adding more racks.
4. Optimized for specific data types such as document, graph, table, key values, streams and multimedia.

MapReduce, message passing, directed acyclic graph, workflow, bulk synchronous parallel and SQL-like are the standard programming models for analyzing large dataset.

3.3.5 Parallel Programming

Data analysis with parallel programming is used for analyzing data using parallel processes that run concurrently on multiple computers. Here, the entire dataset is divided into smaller chunks and sent to workers, where a similar parallel computation takes place on those chunks of data, the results are further accrued back, and finally the result is computed. Map and Reduce applications are generally linearly scalable to thousands of nodes because Map and Reduce functions are designed to facilitate parallelism [26].

Map and Reduce

The Map and Reduce is a new parallel processing framework with two main functions, namely (i) Map function and (ii) Reduce function in functional programming. In the Map Phase, the Map function '*f*' is applied to every data 'chunk' that produces an intermediate key–value pair. All intermediate pairs are then grouped based on a common intermediate key and passed into the Reduce function. In the Reduce Phase, the Reduce function '*g*' is applied once to all values with the same key. Hadoop is its open-source implementation on a single computing node or on cluster of nodes. The structure of Map and Reduce is given in Fig. 3.24.

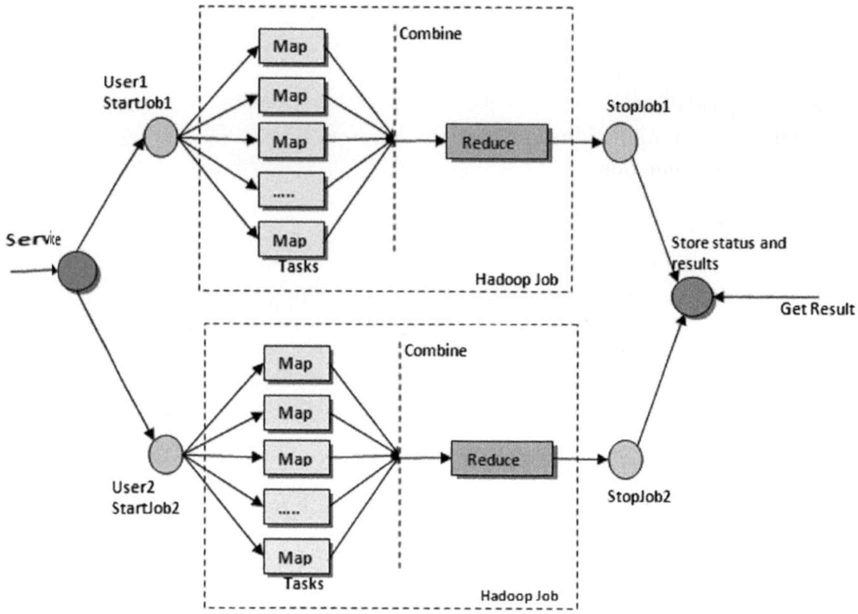

Fig. 3.24 Map and Reduce structure [29]

3.3.6 Functional Programming

In functional programming, the computation is considered as an elegance of building a structure and elements including programming interfaces and calculation of functions that applied on input data sources [27]. Here, the programming provides framework for declarations or expressions and functional inputs, instead of writing computing statements. Functional programs are simpler to write, understand, maintain and also easy to debug. Compared to OOP, functional programs are more compact and in-built for data-driven applications. Spark, Flink and SQL-like are the example framework for functional programming.

3.3.7 Distributed Programming

Analytics using big data technologies enable to make better human- and machine-level decisions in a fast and predictable way. Distributed computing principles are the keys to big data technologies and analytics that associated with predictive modeling, storage, access, transfer and visualization of data. Hadoop supports distributed computing using MapReduce that uses HDFS—Hadoop Distributed File System that is aimed for job distribution, balancing, recovery, scheduler, etc.

3.4 Conclusion

The art of modeling involves selecting the right datasets, variables, algorithms and appropriate techniques to format data for solving any kind of business problems in an efficient way. An analytical model estimates or classifies data values by essentially drawing a line of conclusion through data points. When the model is applied to new data, it predicts outcomes based on historical patterns. The applications of Big Data Analytics assist all analytics professionals, statisticians and data scientists to analyze dynamic and large volumes of data that are often left untapped by conventional business intelligence and analytics programs. Use of Big Data Analytics can produce big value to business by reducing complex datasets to actionable business intelligence by making more accurate business decisions.

3.5 Review Questions

1. Compare and contrast four types of analytics.
2. Discuss the biggest dataset you have worked with, in terms of training set size, algorithm implemented in production mode to process billions of transactions.
3. Explain any two data structures to handle big data with example.
4. Differentiate count–min sketch and MinHash data structures.
5. Why feature selection is important in data analysis?
6. Suppose you are using a bagging-based algorithm say a random forest in model building. Describe its process of generating hundreds of trees (say $T1$, $T2$ ... Tn) and aggregating the results.
7. Illustrate with example any two ensemble learning techniques.

References

1. H. Kalechofsky, A Little Data Science Business Guide (2016). http://www.msquared.com/wp-content/uploads/2017/01/A-Simple-Framework-for-Building-Predictive-Models.pdf
2. T. Maydon, The Four types of Data Analytics (2017). https://www.kdnuggets.com/2017/07/4-types-data-analytics.html
3. T. Vlamis, The Four Realms of Analytics (2015). http://www.vlamis.com/blog/2015/6/4/the-four-realms-of-analytics.html
4. Dezire, Types of Analytics: descriptive, predictive, prescriptive analytics (2016). https://www.dezyre.com/article/types-of-analytics-descriptive-predictive-prescriptive-analytics/209
5. I. Scholtes, Understanding Complex Systems: When Big Data meets Network Science. Information Technology, de Gruyter Oldenbourg (2015). https://pdfs.semanticscholar.org/cb41/248ad7a30d8ff1ddacb3726d7ef067a8d5db.pdf
6. Y. Niu, Introduction to Probabilistic Data Structures (2015). https://dzone.com/articles/introduction-probabilistic-0
7. C. Low, Big Data 101: Intro to Probabilistic Data Structures (2017). http://dataconomy.com/2017/04/big-data-101-data-structures/

8. T. Treat, Probabilistic algorithms for fun and pseudorandom profit (2015). https://bravenewgeek.com/tag/hyperloglog/
9. A.S. Hassan, Probabilistic Data structures: Bloom filter (2017). https://hackernoon.com/probabilistic-data-structures-bloom-filter-5374112a7832
10. S. Kruse et al., Fast Approximate Discovery of Inclusion Dependencies. Conference: Conference on Database Systems for Business, Technology, and Web at: Stuttgart, Germany. Lecture Notes in Informatics (LNI), pp. 207–226 (2017). https://www.researchgate.net/publication/314216122_Fast_Approximate_Discovery_of_Inclusion_Dependencies/figures?lo=1
11. B. Trofimoff, Audience Counting (2015). https://www.slideshare.net/b0ris_1/audience-counting-at-scale
12. I. Haber, Count Min Sketch: The Art and Science of Estimating Stuff (2016). https://redislabs.com/blog/count-min-sketch-the-art-and-science-of-estimating-stuff/
13. J. Lu, Data Sketches (2016). https://www.cs.helsinki.fi/u/jilu/paper/Course5.pdf
14. T. Roughgarden, G. Valiant, CS168: The Modern Algorithmic Toolbox Lecture #2: Approximate Heavy Hitters and the Count-Min Sketch (2015). http://theory.stanford.edu/~tim/s15/l/l2.pdf
15. A. Rajaraman, Near Neighbor Search in High Dimensional Data (nd). https://web.stanford.edu/class/cs345a/slides/04-highdim.pdf
16. R. Motwani, J. Ullman, Finding Near Duplicates (nd). https://web.stanford.edu/class/cs276b/handouts/minhash.pdf
17. Online: https://www.geeksforgeeks.org/b-tree-set-1-introduction-2/
18. Online: https://www.cs.cmu.edu/~ckingsf/bioinfo-lectures/kdtrees.pdf
19. Online: https://www.geeksforgeeks.org/k-dimensional-tree-set-3-delete/
20. R. Biswas, Processing of heterogeneous big data in an atrain distributed system (ADS) using the heterogeneous data structure r-Atrain. Int. J. Comput. Optim. **1**(1), 17–45 (2014). http://www.m-hikari.com/ijco/ijco2014/ijco1-4-2014/biswasIJCO1-4-2014.pdf
21. P. Rajapaksha, Analysis of Feature Selection Algorithms (2014). https://www.slideshare.net/parindarajapaksha/analysis-of-feature-selection-algorithms
22. Wikipedia, Feature Learning (2018). https://en.wikipedia.org/wiki/Feature_learning
23. Z.-H. Zhou, Ensemble Learning (nd). https://cs.nju.edu.cn/zhouzh/zhouzh.files/publication/springerEBR09.pdf
24. T. Srivastava, Basics of Ensemble Learning Explained in Simple English (2015). https://www.analyticsvidhya.com/blog/2015/08/introduction-ensemble-learning/
25. Datafloq, 3 Data Science Methods and 10 Algorithms for Big Data Experts (nd). https://datafloq.com/read/data-science-methods-and-algorithms-for-big-data/2500
26. L. Belcastro, F. Marazzo, Programming models and systems for Big Data analysis (2017). https://doi.org/10.1080/17445760.2017.1422501. https://www.tandfonline.com/doi/abs/10.1080/17445760.2017.1422501
27. D. Wu, S. Sakr, L. Zhu, Big Data Programming Models (2017). https://www.springer.com/cda/content/document/cda_downloaddocument/9783319493398-c2.pdf%3FSGWID%3D0-0-45-1603687-p180421399+&cd=1&hl=en&ct=clnk&gl=in
28. E. Lutins, Ensemble Methods in Machine Learning: What are They and Why Use Them? (2017). Available in: https://towardsdatascience.com/ensemble-methods-in-machine-learning-what-are-they-and-why-use-them-68ec3f9fef5f
29. Wikispace, Map-Reduce. Cloud Computing—An Overview (nd). http://map-reduce.wikispaces.asu.edu/

Chapter 4
Big Data Tools—Hadoop Ecosystem, Spark and NoSQL Databases

4.1 Introduction

In Chap. 1, we have surveyed in brief the total overview for Big Data and Hadoop.

In this chapter, we will delve deep into some of the modules of the Hadoop ecosystem, in the following sections.

4.1.1 Hadoop Ecosystem

The Hadoop Ecosystem comprises:

(1) [1]**HDFS** (Hadoop Distributed File System) which simply stores data files as close to the original format as possible.
(2) **HBase** is a Hadoop database management system and compares well with RDBMS. It supports structured data storage for large tables.
(3) **Hive** enables analysis of large data with a language similar to SQL, thus enabling SQL-type processing of data in a Hadoop cluster.
(4) **Pig** is an easy-to-understand data flow language, helpful in analyzing Hadoop-based data. Pig scripts are automatically converted to MapReduce jobs by the Pig Interpreter, thus enabling SQL-type processing of Hadoop data.
(5) **ZooKeeper** is a coordinator service for distributed applications.
(6) **Oozie** is a workflow schedule system to manage Apache Hadoop Jobs.
(7) **Mahout** is a scalable machine learning and data mining library.
(8) **Chukwa** is a data collection system for managing large distributed systems.
(9) **Sqoop** is used to transfer bulk data between Hadoop and as structured data management systems such as relational databases.
(10) **Ambari** web-based tool for provisioning, managing and monitoring Apache Hadoop clusters (Fig. 4.1).

[1] stackoverflaw.com.

C. S. R. Prabhu et al., *Big Data Analytics: Systems, Algorithms, Applications*,
https://doi.org/10.1007/978-981-15-0094-7_4

Data Management	Data Access	Data Processing	Data Storage
Oozie (Workflow Monitoring)	Hive (SQL)	MapReduce (Cluster Management)	HDFS (Distributed File System)
Chukwa (Monitoring)	Pig (Data Flow)	Yarn (Cluster & Resource Management)	HBase (Column DB Storage)
Flume (Monitoring)	Mahout (Machine Learning)		
Zookeeper (Management)	Avio (RPC Serialization)		
	Sqoop (RDEMS Connector)		

Fig. 4.1 Hadoop ecosystem elements at various stages of data processing

4.1.2 HDFS Commands [1]

The Hadoop Distributed File System is a distributed file system designed for storing and managing huge volume of data residing on commodity hardware. It is a scalable, effective and fault tolerant.

HDFS was developed using distributed file system approach. It holds very large amount of data files, and these files are stored across multiple machines.

Commands:

1. **Starting HDFS**

The following command will start the name node as well as the data nodes as cluster.

$ start-all.sh

2. **Checking the Nodes**

The following command checks whether name node, data node, Task Tracker, job tracker, etc., are working or not.

$ jps

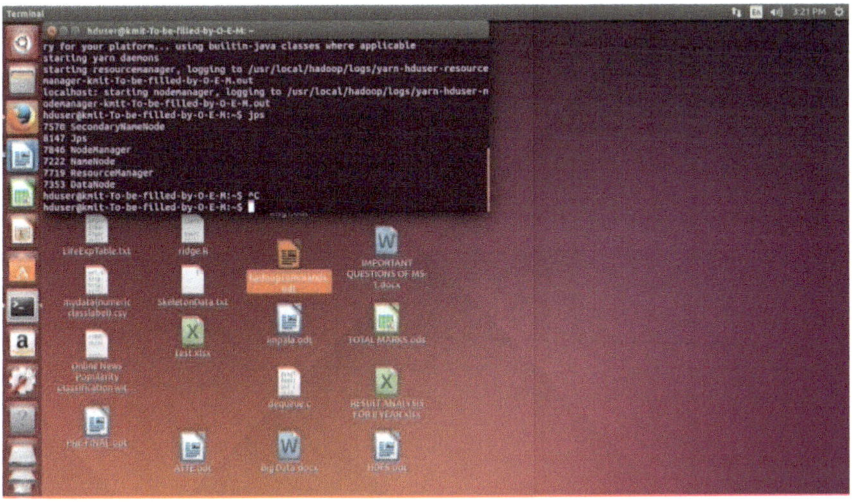

3. **Listing Files in HDFS**

The following command lists the files in a directory. We can find the list of files in a directory, status of a file, using 'ls'.

$ hdfs dfs -ls /

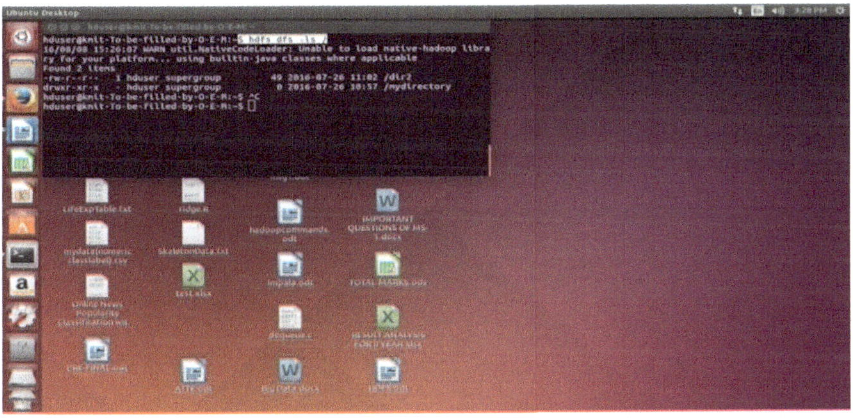

Given below is the syntax of ls that you can pass to a directory or a filename as an argument.

$ hdfs dfs -ls /dir2

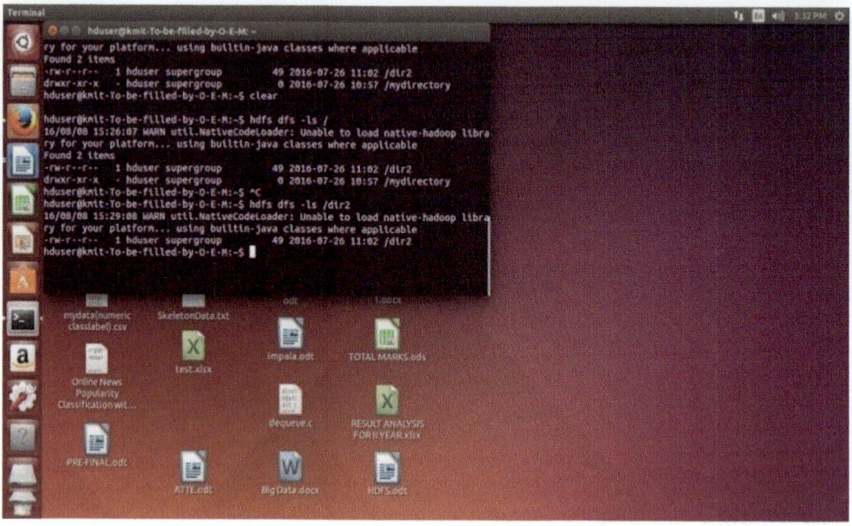

4. **Create a Directory**

The following command is used for creating a directory under HDFS

$ hdfs dfs -mkdir /testdir

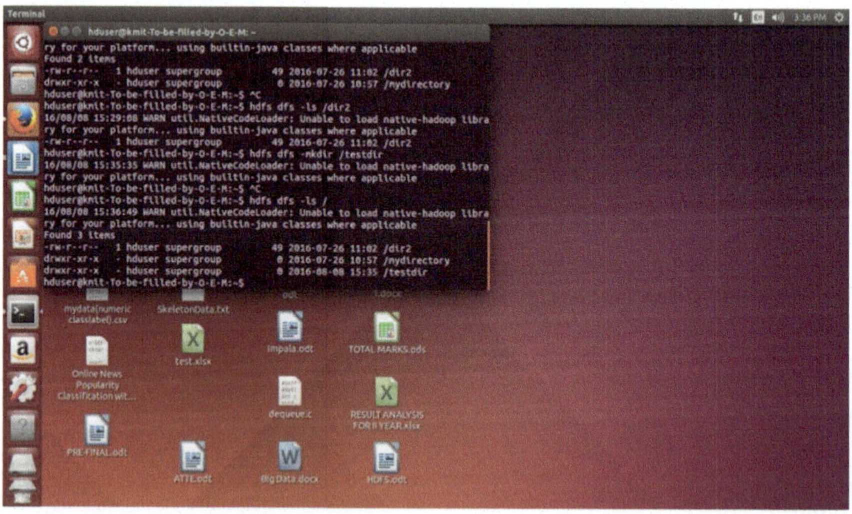

5. **Put Command**

The following command is used to copy the file from local file systems to HDFS file system.

$ hdfs dfs -put sample1 /testdir/sample.txt

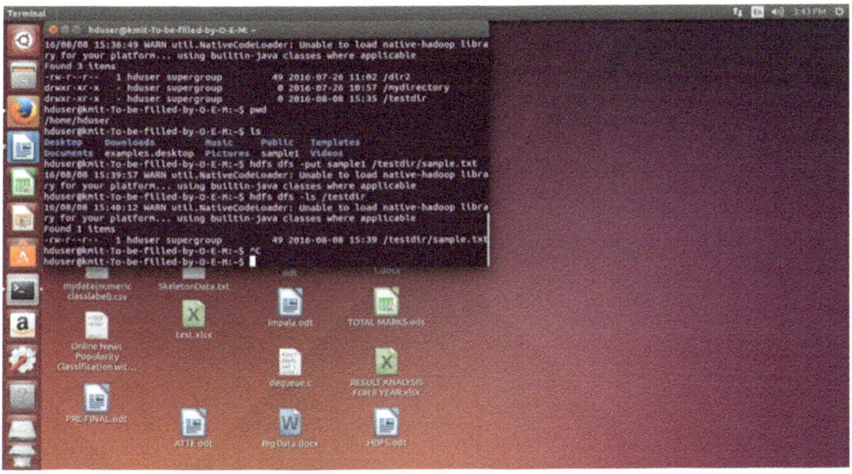

6. **lSR Command**

The following command is a recursive version of ls.

$ hdfs dfs -lsr /

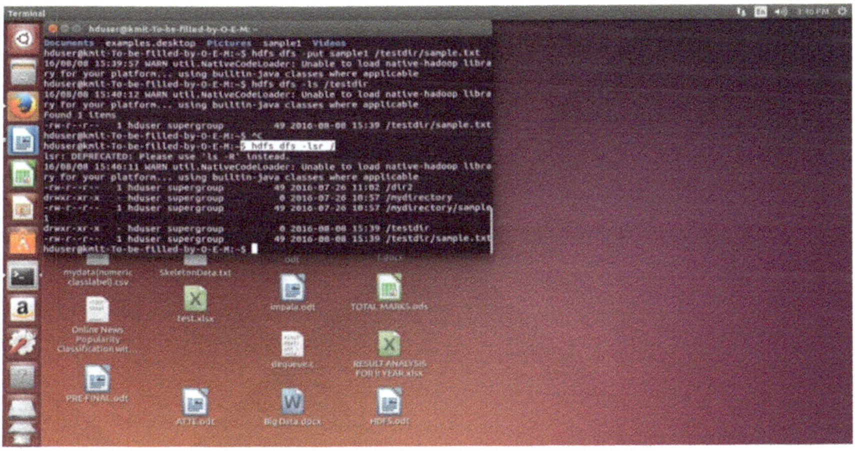

7. **RM command**

The following command is used to delete the files/directories.

$ hdfs dfs -lsr /

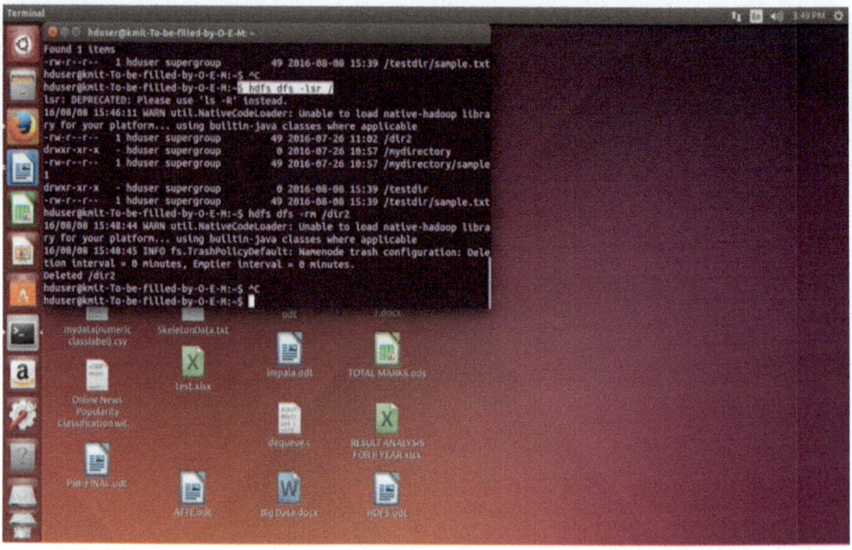

8. **Cat Command**

The following command used to view the data in the files.

$ hdfs dfs -cat /testdir/sample.txt

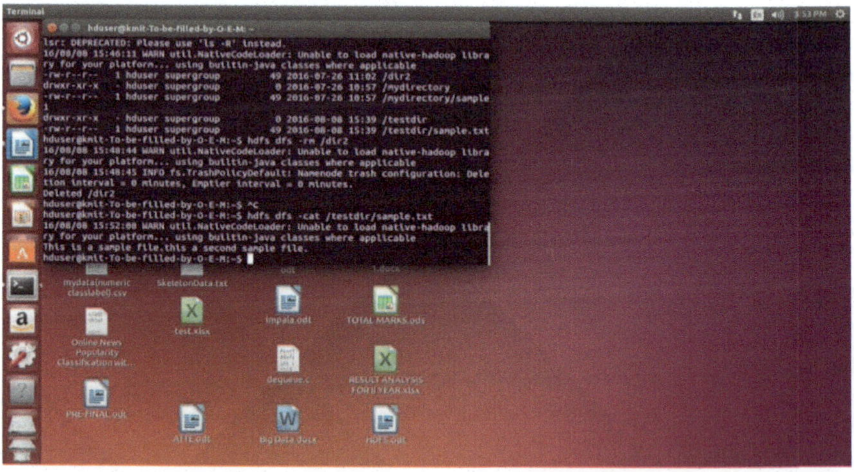

9. **Copy from Local**

This command is similar to put command, except that the source is restricted to a local file reference.

$ hdfs dfs -copyFromLocal /home/hduser/sample1 /mydirectory/sample2.txt

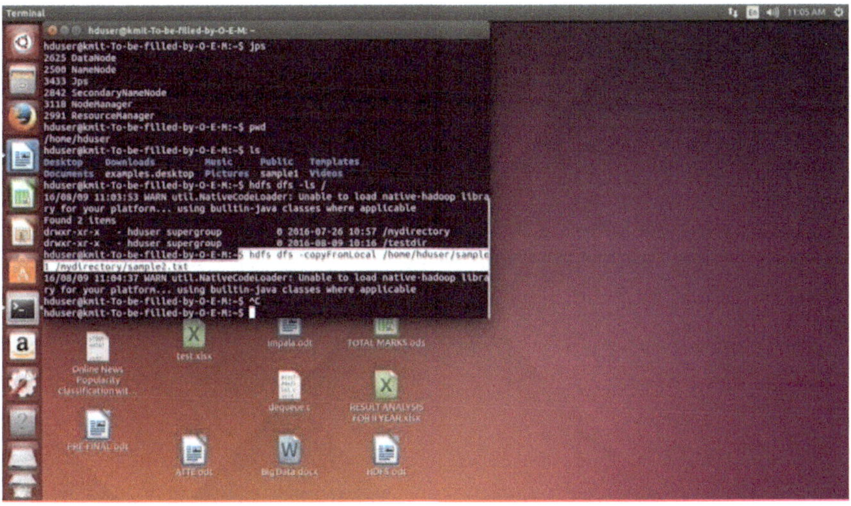

10. **DU Command**

This command shows the amount of space, in bytes used by the files.

$ hdfs dfs -du /

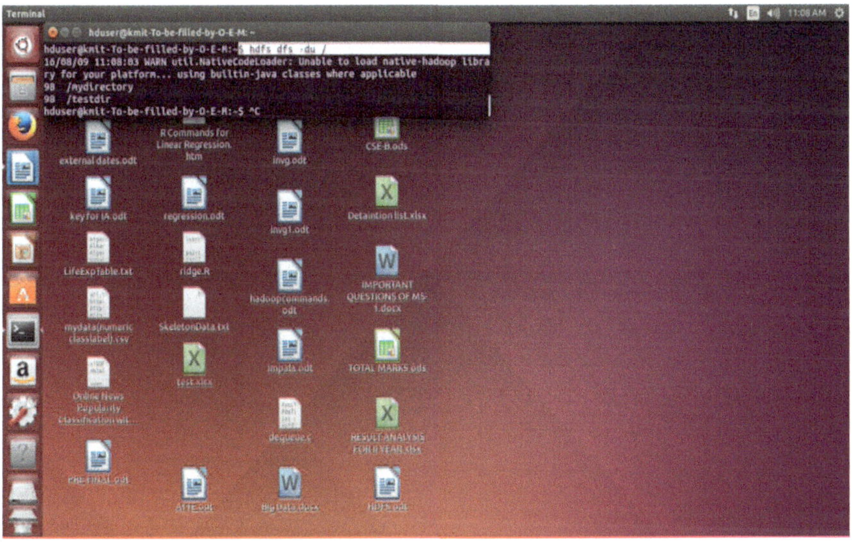

11. **Count Command**

This command counts the number of directories, files and bytes.

$ hdfs dfs -count /

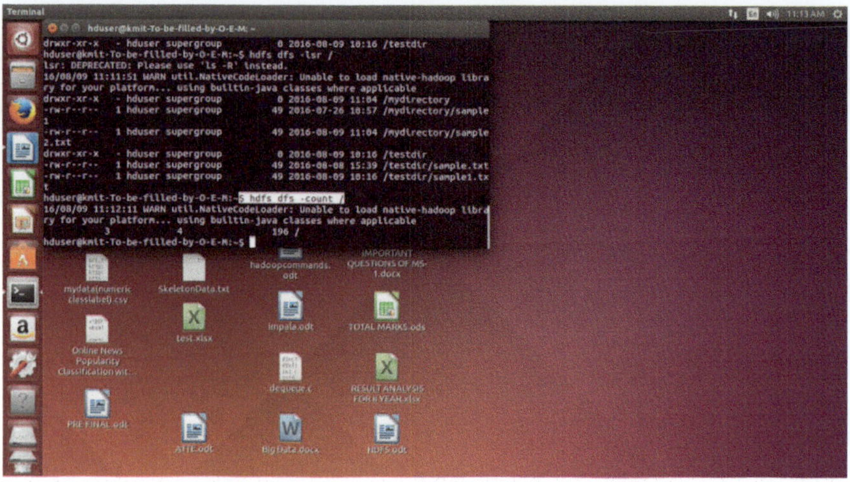

12. **Text Command**

This command takes a source file and outputs the content of the file in text format.

$ hdfs dfs -text /testdir/sample.txt

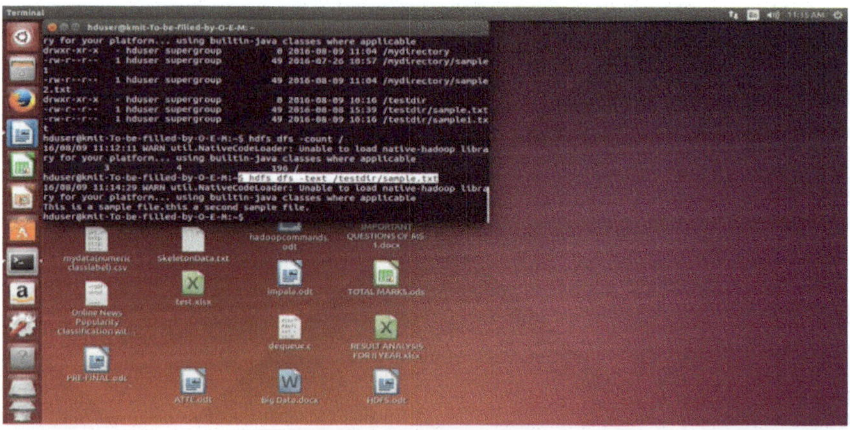

13. **mv Command**

This command is used to move files from source to destination.

$ hdfs dfs -mv /mydirectory/sample1 /testdir/samp1.txt

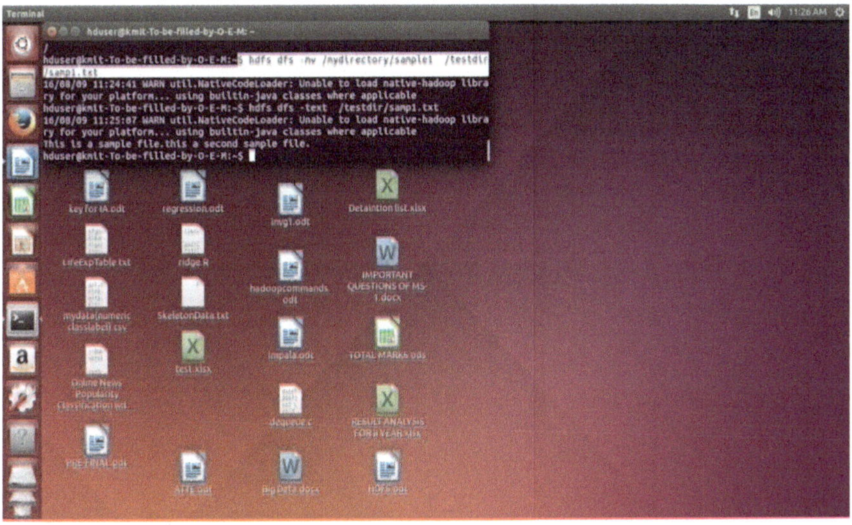

14. **rmr command**

This command is a recursive version of delete command.

$ hdfs dfs -rmr /testdir

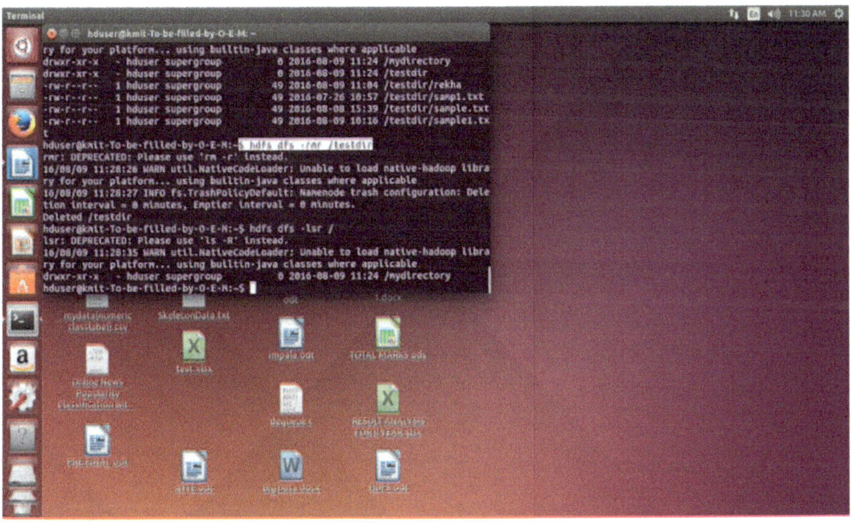

15. **copyToLocal**

This command is used to copy the content from HDFS to local file system.

$ hdfs dfs -copyToLocal /mydirectory/sample2.txt /home/hduser/sample2.txt

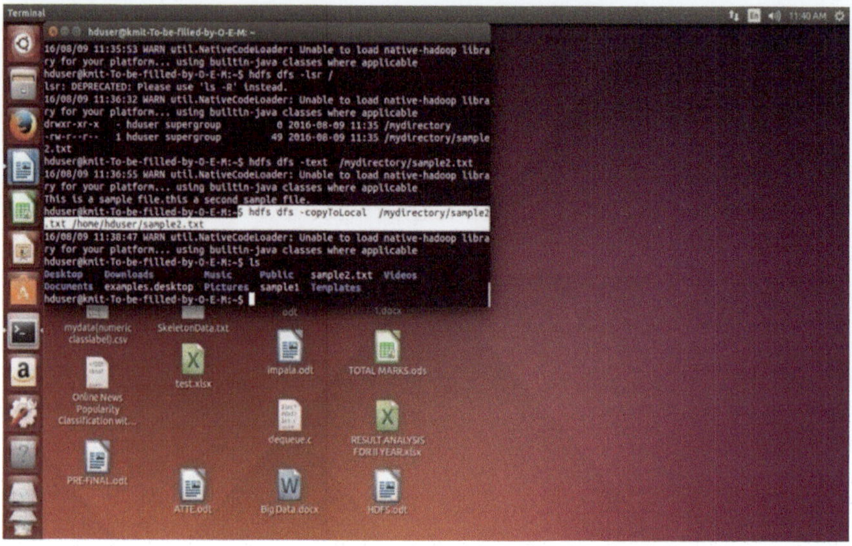

16. **cp Command**

This command is used to copy the files from source to destination.

$ hdfs dfs -cp /mydirectory/sample2.txt /testdir

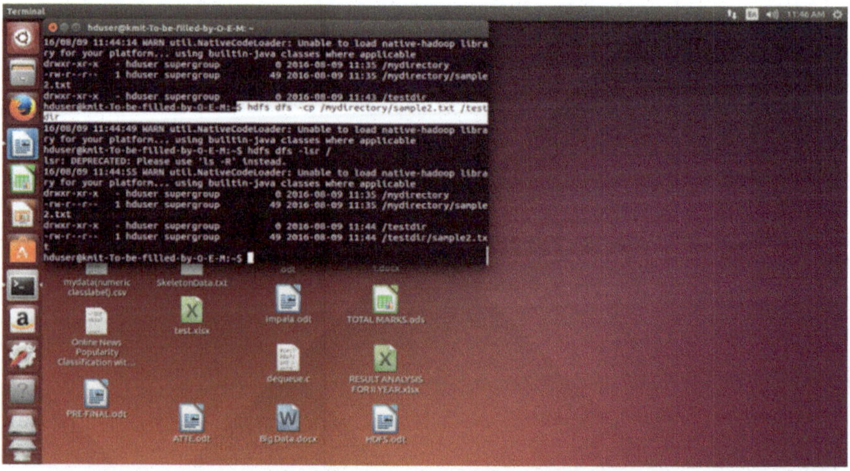

17. stop Command

The following command will stop the name node as well as the data nodes.

$ stop-all.sh

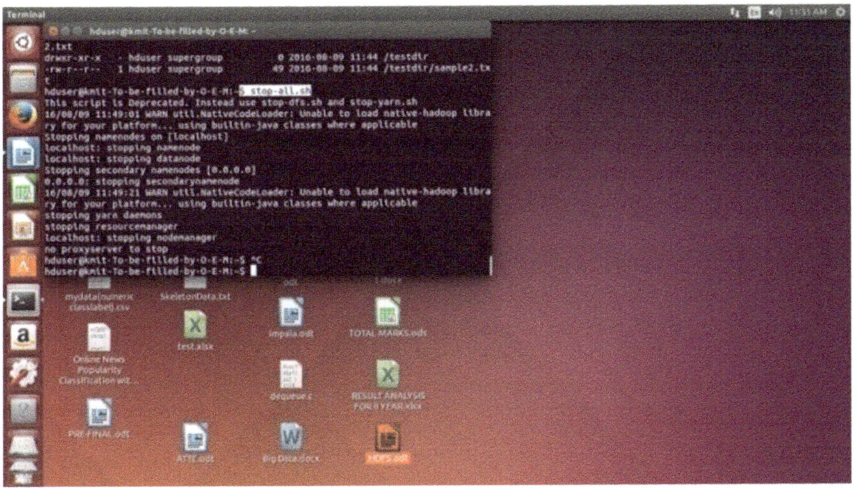

4.2 MapReduce

MapReduce programs are designed to compute large volumes of data in a parallel fashion. This requires dividing the workload across a large number of machines.

In MapReduce, there are two components: One is Mapper, and the other is Reducer.

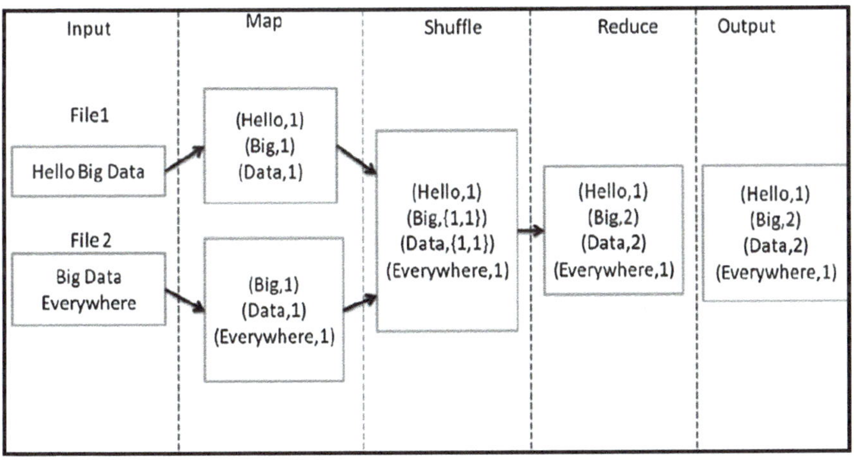

(Key value) pairs—a key is associated with a value

[2]A reducing function turns a large list of values into one (or a few) output value. In MapReduce, all of the output values are not usually reduced together. All of the values *with the same key* are presented to a single Reducer together. This is performed independently of any reduce operations occurring on other lists of values, with different keys attached.

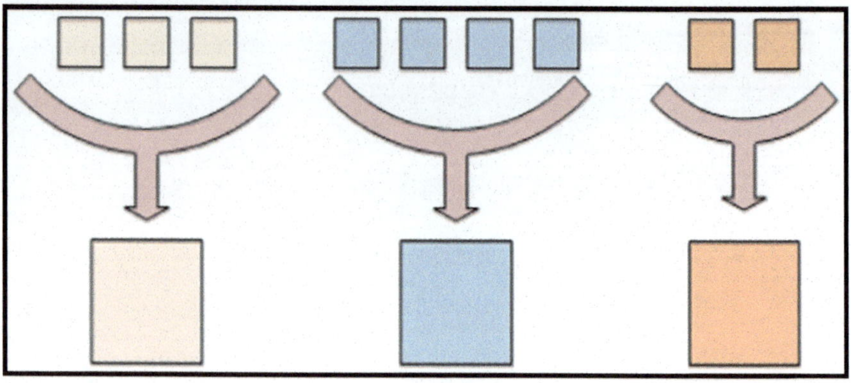

Different colors represent different keys. All values with the same key are presented to a single reduce task.

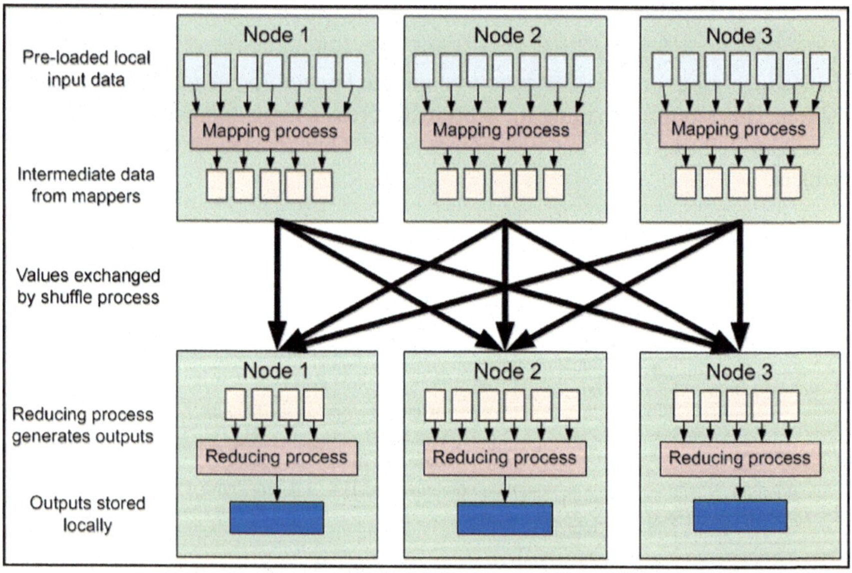

Input to MapReduce comes from Input files in HDFS cluster. In order to run MapReduce, we have to run mapping tasks on (all or many) nodes of the HDFS

[2]sercanbilgic.com.

cluster. All mapping tasks are equivalent. Mappers do not have particular identities associated with them. A Mapper will load the relevant set of files local to itself and then process the same.

The nodes of Reducer task are the same as the nodes of Mapper. In MapReduce, this is the only communication step. Information exchange is not done between individual map tasks and also between different reduce tasks. All data transferred is done automatically.

All values with same key are required to be sent to the same (single) Reducer. Therefore, after mapping the intermediate (key value) pairs are required to be appropriately/accordingly exchanged between modules.

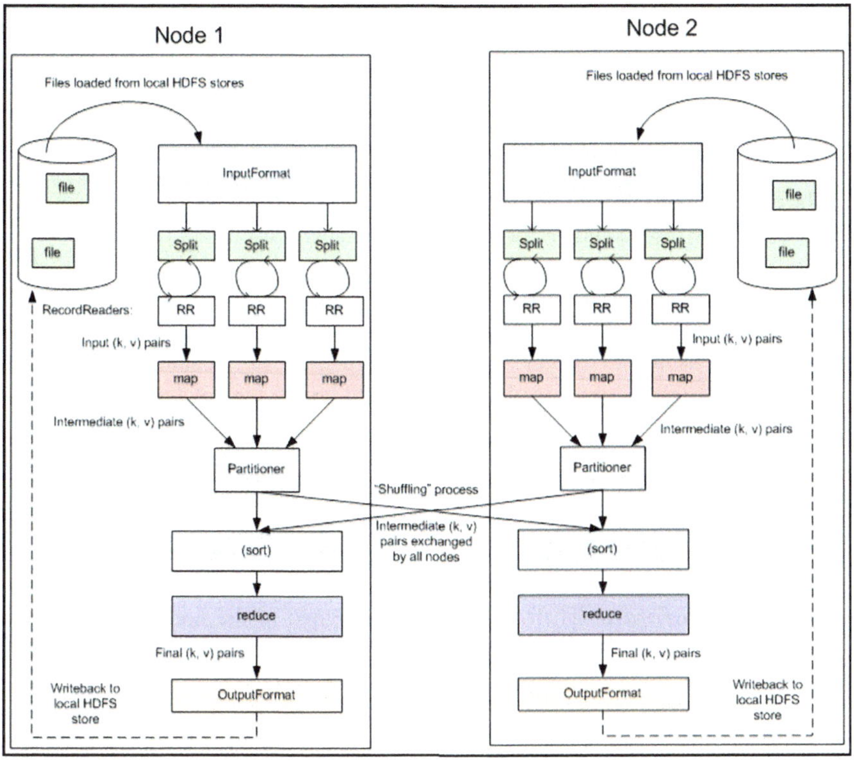

We can see and show in the above diagram, the Mapper and Reducer for the 'WordCount' application.

Even though only two nodes are shown above, the same is applicable to a large number of nodes.

Input Files: Large Input files contain the data for a MapReduce task. Input files can also reside in HDFS (the size of input files can go up to gigabytes).

Input Format: Multiple formats are possible. They inherit from class *FileInput-Format.* While starting a Hadoop job, the above is provided with a path containing files to read. '*FileInputFormat*' can read all files in the directory. Then it splits each file into one or more Input splits each.

Use *TextInputFormat,* the default, unless *job.setInputFormatClass* is used where job is the object that defines the Job. A table of standard *InputFormat* is given below.

InputFormat	Description	Key	Value
TextInputFormat	Default format; reads lines of text files	The byte offset of the line	The line contents
KeyValueInputFormat	Parses lines into key, val pairs	Everything up to the first tab character	The remainder of the line
SequenceFileInputFormat	A Hadoop-specific high-performance binary format	User-defined	User-defined

The default InputFormat is the TextInputFormat. This treats each line of each input file as a separate record and performs no parsing. This is useful for unformatted data or line-based records like log files. A more interesting input format is the *KeyValueInputFormat.* This format also treats each line of input as a separate record. While the TextInputFormat treats the entire line as the value, the *KeyValueInput* Format breaks the line itself into the key and value by searching for a tab character.

SequenceFileInputFormat reads all special and binary files (which are specific to Hadoop). Data from these files can be read into Mappers of Hadoop.

InputSplit: A single map task is a *MapReduce* program called *InputSplit.* A job is a *MapReduce* program applied to a dataset. It is made of a (large) number of tasks. A map task may read a whole or only a part of file, say a chunk of 64 MB as the file may be broken up. Thus, several map tasks operate on a single file in parallel.

The list of tasks corresponding to the single task comprising mapping phase is available in *InputFormat.* Then the tasks are assigned to the nodes in the system according to the actual locations of chunks of input file. As a result, it is possible that all individual node has a number of tasks (multiple of tens also) assigned to it.

RecordReader: The *RecordReader* class loads the data from its source and converts into (key, value) pairs. Mapper reads them. The InputFormat defines the *RecordReader* instance. Each line of input value is treated as a new value by *LineRecordReader* which is provided by that *TextInputFormat.* The invocation of *RecordReader* with corresponding call to Mapper is done repeatedly on the input until all input is completely consumed.

Mapper: The Mapper performs the concerned work (of the user-defined table) of the first phase of MapReduce program. For every given inputs of key and value, the Map() method produces an output key value pairs. Then these are forwarded to Reducers. For each map task for each new instance of a Mapper, a separate Java process is created as an instance.

Shuffling and Partitioning: After first mapping is completed, many more mapping tasks may be continued being processed by the nodes. Intermediate outputs may be exchanged from map tasks to wherever they are required to reach by the reducers. Intermediate key spaces assigned to each node in Reducer subsets are called partitions, and they are inputs to reduce tasks. The partition decides to which partition a given key/value pair will go. A hash value is computed for the key and partition concerned based on the hash value computed for the key.

Sorting: The set of intermediate keys on a single node is sorted by Hadoop before presenting the same to Reducer.

Reduce: For each reduce task, a Reducer instance is created. This instance is the code provided by the user to perform the job-specific work. Every time a key assigned to Reduce, the Reducers reduce method is invoked once. The key and the iterator over all the values associated with the key are received by it.

OutputFormat: All (key[3], value) pairs provided to this OutputCollector are then written to output files. The way they are written is governed by the *OutputFormat*. The OutputFormat functions much like the InputFormat class described earlier. The instances of OutputFormat provided by Hadoop write to files on the local disk or in HDFS; they all inherit from a common *FileOutputFormat*. Each Reducer writes a separate file in a common output directory. These files will typically be named part-*nnnnn*, where *nnnnn* is the partition id associated with the reduce task. The output directory is set by the *FileOutputFormat.setOutputPath*() method.

OutputFormat:	Description
TextOutputFormat	Default; writes lines in "key \t value" form
SequenceFileOutputFormat	Writes binary files suitable for reading into subsequent MapReduce jobs
NullOutputFormat	Disregards its inputs

The *TextOutputFormat* instance writes (key, value) pairs on individual lines in a text file. This text file can be read by humans or by a later MapReduce task (by deploying *KeyValueInputFomat* class). *SequenceFileInputFormat* is an intermediate format for use by multiple MapReduce jobs. *NullOutputFormat* can produce output files and all (key value) pairs input to it.

RecordWriter: *RecordWriter* writes the individual records to the files as directed by the *OuputFormat*. The output files (written by reducers) are kept in HDFS for their subsequent use either by another *MapReduce* job or by a separate program or for human usage.

- WordCount Source Code[4]

[3]bigdataprojects.org.

[4]cad.kpi.ua.

```
package org.myorg;
import java.io.IOException;
import java.util.regex.Pattern;
import org.apache.hadoop.conf.Configured;
import org.apache.hadoop.util.Tool;
import org.apache.hadoop.util.ToolRunner;
import org.apache.log4j.Logger;
import org.apache.hadoop.mapreduce.Job;
import org.apache.hadoop.mapreduce.Mapper;
import org.apache.hadoop.mapreduce.Reducer;
import org.apache.hadoop.fs.Path;
import org.apache.hadoop.mapreduce.lib.input.FileInputFormat;
import org.apache.hadoop.mapreduce.lib.output.FileOutputFormat;
import org.apache.hadoop.io.IntWritable;
import org.apache.hadoop.io.LongWritable;
import org.apache.hadoop.io.Text;
public class WordCount extends Configured implements Tool {
        private static final Logger LOG =
Logger.getLogger(WordCount.class);
        public static void main(String[] args) throws Exception {
                int res = ToolRunner.run(new WordCount(), args);
                System.exit(res);
        }
        public int run(String[] args) throws Exception {
                Job job = Job.getInstance(getConf(), "wordcount");
                job.setJarByClass(this.getClass());
        // Use TextInputFormat, the default unless
job.setInputFormatClass is used
        FileInputFormat.addInputPath(job, new Path(args[0]));
        FileOutputFormat.setOutputPath(job, new Path(args[1]));
                job.setMapperClass(Map.class);
                job.setReducerClass(Reduce.class);
                job.setOutputKeyClass(Text.class);
                job.setOutputValueClass(IntWritable.class);
                return job.waitForCompletion(true) ? 0 : 1;
        }
public static class Map extends Mapper<LongWritable, Text, Text,
IntWritable> {
private final static IntWritable one = new IntWritable(1);
private Text word = new Text();
private long numRecords = 0;
private static final Pattern WORD_BOUNDARY =
Pattern.compile("\\s*\\b\\s*");
public void map(LongWritable offset, Text lineText, Context
context) throws IOException, InterruptedException {
                        String line = lineText.toString();
                        Text currentWord = new Text();
                        for (String word :
WORD_BOUNDARY.split(line)) {
                                if (word.isEmpty()) {
                                        continue;
                                }
                                currentWord = new Text(word);
                                context.write(currentWord, one);
                        }
                }
        }
```

```
public static class Reduce extends Reducer<Text, IntWritable,
Text, IntWritable> {
        @Override
public void reduce(Text word, Iterable<IntWritable> counts,
Context context)
                    throws                    IOException,
InterruptedException {
                int sum = 0;
                for (IntWritable count : counts) {
                    sum += count.get();
                }
                context.write(word, new IntWritable(sum));
        }
    }
}
```

The only standard Java classes you need to import are IOException and regex.Pattern. You useregex.Pattern to extract words from input files.

```
import java.io.IOException;
import java.util.regex.Pattern;
```

This application extends the class Configured and implements the Tool utility class. You tell Hadoop what it needs to know to run your program in a configuration object. Then, you use ToolRunner to run your MapReduce application.

```
import org.apache.hadoop.conf.Configured;
import org.apache.hadoop.util.Tool;
import org.apache.hadoop.util.ToolRunner;
```

The Logger class sends debugging messages from inside the Mapper and Reducer classes. When you run the application, one of the standard INFO messages provides a URL you can use to track the job's success. Messages you pass to Logger are displayed in the map or reduce logs for the job on your Hadoop server.

```
import org.apache.log4j.Logger;
```

You need the Job class to create, configure and run an instance of your MapReduce application. You extend the Mapper class with your own Mapclass and add your own processing instructions. The same is true for the Reducer: You extend it to create and customize your own Reduce class.

```
import org.apache.hadoop.mapreduce.Job;
import org.apache.hadoop.mapreduce.Mapper;
import org.apache.hadoop.mapreduce.Reducer;
```

Use the Path class to access files in HDFS. In your job configuration instructions, you pass required paths using the FileInputFormat and FileOutputFormat classes.

```
import org.apache.hadoop.fs.Path;
import org.apache.hadoop.mapreduce.lib.input.FileInputFormat;
import org.apache.hadoop.mapreduce.lib.output.FileOutputFormat;
```

Writable objects have convenience methods for writing, reading and comparing values during map and reduce processing. You can think of the Text class as *String-Writable*, because it performs essentially the same functions as those for integer (IntWritable) and long integer (LongWritable) objects.

```
import org.apache.hadoop.io.LongWritable;
import org.apache.hadoop.io.Text;
import org.apache.hadoop.io.IntWritable;
```

WordCount includes main and run methods, and the inner classes Map and Reduce. The class begins by initializing the logger.

```
public class WordCount extends Configured implements Tool {
  private static final Logger LOG = Logger.getLogger(WordCount.class);
```

The main method invokes ToolRunner, which creates and runs a new instance of WordCount, passing the command-line arguments. When the application is finished, it returns an integer value for the status, which is passed to the System object on exit.

```
public static void main(String[] args) throws Exception {
    int res = ToolRunner.run(new WordCount(), args);
    System.exit(res);
}
```

The run method configures the job (which includes setting paths passed in at the command line), starts the job, waits for the job to complete, and then returns an integer value as the success flag.

```
public int run(String[] args) throws Exception {
```

Create a new instance of the Job object. This example uses the Configured.getConf() method to get the configuration object for this instance of WordCount and names the job object *wordcount*.

```
Job job = Job.getInstance(getConf(), "wordcount");
```

Set the JAR to use, based on the class in use.

```
job.setJarByClass(this.getClass());
```

Set the input and output paths for your application. You store your input files in HDFS, and then pass the input and output paths as command-line arguments at runtime.

```
FileInputFormat.addInputPaths(job, new Path(args[0]));
FileOutputFormat.setOutputPath(job, new Path(args[1]));
```

Set the map class and reduce class for the job. In this case, use the Map and Reduce inner classes defined in this class.

```
job.setMapperClass(Map.class);
job.setReducerClass(Reduce.class);
```

Use a Text object to output the key (in this case, the word being counted) and the value (in this case, the number of times the word appears).

```
job.setOutputKeyClass(Text.class);
job.setOutputValueClass(IntWritable.class);
```

Launch the job and wait for it to finish. The method syntax is waitForCompletion(boolean verbose). When true, the method reports its progress as the Map and Reduce classes run. When false, the method reports progress up to, but not including, the Map and Reduce processes. In Unix, 0 indicates success, and anything other than 0 indicates a failure. When the job completes successfully, the method returns 0. When it fails, it returns 1.

```
return job.waitForCompletion(true) ? 0 : 1;
}
```

The Map class (an extension of Mapper) transforms key/value input into intermediate key/value pairs to be sent to the Reducer. The class defines several global variables, starting with an IntWritable for the value 1, and a Text object is used to store each word as it is parsed from the input string.

```
[5]public static class Map extends Mapper<LongWritable, Text, Text, IntWritable>{
private final static IntWritable one = new IntWritable(1);
private Text word = new Text();
```

Create a regular expression pattern you can use to parse each line of input text on word boundaries ("\b"). Word boundaries include spaces, tabs and punctuation.

```
private static final Pattern WORD_BOUNDARY =
Pattern.compile("\\s*\\b\\s*");
```

Hadoop invokes the map method once for every key/value pair from your input source. This does not necessarily correspond to the intermediate key/value pairs output to the Reducer. In this case, the map method receives the offset of the first character in the current line of input as the key, and a Text object representing an

[5]cad.kpi.ua

entire line of text from the input file as the value. It further parses the words on the line to create the intermediate output.
public void map(LongWritable offset, Text lineText, Context context)

```
throws IOException, InterruptedException {
```

Convert the Text object to a string. Create the currentWord variable, which you use to capture individual words from each input string.

```
String line = lineText.toString();
Text currentWord = new Text();
```

Use the regular expression pattern to split the line into individual words based on word boundaries. If the word object is empty (e.g., consists of white space), go to the next parsed object. Otherwise, write a key/value pair to the context object for the job.

```
for (String word : WORD_BOUNDARY.split(line)) {
  if (word.isEmpty()) {
    continue;
  }
  currentWord = new Text(word);
  context.write(currentWord,one);
}
}
```

The Mapper creates a key/value pair for each word, composed of the word and the IntWritable value 1. The Reducer processes each pair, adding one to the count for the current word in the key/value pair to the overall count of that word from all Mappers. It then writes the result for that word to the Reducer context object and moves on to the next. When all of the intermediate key/value pairs are processed, the map/reduce task is complete. The application saves the results to the output location in HDFS.

```
  public static class Reduce extends Reducer<Text, IntWritable, Text,
  IntWritable> {
    @Override public void reduce(Text word, Iterable<IntWritable>counts,
    Context context)
```

```
            throws IOException, InterruptedException {
            int sum = 0;
            for (IntWritable count : counts) {
            sum += count.get();
            }
            context.write(word, new IntWritable(sum));
        }
    }
}
```

- **Running WordCount**

Before you run the sample, you must create input and output locations in HDFS. Use the following commands to create the input directory in HDFS:

```
hduser@kmit-G41MT-S2P:~$ hdfs dfs -mkdir /kmit/wordcount
```

```
hduser@kmit-G41MT-S2P:~$ hdfs dfs -mkdir /kmit
```

```
hduser@kmit-G41MT-S2P:~$ hdfs dfs -mkdir /kmit/wordcount/input
```

Create sample text files to use as input and move them to the/user/cloudera/wordcount/inputdirectory in HDFS.

```
hduser@kmit-G41MT-S2P:/$ cd home/hduser/
hduser@kmit-G41MT-S2P:~$ echo "Hadoop is an elephant . Hadoop is a BigData framework " > file0
hduser@kmit-G41MT-S2P:~$ echo "Hadoop is as yellowish as can be" > file1
hduser@kmit-G41MT-S2P:~$ echo "Oh what a yellow fellow is Hadoop awsome" > file2
```

```
hduser@kmit-G41MT-S2P:~$ hdfs dfs -cat /kmit/wordcount/input/file*
```

```
hduser@kmit-G41MT-S2P:~$ hdfs dfs -put file* /kmit/wordcount/input
```

```
Hadoop is an elephant . Hadoop is a BigData framework
Hadoop is as yellowish as can be
Oh what a yellow fellow is Hadoop awsome
```

Create a JAR file for the WordCount application in eclipse and run the WordCount application from the JAR file, passing the paths to the input and output directories in HDFS.

```
hadoop jar WordCount.jar org.myorg.WordCount  /kmit/wordcount/input /kmit/wordcount/output
hduser@kmit-G41MT-S2P:~$ hdfs dfs -cat /kmit/wordcount/output/*
17/12/07 14:45:09 WARN util.NativeCodeLoader: Unable to load nat
cable
.       1
BigData 1
Hadoop  4
Oh      1
a       2
an      1
as      2
awsome  1
be      1
can     1
elephant        1
fellow  1
framework       1
is      4
what    1
yellow  1
yellowish       1
```

4.3 Pig

Pig is an open-source high-level dataflow system. It provides a simple language called Pig Latin which is used for querying and data manipulation. This Pig Latin script is compiled into MapReduce jobs that run on Hadoop.

Why Pig was Created?

It is an ad hoc way of creating and executing MapReduce jobs on very large datasets. It is useful for rapid development. The developer need not have to know Java programming. It is developed by Yahoo.

Why Should I Go For Pig When There Is MR?

- MapReduce.
- Powerful model for parallelism.
- Based on a rigid procedural structure.
- Provides a good opportunity to parallelize algorithm.

PIG

- Does not require developer to have any Java skills.
- It is desirable to have a higher-level declarative language.
- Similar to SQL query where the user specifies the 'what' and
- Leaves the 'how' to the underlying processing engine.

Who are Using PIG ?

70% MapReduce jobs are written in Pig in Yahoo.

Usage

- Web log processing
- Build user behavior models
- Process images
- Data mining
- Twitter, LinkedIn, eBay, AOL, etc.

Where I Should Use Pig?

Sampling Pig is a data flow language. It is at the top of Hadoop and makes it possible to create complex jobs to process large volumes of data quickly and efficiently.

Case 1—Time-Sensitive Data Loads
Case 2—Processing Many Data Sources
Case 3—Analytic Insight Through Sampling Pig.

[6]**Where not to use PIG?**

Really nasty data formats or completely unstructured data (video, audio, raw human-readable text).

Pig is definitely slow compared to MapReduce jobs.

When you would like more power to optimize your code.

[7]Are there any problems which can only be solved by MapReduce and cannot be solved by PIG? In which kind of scenarios MR jobs will be more useful than PIG?

Let us take a scenario where we want to count the population in two cities. I have a dataset and sensor list of different cities. I want to count the population by using one MapReduce for two cities. Let us assume that one is Bangalore and the other is Noida. So I need to consider key of Bangalore city similar to Noida through which I can bring the population data of these two cities to one Reducer. The idea behind this is somehow I have to instruct MapReduce program—whenever you find city with the name 'Bangalore' and city with the name 'Noida,' you create the alias name which will be the common name for these two cities so that you create a common key for both the cities and it get passed to the same Reducer. For this, we have to write custom partitioner.

In MapReduce when you create a 'key' for city, you have to consider 'city' as the key. So, whenever the framework comes across a different city, it considers it as a different key. Hence, we need to use customized partitioner. There is a provision in MapReduce only, where you can write your custom partitioner and mention if city = bangalore or noida then pass similar hashcode. However, we cannot create custom partitioner in Pig. As Pig is not a framework, we cannot directly execute engine to customize the partitioner. In such scenarios, MapReduce works better than Pig.

---

[6]doctuts.com.

[7]hakimshabir.blogspot.com.

How Yahoo Uses PIG?

Pig is best suited for the data factory.

Data Factory Contains

Pipelines
Pipelines bring logs from Yahoo!'s web servers. These logs undergo a cleaning step where bots, company internal views and clicks are removed.

Research
Researchers want to quickly write a script to test a theory. Pig integration with streaming makes it easy for researchers to take a Perl or Python script and run it against a huge dataset.

Use Case in Healthcare
Personal health information of a person in healthcare data is confidential and is not supposed to expose to others. There is a need to mask this information. The data associated with health care is huge, so identifying the personal health information and removing it is crucial.

Problem Statement
De-identify personal health information.

Challenges
Huge amount of data flows into the systems daily, and there are multiple data sources that we need to aggregate data from. Crunching this huge data and de-identifying it in a traditional way had problems.

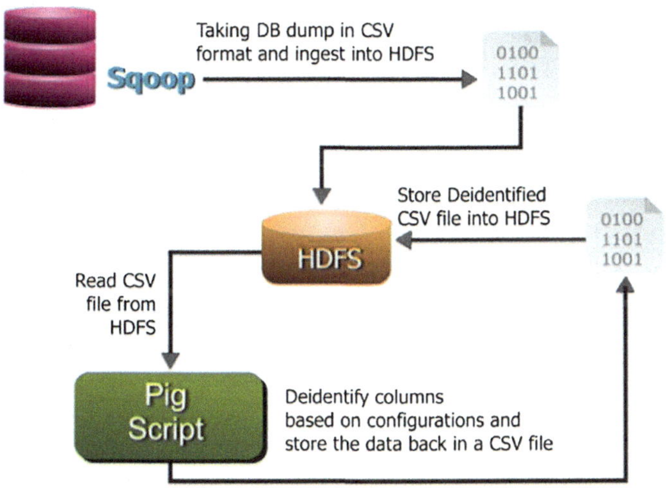

To de-identify health information, Pig can be used. Sqoop helps to export/import from relational database to HDFS and vice versa. Take a database dump into HDFS using Sqoop and then de-identify columns using Pig script. Then store the de-identified data back into HDFS.

Pig Components

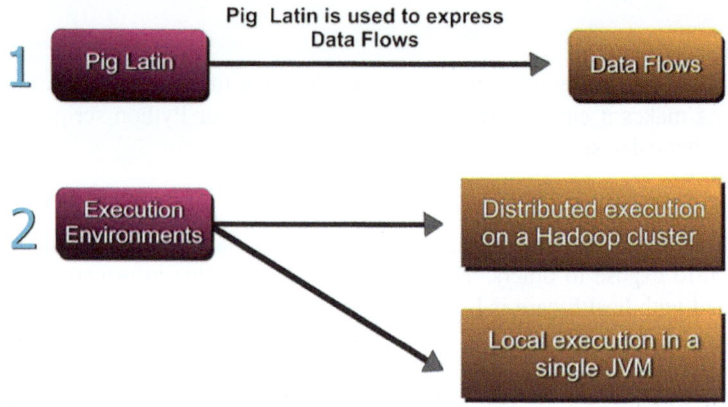

Pig is made up of Pig Latin and execution environment. Execution can be done in a single JVM which is called local execution, and the other way is to execute in distributed environment.

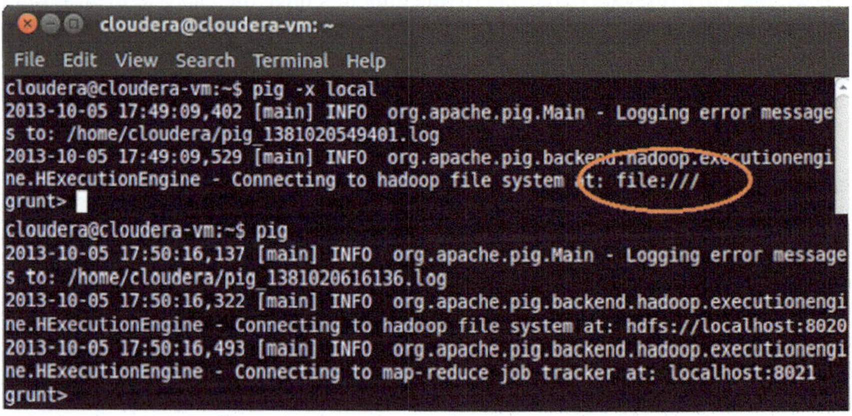

Pig Latin Program

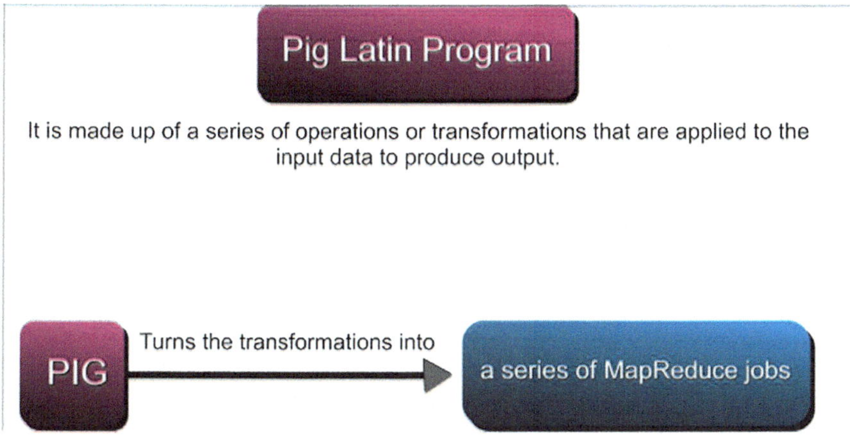

Pig Execution

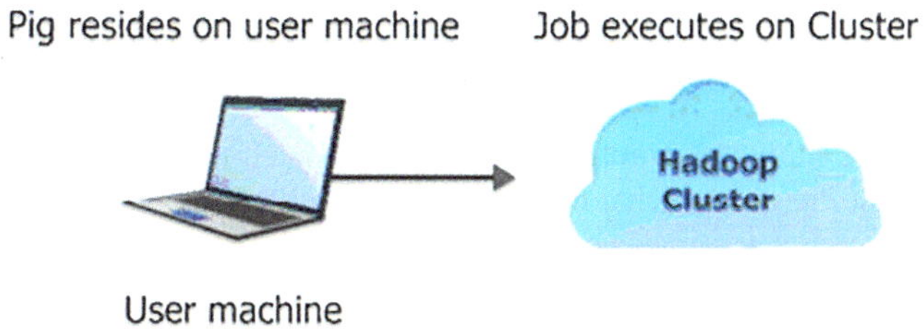

Pig installation can be on the user machine, and the job executes on cluster.

Data Models

Supports Four Basic Types

- **Atom**: a simple atomic value (int, long, double, string), e.g., Edureka.
- **Tuple**: a sequence of fields that can be any of the data types, e.g., (Edureka., Bangalore.)
- **Bag**: a collection of tuples of potentially varying structures, e.g., {(„Educomp), („Edureka, „Bangalore.)}

A bag is one of the data models present in Pig. It is an unordered collection of tuples with possible duplicates. Bags are used to store collections while grouping. The size of bag is the size of the local disk; this means that the size of the bag is limited. When the bag is full, then Pig will spill this bag into local disk and keep only some parts of the bag in memory. There is no necessity that the complete bag should fit into memory. We represent bags with '{}'.

- **Map**: an associative array, the key must be a char array but the value can be any type, e.g., {Name:.Edureka.}

Pig Data Types

Pig Data Type	Implementing Class
Bag	org.apache.pig.data.DataBag
Tuple	org.apache.pig.data.Tuple
Map	java.util.Map<Object, Object>
Integer	java.lang.Integer
Long	java.lang.Long
Float	java.lang.Float
Double	java.lang.Double
Chararray	java.lang.String
Bytearray	byte[]

Does Pig give any warning when there is a type mismatch or missing field?

No, Pig will not show any warning if there is no matching field or a mismatch. If you assume that Pig gives such a warning, then it is difficult to find in log file. If any mismatch is found, it assumes a null value in Pig.

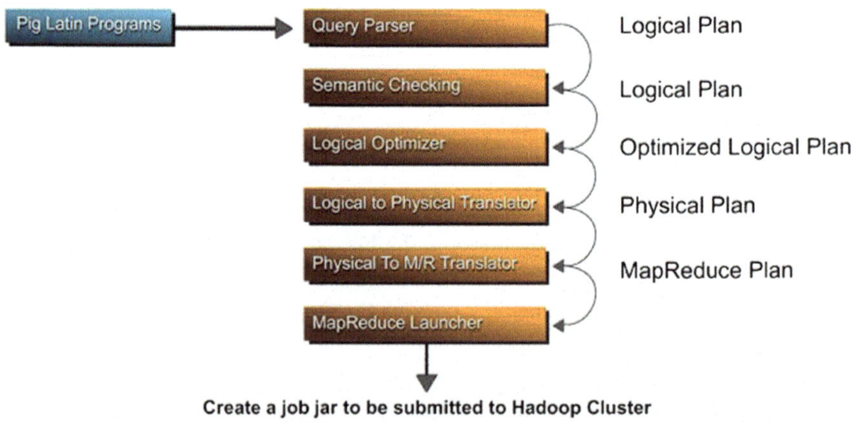

Pig undergoes some steps when a Pig Latin script is converted into MapReduce jobs. After performing the basic parsing and semantic checking, it produces a logical plan. The logical plan describes the logical operators that have to be executed by Pig during execution. After this, Pig produces a physical plan. The physical plan describes the physical operators that are needed to execute the script.

Pig Scripts

Step 1: Input file a

```
0,1,2
1,3,4
```

Step 2: Create a file by namepigex1.pig and add the below code to it

```
/*
* pigex1.pig
* Another line of comment
*/
log = LOAD '$input' using PigStorage AS (a1:int,a2:int,a3:int);
lmt = LIMIT log $size;
DUMP lmt;
--End of program
```

Step 3:

```
cloudera@cloudera-vm:/var/lib/hadoop-0.20/inputs/pigexamples$ pig -param input=/pigexamples/inputs/a -
param size=2 /var/lib/hadoop-0.20/inputs/pigexamples/pigex1.pig
```

(or)

Step 4: Using parm file paramspig and add the below parms in it Input=/var/lib/hadoop-0.20/inputs/pigexamples/pigex1.pig size = 2

```
cloudera@cloudera-vm:/var/lib/hadoop-0.20/inputs/pigexamples$ pig -param_file /var/lib/hadoop-0.20/inp
uts/pigexamples/paramspig /var/lib/hadoop-0.20/inputs/pigexamples/pigex1.pig
```

JOINS Overview

Critical tool for data processing.

[8]Pig supports

- Inner JOINS
- Outer JOINS
- Full JOINS

How to JOIN in Pig

JOIN Steps

1. Load records into a bag from input #1
2. Load records into a bag from input #2
3. JOIN the datasets (bags) by provided JOIN key.

Default JOIN is Inner JOIN

- Rows are joined where the keys match
- Rows that do not have matches are not included in the result.

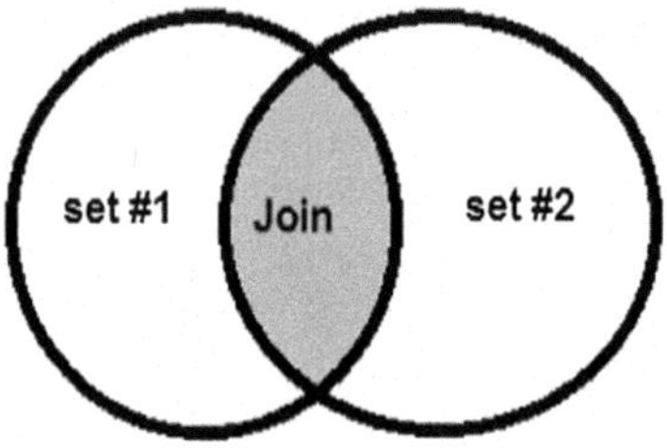

Inputs:

pigjoin1.txt:
user1,Funny story,1343182026191
user2,Cool deal,1343182022222
user5,Yet another blog,1343182044444
user4,Interesting post,1343182011111

pigjoin2.txt:
user1,12,1343182026191
user2,7,1343182021111
user3,0,1343182023333
user4,50,1343182027777

[8]courses.coreservelets.com.

Code:

```
grunt> aa = LOAD '/var/lib/hdoop-0.20/inputs/pigexamples/pigjoin1.txt' using PigStorage(',') as (user:
chararray,post:chararray,date:long);
grunt> bb = LOAD '/var/lib/hdoop-0.20/inputs/pigexamples/pigjoin2.txt' using PigStorage(',') as (user:
chararray,like:int,date:long);
grunt> userinfo = join aa by user,bb by user;
```

Output:

```
(user1,funny story,111222333,user1,12,444555666)
(user2,interesting story,444555666,user2,10,333322211)
(user5,yet another blog,777888999,user5,32,111222333)
```

Inner JOIN Schema:

JOIN reuses the names of the input fields and prepends the name of the input bag.

```
grunt> describe userinfo;
userinfo: {aa::user: chararray,aa::post: chararray,aa::date: long,bb::user: chararray,bb::post: int,bb
::date: long}
```

Inner JOIN with Multiple Keys:

Userinfo = JOIN aa By (user,date), bb BY (user,date);

Outer JOIN

Records which will not JOIN with the other record set are still included in the result.

Left Outer:

Records from the first dataset are included whether they have a match or not. Fields from the unmatched bag are set to null. Use the datasets from inner JOIN tutorial.

Code:

```
grunt> aa = LOAD '/var/lib/hdoop-0.20/inputs/pigexamples/pigjoin1.txt' using PigStorage(',') as (user:
chararray,post:chararray,date:long);
grunt> bb = LOAD '/var/lib/hdoop-0.20/inputs/pigexamples/pigjoin2.txt' using PigStorage(',') as (user:
chararray,like:int,date:long);
grunt> userinfo = join aa by user,bb by user;
```

Userinfo_leftouter = JOIN aa BY user left outer, bb BY user;

Output:

```
(user1,funny story,111222333,user1,12,111222333)
(user2,interesting story,444555666,user2,10,333322211)
(user4,new story,999888,,,)
(user5,yet another blog,777888999,user5,32,111222333)
```

User 1, user 2 and user 5 have matches; it shows concatenated results. User 4 does not match in bb, so the bb part has null values. User 3 in bb is not displayed at all.

Right Outer:

The opposite to the left outer JOIN. Records from the second dataset are included no matter what. Fields from unmatched bag are set to null.

Code:

Userinfo_leftouter = JOIN aa BY user right outer, bb BY user;

Output:

```
(user1,funny story,111222333,user1,12,111222333)
(user2,interesting story,444555666,user2,10,333322211)
(,,,user3,0,777555999)
(user5,yet another blog,777888999,user5,32,111222333)
```

Full Outer:

Records from both sides are included. For unmatched records, the fields from the other bag are set to NULL.

Code:

Userinfo_leftouter = JOIN aa BY user full outer, bb BY user;

Output:

```
(user1,funny story,111222333,user1,12,111222333)
(user2,interesting story,444555666,user2,10,333322211)
(,,,user3,0,777555999)
(user4,new story,999888,,,)
(user5,yet another blog,777888999,user5,32,111222333)
```

Summary:

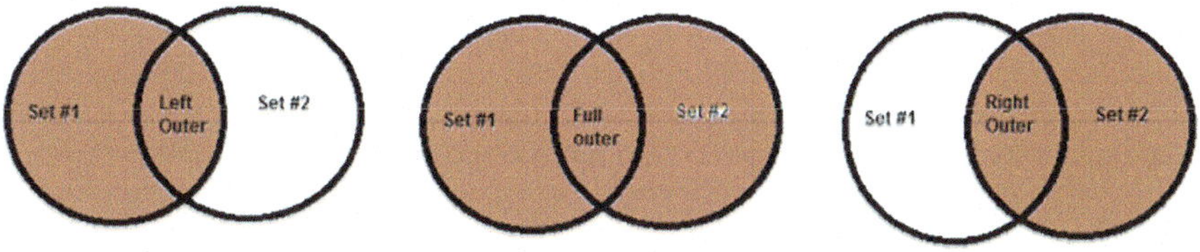

Input:

```
(a,{(a,22,hello1),(a,21,hello1),(a,24,hello1),(a,23,hello1)})
(b,{(b,34,hello2),(b,22,hello2),(b,35,hello2),(b,33,hello2)})
(c,{(c,23,hello3)})
(d,{(d,24,hello4)})
(e,{(e,25,hello5)})
(f,{(f,26,hello6)})
(g,{(g,27,hello7)})
(h,{(h,28,hello8,)})
(i,{(i,29,hello9)})
```

Step 1: Enter into grunt shell
Pig—x local

```
cloudera@cloudera-vm:~$ pig -x local
2013-10-05 22:39:59,579 [main] INFO  org.apache.pig.Main - Logging error message
s to: /home/cloudera/pig_1381037999575.log
2013-10-05 22:40:04,196 [main] INFO  org.apache.pig.backend.hadoop.executionengi
ne.HExecutionEngine - Connecting to hadoop file system at: file:///
grunt>
```

Step 2: Load data
log = LOAD '/var/lib/hadoop-0.20/inputs/pigfile1' AS (user, id, welcome);
On loading data and on executing dump command on the above log, data is stored as shown below.

```
(a,21,hello1)
(b,22,hello2)
(c,23,hello3)
(d,24,hello4)
(e,25,hello5)
(f,26,hello6)
(g,27,hello7)
(h,28,hello8,)
(i,29,hello9)
(a,22,hello1)
(a,23,hello1)
(a,24,hello1)
(b,33,hello2)
(b,34,hello2)
(b,35,hello2)
```

Step 3: Group the log by user id
grpd = GROUP log BY user;
On dumping grpd, grpd contains the below content.

```
(a,{(a,22,hello1),(a,21,hello1),(a,24,hello1),(a,23,hello1)})
(b,{(b,34,hello2),(b,22,hello2),(b,35,hello2),(b,33,hello2)})
(c,{(c,23,hello3)})
(d,{(d,24,hello4)})
(e,{(e,25,hello5)})
(f,{(f,26,hello6)})
(g,{(g,27,hello7)})
(h,{(h,28,hello8,)})
(i,{(i,29,hello9)})
```

Step 4:
cntd = FOREACH grpd GENERATE group, COUNT(log);

```
(a,4)
(b,4)
(c,1)
(d,1)
(e,1)
(f,1)
(g,1)
(h,1)
(i,1)
```

Step 5: Store the output to a file
STORE cntd INTO '/var/lib/hadoop-0.20/inputs/pigfile1output2';

```
a	4
b	4
c	1
d	1
e	1
f	1
g	1
h	1
i	1
```

The above is the final output.

Input Data:

```
a    21    hello1
b    22    hello2
c    23    hello3
d    24    hello4
e    25    hello5
f    26    hello6
g    27    hello7
h    28    hello8
i    29    hello9
a    22    hello1
a    23    hello1
a    24    hello1
b    33    hello2
b    34    hello2
b    35    hello2
```

Commands:

```
grunt> log = LOAD '/var/lib/hadoop-0.20/inputs/pigfile1' AS (user,id,welcome);
grunt> grpd = GROUP log BY user;
grunt> cntd = FOREACH grpd GENERATE group,COUNT(log) as cnt;
grunt> fltr = FILTER cntd BY cnt>1;
grunt> STORE fltr INTO '/var/lib/hadoop-0.20/inputs/pigfile1outfilt';
```

Output:

a 4
b 4

Input:

```
a    21    hello1
b    22    hello2
c    23    hello3
d    24    hello4
e    25    hello5
f    26    hello6
g    27    hello7
h    28    hello8
i    29    hello9
a    22    hello1
a    23    hello1
a    24    hello1
b    33    hello2
b    34    hello2
b    35    hello2
```

Commands:

```
grunt> loadorder = LOAD '/var/lib/hadoop-0.20/inputs/piginputfile' AS (user,id,m
sg);
grunt> grpd = GROUP loadorder BY user;
grunt> cntd = FOREACH grpd GENERATE group,COUNT(loadorder) as cnt;
grunt> fltr = FILTER cntd BY cnt>0;
grunt> srtd = ORDER fltr BY cnt;
grunt> STORE srtd INTO '/var/lib/hadoop-0.20/inputs/pigorderout1';
```

Output:

```
c    1
e    1
f    1
g    1
h    1
i    1
d    3
a    4
b    4
```

Dump Command:

DUMP command is used for development only.
If you DUMP an alias, the content is small enough to show on the screen.

lmt = LIMIT log 4;
DUMP lmt;

Describe Command:

```
grunt> loadorder = LOAD '/var/lib/hadoop-0.20/inputs/piginputfile' AS (user,id,m
sg);
grunt> describe loadorder;
loadorder: {user: bytearray,id: bytearray,msg: bytearray}
```

Illustrate Command:

ILLUSTRATE does a step-by-step process on how Pig will compute the relation

- Illustrate cntd

[9]**Union**:

Pig Latin provides union to put two datasets together by concatenating them instead of joining them. Unlike union in SQL, Pig does not require that both inputs share the same schema.

If both do share the same schema, the output of the union will have that schema. If one schema can be produced from another by a set of implicit casts, the union will

[9]chimera.labs.oreilly.com.

have that resulting schema. If neither of these conditions hold, the output will have no schema (i.e., different records will have different fields).

Example 4: Union and Split

Inputs:

File A:File B

(0,1,2)(0,5,2)
(1,3,4)(1,7,8)

Problem: Group all rows starting with 0 and starting with 1 separately

Code

COGROUP

COGROUP is a generalization of group. Instead of collecting records of one input based on a key, it collects records of *n* inputs based on a key. The result is a record with a key and one bag for each input. Each bag contains all records from that input that have the given value for the key:

A = load 'input1' as (id:int, val:float);
B = load 'input2' as (id:int, val2:int);
C = COGROUP A by id, B by id;
describe C;
C: {group: int,A: {id: int,val: float},B: {id: int,val2: int}}

COGROUP is a group of one dataset. But in the case of more than one dataset, COGROUP will group all the datasets and JOIN them based on the common field. Hence, we can say that COGROUP is a group of more than one dataset and JOIN of that dataset as well

Inputs:

File a:File b
0,1,20,5,2
1,3,41,7,8

Code:

```
grunt> aa = LOAD 'a' using PigStorage(',') as (a1:int,a2:int,ab:int);
grunt> bb = LOAD 'b' using PigStorage(',') as (b1:int,b2:int,ab:int);
grunt> cc =COGROUP aa BY ab,bb BY ab;
grunt> describe cc;
cc: {group: int,aa: {a1: int,a2: int,ab: int},bb: {b1: int,b2: int,ab: int}}
```

Output:

```
(2,{(0,1,2)},{(0,5,2)})
(4,{(1,3,4)},{})
(8,{},{(1,7,8)})
```

Example 5: Word Count example

Input:

```
orange mange banana
orange mange banana
orange mange banana
orange mange banana
orange mange banana
orange mange banana
orange mange banana
orange mange banana
orange mange banana
orange mange banana
```

Code:

```
grunt> aa = LOAD '/var/lib/hadoop-0.20/inputs/wordcountex/pigwc';
grunt> bb = FOREACH aa GENERATE FLATTEN(TOKENIZE((chararray)$0)) as word;
grunt> cc = group bb by word;
grunt> dd = FOREACH cc GENERATE group,COUNT(bb);
grunt> store ee INTO '/var/lib/hadoop-0.20/inputs/wordcountex/pigwcout';
```

Output:

Orange 10
Banana 10
Mango 10

Notes

A relation has a tuple {a, {(b,c), (d,e)}}. Giving the command GENERATE $0, flatten ($1) to this tuple, the tuples (a, b, c) and (a, d, e) are created.

Dump aa:

```
(orange mange banana)
(orange mange banana)
(orange mange banana)
(orange mange banana)
(orange mange banana)
(orange mange banana)
(orange mange banana)
(orange mange banana)
(orange mange banana)
(orange mange banana)
```

Dump bb:

```
(mange)
(banana)
(orange)
(mange)
(banana)
(orange)
(mange)
(banana)
(orange)
(mange)
(banana)
(orange)
(mange)
(banana)
(orange)
(mange)
(banana)
(orange)
(mange)
(banana)
(orange)
(mange)
(banana)
```

Dump cc:

```
(mange,{(mange),(mange),(mange),(mange),(mange),(mange),(mange),(mange),(mange),
(mange)})
(banana,{(banana),(banana),(banana),(banana),(banana),(banana),(banana),(banana)
,(banana),(banana)})
(orange,{(orange),(orange),(orange),(orange),(orange),(orange),(orange),(orange)
,(orange),(orange)})
```

Dump dd:

```
(mange,10)
(banana,10)
(orange,10)
```

Parallelism can be incorporated by having multiple reducers. The number of reducers can be set explicitly using parallel keyword.

The number of Mappers depends upon the number of input splits.

Default

- Sets the number of reducer according to the size of the file
- 1 reducer per 1 GB of input, upto a max of 999 reducers
- pig.exec.reducers.bytes.per.reducer & pig.exec.reducers.max
- #reducers = MIN (pig.exec.reducers.max, total input size (in bytes) / bytes per reducer)

Explicit

- set default_parallel 2
- set parallel 2

PARALLEL clause can be included with any operator that starts a reduce phase

- COGROUP
- CROSS,
- DISTINCT
- GROUP,
- JOIN
- ORDER.

B = GROUP A BY country PARALLEL 4;

[10]UDF

There are times when pigs built-in operators and functions will not suffice. Pig provides the ability to implement your own

1. **Filter**:
   e.g., res = FILTER bag BY udfFilter(post)
2. **Load Function**:
   e.g., res = load 'file.txt' using udfload();
3. **Eval**:
   e.g., res = FOREACH bag GENERATE udfEval($1)

Implement Custom Eval Function:
Eval is the most common type of function. It looks like as below

```
Public abstract class EvalFunc<T>{
Public abstract T exec(Tuple Input) throws IOException;
}
```

[10]chimera.labs.oreilly.com.

Input:

Input File (student_data)		
Ranjit	35	4.5
Nisha	26	4
Amar	25	4.9
Mohan	45	4.3

Convert the input data into capitals.

Code:

```
import java.io.IOException;
import org.apache.pig.EvalFunc;
import org.apache.pig.data.Tuple;
public class UPPER extends EvalFunc<String>
{
public String exec(Tuple input) throws IOException {
if (input == null || input.size() == 0)
return null; 11 try{
String str = (String)input.get(0);
return str.toUpperCase();
}
catch(Exception e){
 throw new IOException("Caught exception processing input row ", e);
16 }
}
}
```

Write the above java code and generate a jar out of it.

Write a Pig script myscript.pig as shown below:

```
REGISTER myUDF.jar
A = LOAD 'student_data' as (nme:chararray,age:int,gpa:float);
B = FOREACH A GENERATE UPPER(name);
Dump B;
```

Implement custom filter function:

[11]**Write a custom filter function which will remove records with the provided value of more than 15 characters**.

---

[11]chimera.labs.oreilly.com.

Filtered = FILTER posts BY isShort(post);

Steps to implement a custom filter:

1. Extend FilterFunc class and implement exec method
2. Register Jar with pig script
3. Use custom filter function in the pig script

Java Code

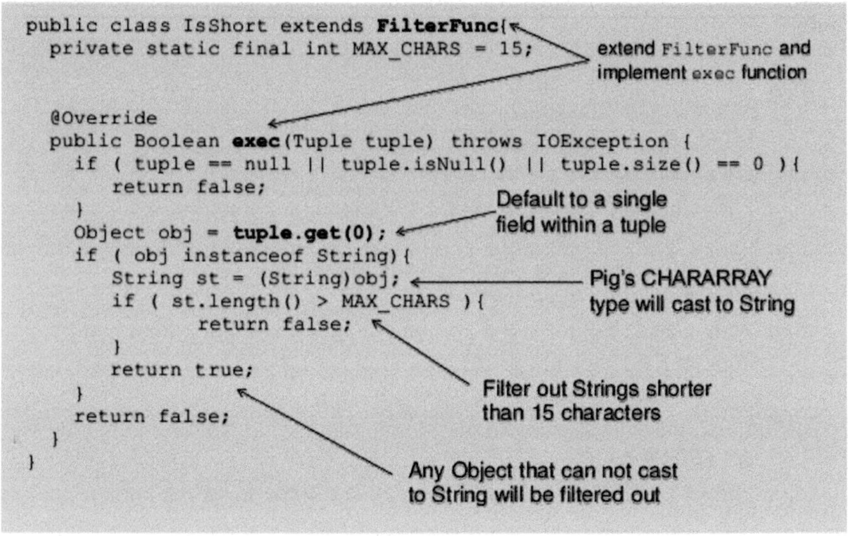

[12]Compile the Java code with filter function and package it into a jar file.

Register the jar file in Pig script

REGISTER Hadoopsamples.jar
Path of the jar file can be either absolute or relative to the execution path.
Path must not be wrapped with quotes.
Add JAR file to the Java's classpath.
Pig locates functions by looking on classpath for fully qualified class name.
Filtered = FILTER posts BY pig.IsShort(post);
Alias can be added to the function name using DEFINE operator. DEFINE isshort pig.isShort();

[12]chimera.labs.oreilly.com.

Pig Script

```
--customfilter.pig
REGISTER Hadoopsamples.jar
DEFINE isShort pig.IsShort();
Posts = LOAD '/training/data/user-post.txt' USING PigStorage(',')
AS (user:chararray,post:chararray,date:long);
Filtered = FILTER posts BY isShort(post);
Dump filtered;
```

Data hierarchy in Pig

Atoms → Tuples → Relations/Bag

Running in Local mode

```
bin/pig -x local -e 'explain -script /home/vm4learning/Desktop/max-temp.pig'
```

Running from a file

```
bin/pig script.pig
```

Executing the Pig script to find the maximum temperature for a particular year copy the files to HDFS

Copy the files to HDFS

```
cd $HADOOP_HOME
bin/hadoop dfs -rmr /user/vm4learning/pig-input1
bin/hadoop dfs -mkdir /user/vm4learning/pig-input1
bin/hadoop fs -put /home/vm4learning/PublicDataSets/ WeatherSimple/Simple.txt
pig input1/
Simple.txt
```

Start the grunt

```
cd $PIG_INSTALL
bin/pig
```

Execute the below commands

```
records = LOAD 'pig-input1/Simple.txt' AS (year:chararray, temperature:int,quality:int);
filtered_records = FILTER records BY temperature ! = 9999 AND (quality == 0
OR
quality == 1 OR quality == 4 OR quality == 5 OR quality == 9);
```

```
grouped_records = GROUP filtered_records BY year;
max_temp = FOREACH grouped_records GENERATE group,
MAX(filtered_records.temperature);
DESCRIBE max_temp;
```

Equivalent of the above in SQL (can be executed in Hive)

```
SELECT year, MAX(temperature) FROM records WHERE temperature ! = 9999
AND (quality = 0 OR quality = 1 OR quality = 4 OR quality = 5 OR quality = 9)
GROUP BY year;
```

To get the data flow on a subset of data

```
ILLUSTRATE max_temp;
```

To get the plan generated by Pig

```
EXPLAIN max_temp;
```

To trigger the MR job and store the results in max_temp folder instead of the grunt console

```
STORE max_temp into 'max_temp';
```

To come out of grunt

```
Quit
```

[13]**UDF (for converting an atom to an upper case)**

```
package myudfs;
import java.io.IOException;
import org.apache.pig.EvalFunc;
import org.apache.pig.data.Tuple;
import org.apache.pig.impl.util.WrappedIOException;
public class UPPER extends EvalFunc <String>
{
public String exec(Tuple input) throws IOException {
if (input == null || input.size() == 0)
return null;
try{
String str = (String)input.get(0);
return str.toUpperCase();
}catch(Exception e){
throw WrappedIOException.wrap("Caught exception processing input row", e);
}}}
```

[13]www.datastax.com.

Put the file in HDFS

bin/hadoop dfs -rmr /user/vm4learning/students
bin/hadoop dfs -mkdir /user/vm4learning/students
bin/hadoop fs -put /home/vm4learning/PublicDataSets/Students/students.txt students/

Execute the Pig scripts using above UDF

//Prepare the register the jar file
REGISTER /home/vm4learning/Desktop/pig-udf.jar;
A = LOAD '/user/vm4learning/students/students.txt' AS (name: chararray, age: int, zip: int);
//Start using the UDF
B = FOREACH A GENERATE UPPER(name);
DUMP B;

Execute the Pig scripts using above UDF (using projection)

//Prepare the register the jar file
REGISTER /home/vm4learning/Desktop/pig-udf.jar;
//Projection is used, so $0, $1,…. are used instead of column names
A = LOAD '/user/vm4learning/students/students.txt';
B = FOREACH A GENERATE UPPER($0);
DUMP B;

Dynamic Invokers (A Java function can be invoked without creating a UDF)

Copy the data into HDFS

bin/hadoop dfs -rmr /user/vm4learning/students
bin/hadoop dfs -mkdir /user/vm4learning/students
bin/hadoop fs -put /home/vm4learning/PublicDataSets/Students/ students.txt students/

Define the dynamic invoker and invoke it

DEFINE trim InvokeForString('org.apache.commons.lang.StringUtils.trim', 'String');
A = LOAD '/user/vm4learning/students/students.txt' AS (name: chararray, age: int,zip: int);
B = FOREACH A GENERATE trim(name);
DUMP B;

Macros in PIG

Define the macro in a file

vi /home/vm4learning/Desktop/max_temp.macro
DEFINE max_by_group(X, group_key, max_field) RETURNS Y {

A = GROUP $X by $group_key;
$Y = FOREACH A GENERATE group, MAX($X.$max_field);
};

Put the data in HDFS

bin/hadoop dfs -rmr /user/vm4learning/pig-input1
bin/hadoop dfs -mkdir /user/vm4learning/pig-input1
bin/hadoop fs -put /home/vm4learning/PublicDataSets/
WeatherSimple/Simple.txt piginput1/
Simple.txt

Import the macro and invoke it

IMPORT '/home/vm4learning/Desktop/max_temp.macro';
records = LOAD 'pig-input1/Simple.txt' AS (year:chararray, temperature:int,
quality:int);
filtered_records = FILTER records BY temperature ! = 9999 AND (quality == 0
OR
quality == 1 OR quality == 4 OR quality == 5 OR quality == 9);
max_temp = max_by_group(filtered_records, year, temperature);
DUMP max_temp;

The same results without a macro

[14]records = LOAD 'pig-input1/sample.txt' AS (year:chararray, temperature:int,
quality:int);
filtered_records = FILTER records BY temperature ! = 9999 AND (quality == 0
OR
quality == 1 OR quality == 4 OR quality == 5 OR quality == 9);
grouped_records = GROUP filtered_records BY year;
max_temp = FOREACH grouped_records GENERATE group,
MAX(filtered_records.temperature);
DUMP max_temp

Pig latin commands

The below commands operate on the HDFS and run in grunt. More details in 'Hadoop—The Definitive Guide'—Page 380
cat pig-input1/sample.txt
mkdir praveen
ls

Schema on read

If the data type is not defined, then it defaults to byte array.
records = LOAD 'pig-input1/sample.txt' AS (year:chararray, temperature:int,
quality:int);

[14]www.hydpages.com.

describe records;
records = LOAD 'pig-input1/sample.txt' AS (year:chararray,**temperature**, quality:int);
describe records;
records = LOAD 'pig-input1/sample.txt' AS (**year**,**temperature**,**quality**);
describe records;
records = LOAD 'pig-input1/sample.txt';
describe records;

Validations in Pig

Change one of the temperature data to an alphabet and dump the results.
cat /home/vm4learning/Code/hadoop-book-3e/input/ncdc/microtab/sample_corrupt.txt
bin/hadoop fs -put /home/vm4learning/Code/hadoop-book-3e/input/ncdc/micro tab/sample_corrupt.txt pig-input1

If the data types are defined, then the validation is done when a dump is done and the invalid value becomes null.

records = LOAD 'pig-input1/sample_corrupt.txt' AS (year:chararray, temperature:int,
quality:int);
DUMP records;
(1950,0,1)
(1950,22,1)
(1950,,1) <– the field is null
(1949,111,1)
(1949,78,1)

No validations without the data type
records = LOAD 'pig-input1/sample_corrupt.txt' AS (year:chararray,**temperature**, quality:int);
DUMP records;

Operators in Pig

SPLIT Operator

SPLIT records INTO good_records IF temperature is not null, bad_records IF temperature is null;
dump good_records;
dump bad_records;

JOIN and COGROUP Operator

JOIN always gives a flat structure: a set of tuples. The COGROUP statement is similar to JOIN, but instead creates a nested set of output tuples. This can be useful if you want to exploit the structure in subsequent statements.

```
vi /home/vm4learning/Desktop/A.txt
2,Tie
4,Coat
3,Hat
1,Scarf
bin/hadoop fs -put /home/vm4learning/Desktop/A.txt /user/vm4learning
A = load '/user/vm4learning/A.txt' using PigStorage(',');
vi /home/vm4learning/Desktop/B.txt
Joe,2
Hank,4
Ali,0
Eve,3
Hank,2
bin/hadoop fs -put /home/vm4learning/Desktop/B.txt /user/vm4learning
B = load '/user/vm4learning/B.txt' using PigStorage(',');
C = JOIN A BY $0, B BY $1;
DUMP C;
(2,Tie,Joe,2)
(2,Tie,Hank,2)
(3,Hat,Eve,3)
(4,Coat,Hank,4)
D = COGROUP A BY $0, B BY $1;
DUMP D;
(0,{},{(Ali,0)})
(1,{(1,Scarf)},{})
(2,{(2,Tie)},{(Joe,2),(Hank,2)})
(3,{(3,Hat)},{(Eve,3)})
(4,{(4,Coat)},{(Hank,4)})
```

[15]**CROSS**

Pig Latin includes the cross-product operator (also known as the Cartesian product), which JOINS every tuple in a relation to every tuple in a second relation (and with every tuple in further relations if supplied). The size of the output is the product of the size of the inputs, potentially making the output very large:

```
I = CROSS A, B;
dump I;
(2,Tie,Joe,2)
(2,Tie,Hank,4)
(2,Tie,Ali,0)
(2,Tie,Eve,3)
(2,Tie,Hank,2)
(4,Coat,Joe,2)
```

[15]clientrd.com.

(4,Coat,Hank,4)
(4,Coat,Ali,0)
(4,Coat,Eve,3)
(4,Coat,Hank,2)
(3,Hat,Joe,2)
(3,Hat,Hank,4)
(3,Hat,Ali,0)
(3,Hat,Eve,3)
(3,Hat,Hank,2)
(1,Scarf,Joe,2)
(1,Scarf,Hank,4)
(1,Scarf,Ali,0)
(1,Scarf,Eve,3)
(1,Scarf,Hank,2)

UNION

Union combines all the tuples from the relations.

J = UNION A, B;
dump J;
(Joe,2)
(Hank,4)
(Ali,0)
(Eve,3)
(Hank,2)
(2,Tie)
(4,Coat)
(3,Hat)
(1,Scarf)

[16]Two run scenarios are optimized, as explained below: explicit and implicit splits, and storing intermediate results.

Explicit and Implicit Splits
There might be cases in which you want different processing on separate parts of the same data stream.

Example 1

A = LOAD …
…
SPLIT A' INTO B IF …, C IF …
…

[16]docplayer.nat.

STORE B' …
STORE C' …

Example 2

A = LOAD …
…
B = FILTER A' …
C = FILTER A' …
…
STORE B' …
STORE C' …

In prior Pig releases, Example 1 will dump A′ to disk and then start jobs for B′ and C′.

Example 2 will execute all the dependencies of B′ and store it and then execute all the dependencies of C′ and store it. Both are equivalent, but the performance will be different.

Here's what the multiquery execution does to increase the performance:

1. For Example 2, adds an implicit split to transform the query to Example 1. This eliminates the processing of A′ multiple times.
2. Makes the split non-blocking and allows processing to continue. This helps reduce the amount of data that has to be stored right at the split.
3. Allows multiple outputs from a job. This way some results can be stored as a side effect of the main job. This is also necessary to make the previous item work.
4. Allows multiple split branches to be carried on to the combiner/Reducer. This reduces the amount of IO again in the case where multiple branches in the split can benefit from a combiner run.

Store Versus Dump

The datasets which are available in Pig script can be as output in two forms:

1. Dump
2. Store.

Dump evaluates the given relation and all despondent relations and produces the output to standard output.

It is recommended to use Dump command for small datasets only because it consumes a lot of time for evaluating and displaying on the screen.

Store performs the same operation as Dump, but it stores the output on HDFS files. It is recommended for large datasets.

A = load 'emp-data using Pig_storage (1,1)
B = filter A by Sal>5000
Store B into 'newdata'.
Dump B.

Flatten

[17]players = load ‘baseball’ as (name:chararray, team:chararray,
position: bag{t:(p:chararray)}, bat: map[]);
pos = for each players generate name, flatten(position) as position;
bypos = group pos by position;
A for each with a flatten produces a cross-product of every record in the bag with all of the other expressions in the generate statement. Looking at the first record in baseball, we see it is the following (replacing tabs with commas for clarity):
Jorge Posada,New York Yankees,{(Catcher),(Designated_hitter)},...
Once this has passed through the flatten statement, it will be two records:
Jorge Posada,Catcher
Jorge Posada,Designated_hitter
–flatten_noempty.pig
players = load ‘baseball’ as (name:chararray, team:chararray,
position:bag{t:(p:chararray)}, bat:map[]);
noempty = foreach players generate name,
((position is null or IsEmpty(position)) ? {(‘unknown’)} : position)
as position;
pos = foreach noempty generate name, flatten(position) as position;
bypos = group pos by position;

Nested for each

--distinct_symbols.pig
daily = load ‘NYSE_daily’ as (exchange, symbol); – not interested in other fields
grpd = group daily by exchange;
uniqcnt = foreach grpd {
sym = daily.symbol;
uniq_sym = distinct sym;
generate group, COUNT(uniq_sym);
};

Theoretically, any Pig Latin relational operator should be legal inside for each. However, at the moment, only distinct, filter, limit and order are supported.

4.4 Flume

[18]Flume’s functionality is that it collects, aggregates and moves log data which is large sized as soon as it is produced. It enables online analytics applications. Flume

[17]www.commonlounge.com.

[18]www.otnira.com.

collects set of log files on every machine in cluster and accumulates in HDFS as an aggregation. The writing speed rate depends on the destination. HDFS is the destination. Flume offers easy mechanisms for management of output files and output formats. Hadoop and Hive can process the data gathered by Flume.

Flume is intended to solve challenges in safely transferring huge set of data from a node (e.g., log files in company web servers) to data store (e.g., HDFS, Hive, HBase and Cassandra).

For a simple system with relatively small dataset, we usually customize our own solution to do this job, such as to create some script to transfer the log to database.

However, this kind of ad hoc solution is difficult to make it scalable because usually it is created very tailored into our system. It sometimes suffers from problem in manageability, especially when the original programmer or engineer who created the system left the company. It is also often difficult to extend, and furthermore, it may have problem in reliability due to some bugs during the implementation.

And Apache Flume comes into the rescue!!!

Apache Flume is a **distributed data collection service** that gets flows of data (like logs) from their source and aggregates them as they are processed. It is based on the concept of data flow through some Flume nodes. In a node, data come from a *source*, optionally processed by *decorator* and transmitted out to *sink*.

On the node level, Flume is modeled around these concepts:

1. Agents are nodes that received data from the application (such as web server).
2. Processors (optional) are nodes that performed intermediate processing of the data.
3. Collectors are nodes that write to permanent data storage (such as HDFS, HBase and Cassandra).

It has these following goals that will solve aforementioned challenges in previous paragraph:

1. Reliability, by providing three types of failure recovery mechanism: BestEffort, Store on Failure and Retry, and EndtoEnd Reliability
2. Scalability, by applying Horizontally Scalable Data and Control Path, and Load Balancing technique.
3. Extensibility, by providing simple source and sink API, and providing plug-in architecture.
4. Manageability, by providing comprehensive interface to configure Flume and intuitive syntax to configure Flume node.

Flume Components

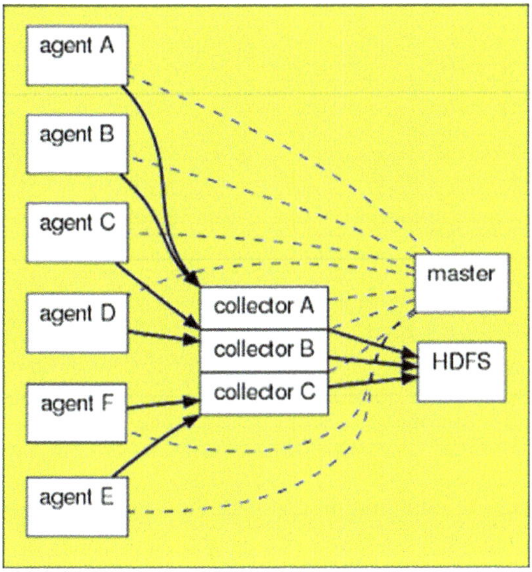

Flume can be configured in many types of topologies such as: (a) single-agent topology, (b) multiagent topology and (c) replicating topology.

In Single-agent topology, a Flume agent can directly send data to its final destination HDFS.

In Multiagent topology, a Flume agent sends data to intermediate Flume agents which may later send to the final destination HDFS (e.g., web services and intermediate agents).

In Replicating Topology, the events can be replicated by Flume. Data can be sent to batch processing systems or archival or storage or even real-time systems.

[19]Many a time, the events from the applications have to be analyzed to know more about the customer behavior for recommendations or to figure any fraudulent use cases. With more data to analyze, it might take a lot of time or sometimes even not possible to process the events on a single machine. This is where distributed systems like Hadoop and others cut the requirement.

Apache Flume and Sqoop can be used to move the data from the source to the sink. The main difference is that Flume is event based, while Sqoop is not. Also, Sqoop is used to move data from structured data stores like RDBMS to HDFS and HBase, while Flume supports a variety of sources and sinks.

[19]www.thecloudavenue.com.

One of the options is to make the application use Log4 J to send the log events to a Flume sink which will store them in HDFS for further analysis.

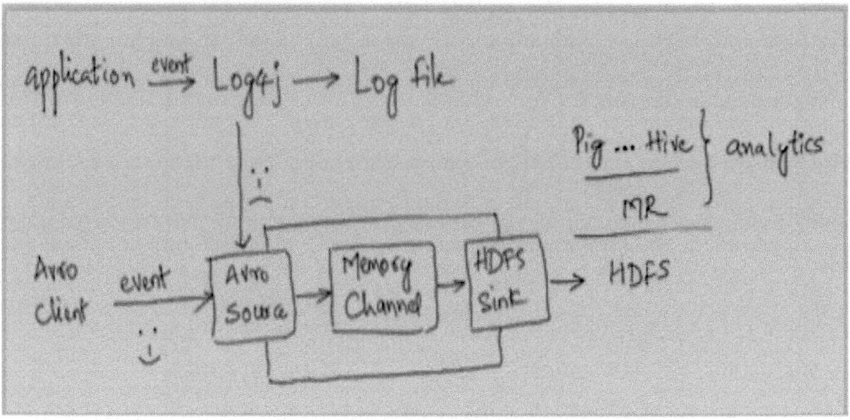

Here are the steps to configure Flume with the Avro Source, Memory Channel, HDFS Sink and chain them together.

cp conf/flumeenv.sh.template conf/flumeenv.sh
Start flume as
bin/fluming agent conf./conf/ fconf/flume.conf Dflume.root.logger = DEBUG,console n agent1
Now, run the Avro client to send message to the Avro source
bin/flum-ngavro-client—conf conf -H localhost-p41414-F /etc./passwd
-Dflume.root.logger = DEBUG,console
Create a project in Eclipse and include all the jars from the <flumeinstallfolder>/lib as dependencies.
include a log4j.properties file in the java project.

4.5 Sqoop

[20]Apache Sqoop

Apache Sqoop is a tool designed for efficiently transferring bulk data in a distributed manner between Apache Hadoop and structured data stores such as relational databases, enterprise data warehouses and NoSQL systems. Sqoop can be used to import data into HBase, HDFS and Hive and out of it into RDBMS, in an automated

[20]hadooped.blogspot.com.

fashion, leveraging Oozie for scheduling. It has a connector-based architecture that supports plug-ins that provide connectivity to new external systems.

Behind the scenes, the dataset being transferred is split into partitions and map-only jobs are launched for each partition with the Mappers managing transferring the dataset assigned to it. Sqoop uses the database metadata to infer the types and handles the data in a type-safe manner.

The following diagrams are from the Apache documentation:

Import process:

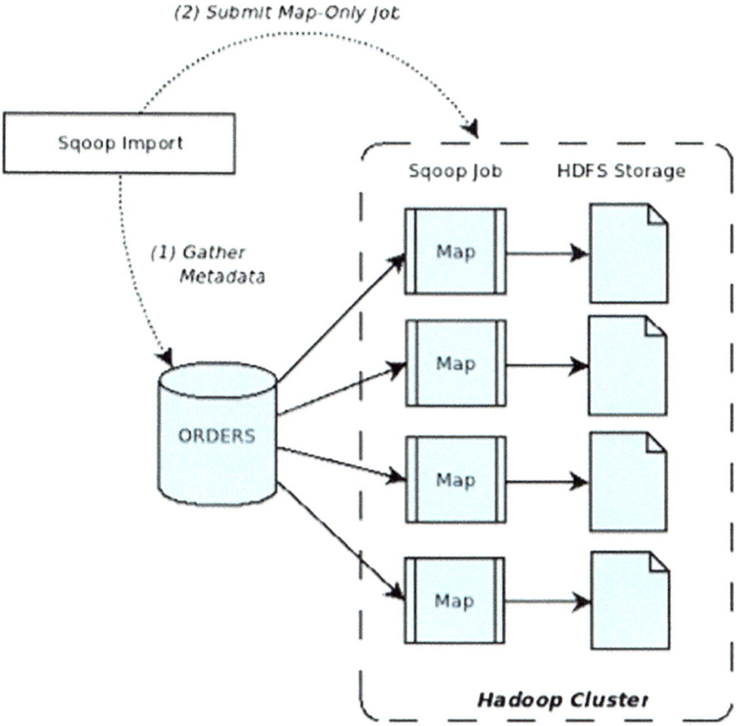

Export process:

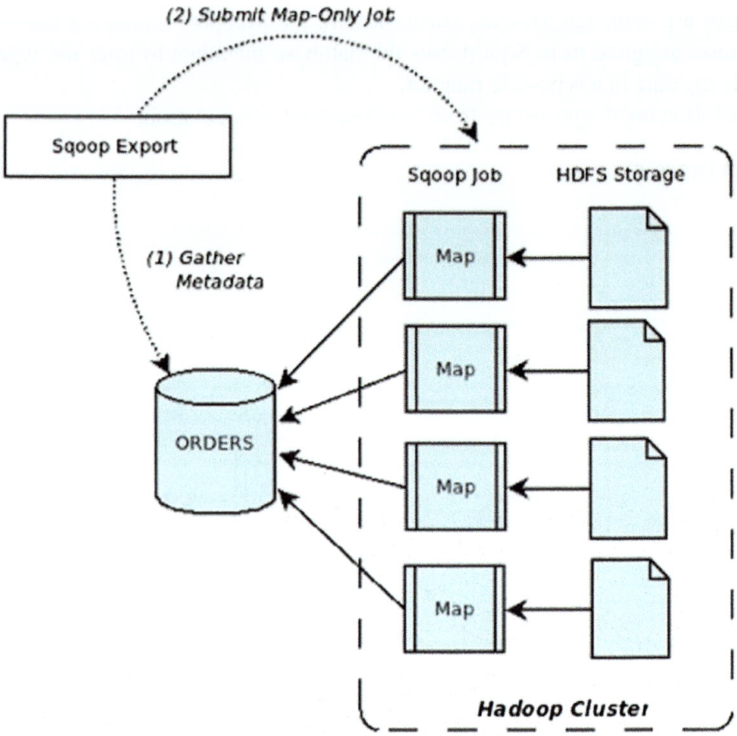

Supported databases:

Database	Version	--direct support	Connect string matches
HSQLDB	1.8.0+	No	jdbc:hsqldb:*//
MySQL	5.0+	Yes	jdbc:mysql://
Oracle	10.2.0+	No	jdbc:oracle:*//
PostgreSQL	8.3+	Yes (import only)	jdbc:postgresql://

```
grant all privileges on hcatalog.* to '%'@'localhost';
Query OK, 0 rows affected (0.04 sec)
mysql> grant all privileges on hcatalog.* to "@'localhost';
Query OK, 0 rows affected (0.00 sec)
-> bin/sqoop list-databases --connect jdbc:mysql://localhost/hcatalog
->bin/sqoop list-tables --connect jdbc:mysql://localhost/hcatalog
->bin/sqoop import --connect jdbc:mysql://localhost/hcatalog --table
TBLS -m 1
-> vi SqoopImportOptions.txt
bin/sqoop import --connect jdbc:mysql://localhost/hcatalog
->bin/sqoop --options-file SqoopImportOptions.txt --table TBLS -m
1
->bin/sqoop import --connect jdbc:mysql://localhost/hcatalog --table
TBLS --where
"SD_ID = 13"
->bin/sqoop import --connect jdbc:mysql://localhost/hcatalog --table
TBLS –P
->bin/sqoop import --connect jdbc:mysql://localhost/hcatalog --
target-dir
/user/vm4learning/manasa --query 'select * from TBLS where
$CONDITIONS' --splitby
TBL_IDd
-> bin/sqoop import --connect jdbc:mysql://localhost/hcatalog --
target-dir
/user/vm4learning/manasa --query 'select * from TBLS where
$CONDITIONS'
--fetch-size=50000 --split-by TBL_IDd
Using a File Format Other Than CSV
-----------------------------------
->bin/sqoop import --connect jdbc:mysql://localhost/hcatalog --table
TBLS --assequencefile
->bin/sqoop import --connect jdbc:mysql://localhost/hcatalog --table
TBLS --asavrodatafile
->bin/sqoop import --connect jdbc:mysql://localhost/hcatalog --table
TBLS
--compress
->bin/sqoop import --connect jdbc:mysql://localhost/hcatalog --table
TBLS --nummappers10
->bin/sqoop import --connect jdbc:mysql://localhost/hcatalog --table
TBLS --nullstring
'\\N'--null-non-string '\\N'
->sqoop import-all-tables --connect
jdbc:mysql://mysql.example.com/sqoop
```

--username sqoop --password sqoop
-> sqoop import-all-tables --connect
jdbc:mysql://mysql.example.com/sqoop
--username sqoop --password sqoop --exclude-tables cities,countries
The eval tool allows users to quickly run simple SQL queries against a database;
results are printed to the console. This allows users to preview their import queries
to ensure they import the data they expect
->bin/sqoop eval --connect jdbc:mysql://localhost/hcatalog --query "select * from
TBLS limit 2"
Incremental Import

->bin/sqoop import --connect jdbc:mysql://localhost/hcatalog --table emp
--incremental append --check-column id --last-value 40 -m 1
bin/sqoop import --connect jdbc:mysql://localhost/hcatalog --table emp
--incremental lastmodified --check-column id --last-value 70
--> Meta store

->bin/sqoop job --create wrap -- \

import --connect jdbc:mysql://localhost/hcatalog --table emp --incremental append
--check-column id \
--last-value 40 -m 1
->sqoop job --exec wrap
--> Custom Boundary Queries
----
bin/sqoop import --connect jdbc:mysql://localhost/hcatalog --target-dir
/user/vm4learning/manasa --query 'select * from TBLS where $CONDITIONS' --splitby
TBL_IDd ----boundary-query "select min(id), max(id) from TBLS"
Importing Data Directly into Hive

Argument	Description
--enclosed-by <char>	Sets a required field enclosing character
--escaped-by <char>	Sets the escape character
--fields-terminated-by <char>	Sets the field separator character
--lines-terminated-by <char>	Sets the end-of-line character
--mysql-delimiters	Uses MySQL's default delimiter set: fields: , lines: \n escaped-by: \ optionally-enclosed-by:
--optionally-enclosed-by <char>	Sets a field enclosing character

--> bin/sqoop import --connect jdbc:mysql://localhost/hcatalog --
table emp --direct
-m 1 --hive-import --create-hive-table --hive-table
departments_mysql --target-dir
/user/hive/warehouse/ --enclosed-by '\"' --fields-terminated-by , --
escaped-by \\

Hive partitioned tables:
--> bin/sqoop import --connect jdbc:mysql://localhost/hcatalog –
query 'select * from
emp where gender = "M" and $CONDITIONS' --direct -m 6 --hive-
import --createhive-
table --hive-table employees_import_parts --target-dir
/user/hive/warehouse/employee-parts --hive-partition-key gender --
hive-partitionvalue 'M' .
--bin/sqoop import --connect jdbc:mysql://localhost/hcatalog –query
'select *
from emp where gender = "F" and $CONDITIONS' --direct -m 6 --
hive-import
--create-hive-table --hive-table employees_import_parts --target-dir

/user/hive/warehouse/employee-parts_F --hive-partition-key gender
--hivepartition-value 'F' .
--> bin/sqoop import --connect jdbc:mysql://localhost/hcatalog --
table emp --hiveimport
--> bin/sqoop import --connect jdbc:mysql://localhost/hcatalog --
table emp --hiveimport--hive-partition-key day --hive-partition-
value "2013-05-22"
HBASE
-----
bin/sqoop import --connect jdbc:mysql://localhost/dbsit --table emp
--hbase-tablemytable --column-family cf1
Export :
bin/sqoop export \
-- connect jdbc:mysql://localhost/hcatalog
--table employees_export \
--staging-table employees_exp_stg \
--clear-staging-table \
-m 4 \
--export-dir /user/vm4learning/sqoop-mysql/Employees

4.6 Mahout, The Machine Learning Platform from Apache

Mahout (http://mahout.spark.org) is a library of Java code of scalable machine learning and data mining algorithms in four categories:

1. Recommendations or collective filtering
2. Classification, categorization clustering and
3. Frequent item set mining and parallel frequent pattern mining.

All the algorithms in Mahout Library can be executed in a distributed manner and have been executable in MapReduce.

Features:

1. Mahout offers a ready-to-use framework to the programmer for executing machine language and data mining tasks on large volume datasets which is a fast manner.
2. Mahout is written on top of the Apache Hadoop, so that it performs well in distributed environment. It uses the Library of Apache Hadoop to scale in the Cloud.
3. Applications that are required to use large quantities of data can use Mahout.
4. Mahout includes many MapReduce-based implementations of data mining algorithms such as k-means, fuzzy k-means, Canopy, Dirichlet and Meanshift.
5. Mahout supports distributed Naïve Bayes, Complementary Naïve Bayes classification implementation.
6. Mahout includes matrix-to-vector libraries.

Applications of Mahout:

Big Data companies such as Adobe, Facebook, LinkedIn, Foursquare, Twitter and Yahoo all use Mahout:

1. Foursquare helps its users as general public to locate places, food and entertainment available in a particular area. The recommended engine of Mahout is deployed for this purpose.
2. Twitter uses Mahout for interest modeling.
3. Yahoo uses Mahout for pattern mining.

4.7 GANGLIA, The Monitoring Tool

Ganglia (http://ganglia.sourceforge.net) is a monitoring system for high-performance computing (HPC) systems such as clusters and grids. Ganglia is open-source (BSD licensed) project coming out of Millennium Project of University of California, Berkeley. It is based on a hierarchical design targeted at federation of clusters. It uses XML for data representation, XDR for compact, portable data transport and RRD tool for data storage and virtualization. It is designed with efficiency as a goal. It has been ported to many operating system environments with robust implementations. It can be scaled up even up to 2000 node clusters. Ganglia is currently in use in thousands of clusters across the globe in the University Campuses. Ganglia WEB 3.5.12 of 2014 under GSOC (Google Summer of Code) is used in top campuses as University of California, Berkeley, MIT, Research Organizations as NASA, NIH, CERN, enterprises such as Bank of America and Lockheed Martin.

4.8 Kafka, The Stream Processing Platform (http://kafka.apache.org)

Apache Kafka is a distributed streaming platform. Kafka can be used to perform the following tasks:

1. [21]Publish and subscribe to streams of records, as a message queue or a messaging system.
2. Build real-time streaming applications that transform or react to the input data.
3. Build real-time streaming applications that can get and process the data between systems or applications.

Kafka runs on a cluster of servers, which can also be spanning across data centers.
Kafka cluster stores streams of records in categories called topics.
Each record consists of a key, a value and a time stamp.

Kafka APIs

1. Producer API: To produce and publish records to one or more topics.
2. Consumer API: This API allows an application to subscribe to one or more topics and process those streams of records.
3. Stream API: This API allows an application to act as a stream process, consuming an input stream and produce an output stream either of them being from one or more input/output topics, effectively transforming the input streams to output streams.
4. Connector API: This API enables building and executing reusable producers or consumers which connect Kafka topics to existing applications or data systems (e.g., a connection to a relational database will capture all transactions on the database).

Connection between clients and servers is performed through a simple TCP protocol. Java client is provided by Apache for Kafka (but many language clients are produced and are made available for users).

[21]26 reference.

What is a Topic?

Topic is a category or feed name to which records are published. Topics can be multisubscriber in nature. A topic can have any number of consumers.

The partitions of log are distributed over the servers in a Kafka cluster with each server handling data and request for a share in the partition. Each partition is replicated from 'leader' node into 'follower' node. All the read/write transactions in the leader get replicated in the follower.

'Mirror maker': The messages are replicated across multiple data centers or multiple cloud regions. This can be harnessed for backup message recovery plans.

Who are the Producers?

Publishers publish the data to the topics of their choice.

Who is a Consumer?

Consumers have a group name, and each record published to a topic is delivered to one consumer instance within each consuming group.

Stream Processing in Kafka

The objective of Kafka is to read, write, store and also process the streams of data in real time. If we take an example retail application, it has to take input stream data about sales and shipments, process the input data to produce an output of streams of records and price adjustments computed on the input data.

While simple Consumer API and Producer API are adequate at minimum level, Kafka also offers 'Stream API' which enables building applications that can perform non-trivial processing on streams or even perform joining of streams together.

To summarize, Kafka is a successful streaming platform offering the combination of messaging, storage and stream processing. Kafka combines the static distributed file system features of HDFS on one hand and the traditional enterprise messaging, stream pipelining for processing messages that will arrive in future. Kafka combines all these features. Kafka enables user applications to process very low-latency pipelines, store the data when required for critical delivery of output after the requisite processing and also ensure the integration with offline systems that load the data periodically (or shutdown for long time for maintenance). The stream processing facilitates the processing of data as it arrives.

4.9 Spark

Spark is an open-source cluster computing framework with in-memory data processing environment. It was developed in University of California, Berkeley. Later, I was donated to Apache. Since it uses in-memory operations, it performs well compared to other Big Data tools. Spark is compatible with Hadoop.

Hadoop is the solution for Big Data management. It uses MapReduce (framework for parallel processing) as the processing environment. MapReduce uses batch processing for handling multiple tasks which were resolved by introducing Yarn.

Yarn is a resource management and scheduling tool which maintains multiple jobs on a cluster.

Spark Architecture contains master and workers. Master holds the driver and workers contain executers. Spark Context is the entry point for spark application which is the part of the driver. The executer is responsible for executing distributed process.

Spark application can be deployed and executed in various modes: Stand-alone, Cluster and Client. In Stand-alone mode, the driver and worker are initiated and executed on the same machine. In Cluster mode, the driver will be placed in a node and workers will be initiated on various machines. In Client Mode, the driver program will be on client machine and workers will be placed on cluster.

The key concept of spark is Resilient Distributed Dataset (RDD). RDD is the distributed in-memory computing environment. It resides the data in memory as much as possible in distributed manner to make it available with high speed. In general, the RDD is immutable meaning that read only data structure that enables to create new RDD by applying transactions. RDDs are lazy-evaluated environment meaning that it evaluates the RDD whenever it required for an action. It also allows the evaluated RDDs to cache in memory/disk to improve the process performance.

Creating RDDs:

Spark Context provides the environment to create RDDs. It is initialized in spark-shell by default with the name 'sc'.

There are two ways to create RDD: parallelizing the existing collection and referencing the dataset from the existing storage (HDFS/HBase/S3).

Parallelizing Existing Collection:

We can convert the existing collection into RDD by using 'parallelize' method from 'sc'.

```
scala> val a=Array(1,2,3,4,5,6,7,8,9)
a: Array[Int] = Array(1, 2, 3, 4, 5, 6, 7, 8, 9)

scala> val ardd= sc.parallelize(a)
ardd: org.apache.spark.rdd.RDD[Int] = ParallelCollectionRDD[115] at parallelize at <c
onsole>:28

scala> ardd.collect.foreach(print)
123456789
scala> 
```

Referencing Data from Existing Storage:

Spark can extract the data from existing storage like HDFS/HBase/S3 by using the various methods like 'textFile' from 'sc'.

```
scala> val frdd=sc.textFile("sample.txt")
frdd: org.apache.spark.rdd.RDD[String] = sample.txt MapPartitionsRDD[119] at textFile
 at <console>:26

scala> frdd.collect.foreach(println)
1001|Network Related
1002|Router Related
1003|Switch Related
1004|Hub Related
1005|Machine IP Related

scala> 
```

Spark supports two types of operations on RDDs:

Transformations and Actions:

Transformations are the operations that take an RDD as input and return new RDD. Actions accept RDDs as input and return the output to the driver or write the result onto HDFS.

Transformations:

Filter: Filters the data which satisfies the given condition as the part of the Lambda function.

```
scala> val a=Array(1,2,3,4,5,6,7,8,9)
a: Array[Int] = Array(1, 2, 3, 4, 5, 6, 7, 8, 9)

scala> val ardd=sc.parallelize(a)
ardd: org.apache.spark.rdd.RDD[Int] = ParallelCollectionRDD[12] at parallelize at <console>:26

scala> val afilter=ardd.filter(x=> x%2==0)
afilter: org.apache.spark.rdd.RDD[Int] = MapPartitionsRDD[13] at filter at <console>:28

scala> afilter.collect.foreach(println)
2
4
6
8
```

map: It applies the given function on all the elements of the dataset.

```
scala> val a=Array(1,2,3,4,5,6,7,8,9)
a: Array[Int] = Array(1, 2, 3, 4, 5, 6, 7, 8, 9)

scala> val ardd=sc.parallelize(a)
ardd: org.apache.spark.rdd.RDD[Int] = ParallelCollectionRDD[15] at parallelize at <console>:26

scala> val amap=ardd.map(x=> x+5)
amap: org.apache.spark.rdd.RDD[Int] = MapPartitionsRDD[16] at map at <console>:28

scala> amap.collect.foreach(println)
6
7
8
9
10
11
12
13
14
```

```
scala> val a=Array(1,2,3,4,5,6,7,8,9)
a: Array[Int] = Array(1, 2, 3, 4, 5, 6, 7, 8, 9)

scala> val ardd=sc.parallelize(a)
ardd: org.apache.spark.rdd.RDD[Int] = ParallelCollectionRDD[19] at parallelize at <console>:26

scala> val afmap=ardd.flatMap(x=>x.to(3))
afmap: org.apache.spark.rdd.RDD[Int] = MapPartitionsRDD[20] at flatMap at <console>:28

scala> afmap.collect.foreach(println)
1
2
3
2
3
3
```

distinct: It returns the elements from the given dataset by eliminating duplicates.

```
scala> val a=Array(1,2,3,4,1,2,3,4,5)
a: Array[Int] = Array(1, 2, 3, 4, 1, 2, 3, 4, 5)

scala> val ardd=sc.parallelize(a)
ardd: org.apache.spark.rdd.RDD[Int] = ParallelCollectionRDD[21] at parallelize at <console>:26

scala> val adist=ardd.distinct
adist: org.apache.spark.rdd.RDD[Int] = MapPartitionsRDD[24] at distinct at <console>:28

scala> adist.collect.foreach(println)
4
2
1
3
5
```

sample: It returns the sample of given percentage from the RDD.

```
scala> val a=Array(1,2,3,4,5,6,7,8,9)
a: Array[Int] = Array(1, 2, 3, 4, 5, 6, 7, 8, 9)

scala> val ardd=sc.parallelize(a)
ardd: org.apache.spark.rdd.RDD[Int] = ParallelCollectionRDD[25] at parallelize at <console>:26

scala> val srdd=ardd.sample(false,0.3)
srdd: org.apache.spark.rdd.RDD[Int] = PartitionwiseSampledRDD[26] at sample at <console>:28

scala> srdd.collect.foreach(println)
1
2
4
6
```

union: It performs the union operation on given two RDDs.

```
scala> val a=Array(1,2,3,4,5)
a: Array[Int] = Array(1, 2, 3, 4, 5)

scala> val b=Array(4,5,6,7,8)
b: Array[Int] = Array(4, 5, 6, 7, 8)

scala> val ardd=sc.parallelize(a)
ardd: org.apache.spark.rdd.RDD[Int] = ParallelCollectionRDD[27] at parallelize at <console>:26

scala> val brdd=sc.parallelize(b)
brdd: org.apache.spark.rdd.RDD[Int] = ParallelCollectionRDD[28] at parallelize at <console>:26

scala> val aub=ardd.union(brdd)
aub: org.apache.spark.rdd.RDD[Int] = UnionRDD[29] at union at <console>:32

scala> aub.collect.foreach(print)
1234545678
```

intersection: It performs the intersection operation on RDDs.

```
scala> val a=Array(1,2,3,4,5)
a: Array[Int] = Array(1, 2, 3, 4, 5)

scala> val b=Array(4,5,6,7,8)
b: Array[Int] = Array(4, 5, 6, 7, 8)

scala> val ardd=sc.parallelize(a)
ardd: org.apache.spark.rdd.RDD[Int] = ParallelCollectionRDD[30] at parallelize at <console>:26

scala> val brdd=sc.parallelize(b)
brdd: org.apache.spark.rdd.RDD[Int] = ParallelCollectionRDD[31] at parallelize at <console>:26

scala> val aintb=ardd.intersection(brdd)
aintb: org.apache.spark.rdd.RDD[Int] = MapPartitionsRDD[37] at intersection at <console>:32

scala> aintb.collect.foreach(print)
45
```

Cartesian: It performs the Cartesian product operation on given RDDs (maps each element of first RDD to each element on second RDD.

```
scala> val a=Array(1,2,3,4,5)
a: Array[Int] = Array(1, 2, 3, 4, 5)

scala> val b=Array(4,5,6,7,8)
b: Array[Int] = Array(4, 5, 6, 7, 8)

scala> val ardd=sc.parallelize(a)
ardd: org.apache.spark.rdd.RDD[Int] = ParallelCollectionRDD[38] at parallelize at <console>:26

scala> val brdd=sc.parallelize(b)
brdd: org.apache.spark.rdd.RDD[Int] = ParallelCollectionRDD[39] at parallelize at <console>:26

scala> val acarb=ardd.cartesian(brdd)
acarb: org.apache.spark.rdd.RDD[(Int, Int)] = CartesianRDD[40] at cartesian at <console>:32

scala> acarb.collect.foreach(print)
(1,4)(1,5)(2,4)(2,5)(1,6)(1,7)(1,8)(2,6)(2,7)(2,8)(3,4)(3,5)(4,4)(4,5)(5,4)(5,5)(3,6)(3,7)(3,8)(4,
6)(4,7)(4,8)(5,6)(5,7)(5,8)
```

Actions:

collect: It retrieves all the elements from distributed set into driver.
It is not recommended for large datasets.

```
scala> val a=Array(1,2,3,4,5,6,7,8,9)
a: Array[Int] = Array(1, 2, 3, 4, 5, 6, 7, 8, 9)

scala> val ardd=sc.parallelize(a)
ardd: org.apache.spark.rdd.RDD[Int] = ParallelCollectionRDD[0] at parallelize at <console>:26

scala> ardd.collect
res0: Array[Int] = Array(1, 2, 3, 4, 5, 6, 7, 8, 9)

scala> a
res1: Array[Int] = Array(1, 2, 3, 4, 5, 6, 7, 8, 9)
```

count: It returns the number of elements in the current RDD.

```
scala> val a=Array(1,2,3,4,5,6,7,8,9)
a: Array[Int] = Array(1, 2, 3, 4, 5, 6, 7, 8, 9)

scala> val ardd=sc.parallelize(a)
ardd: org.apache.spark.rdd.RDD[Int] = ParallelCollectionRDD[1] at parallelize at <console>:26

scala> ardd.count
res2: Long = 9
```

countByValue: It returns the counts of each value in the current RDD.

```
scala> val a=Array(1,2,3,4,5,3,1,3,5)
a: Array[Int] = Array(1, 2, 3, 4, 5, 3, 1, 3, 5)

scala> val ardd=sc.parallelize(a)
ardd: org.apache.spark.rdd.RDD[Int] = ParallelCollectionRDD[3] at parallelize at <console>:26

scala> ardd.countByValue
res4: scala.collection.Map[Int,Long] = Map(5 -> 2, 1 -> 2, 2 -> 1, 3 -> 3, 4 -> 1)
```

take: It returns first n elements for the given n value.

```
scala> val a=Array(1,2,3,4,5,6,7,8,9)
a: Array[Int] = Array(1, 2, 3, 4, 5, 6, 7, 8, 9)

scala> val ardd=sc.parallelize(a)
ardd: org.apache.spark.rdd.RDD[Int] = ParallelCollectionRDD[7] at parallelize at <console>:26

scala> ardd.take(3)
res5: Array[Int] = Array(1, 2, 3)
```

top: It returns the top n(large/big) elements for given n value.

```
scala> val a=Array(1,2,3,4,5,6,7,8,9)
a: Array[Int] = Array(1, 2, 3, 4, 5, 6, 7, 8, 9)

scala> val ardd=sc.parallelize(a)
ardd: org.apache.spark.rdd.RDD[Int] = ParallelCollectionRDD[8] at parallelize at <console>:26

scala> ardd.top(3)
res6: Array[Int] = Array(9, 8, 7)
```

reduce: It performs the given operation (function) on the current RDD and returns its result.

```
scala> val a=Array(1,2,3,4,5,6,7,8,9)
a: Array[Int] = Array(1, 2, 3, 4, 5, 6, 7, 8, 9)

scala> val ardd=sc.parallelize(a)
ardd: org.apache.spark.rdd.RDD[Int] = ParallelCollectionRDD[10] at parallelize at <console>:26

scala> ardd.reduce((x,y)=> x+y)
res7: Int = 45
```

fold: It is similar to reduce by including the initial value for each partition on performing operations.

```
scala> val a=Array(1,2,3,4,5,6,7,8,9)
a: Array[Int] = Array(1, 2, 3, 4, 5, 6, 7, 8, 9)

scala> val ardd=sc.parallelize(a)
ardd: org.apache.spark.rdd.RDD[Int] = ParallelCollectionRDD[11] at parallelize at <console>:26

scala> ardd.fold(5)((x,y)=>x+y)
res8: Int = 60

scala> ardd.fold(1)((x,y)=>x+y)
res9: Int = 48

scala> ardd.fold(0)((x,y)=>x+y)
res10: Int = 45
```

4.10 NoSQL Databases

In 1998, the term 'NoSQL' was first coined (by Carlo Strozzi) to indicate lightweight, open-source, non-relational database management systems which did not expose the standard SQL interface to the users. However, the real deployments came as late as 2009 when NoSQL or Not only SQL databases were conceived and introduced again (by Eric Evans).

NoSQL database management systems have been used in Big Data and other real-time applications where the conventional SQL will not be adequately capable to handle all the demands of size, scaling and variety of data which are characteristic to Big Data.

NoSQL database are used when the conventional relational DBMS cannot be effectively or efficiently deployed: social network data analysis, log analysis, temporal or time-based data (all situations where RDBMS cannot really handle).

What are NoSQL databases?

NoSQL databases or non-relational databases are Not only relational databases that can handle sizes of Big Data. They are non-relational, open-source, distributed databases (distributed across the nodes in a cluster such as Hadoop cluster) which are (1) capable to handle the following types of data (a) structured, (b) semi-structured and (c) less structured, (2) with no fixed table schema, (3) with no JOINS, (4) with no multi document transactions and (therefore) (5) relaxing one or more ACID properties. The ACID properties of Atomicity, Consistency, Isolation and Durability (which are strictly adhered to in RDBS) are not strictly adhered to in NoSQL databases. In contract, NoSQL databases adhere to CAP (Consistency, Availability and Partition tolerance) properties.

NoSQL databases do not have any fixed table schema (schema can be variable).

Types of NoSQL databases

NoSQL database can be broadly classified into (1) Key/value pairs or the big hash table of

(a) Keys and Values (e.g., Key: First name; value : Rama; Key: Last name; value: Prabhu)
(b) Documents: Data is maintained as a collection of documents (e.g., MongoDB, Apache Couch DB, Couchbase, Mark Logic)

Example

{"Book Name" : "object oriented database"
" Publisher" : "Pentice Hall India"
"Year of publication": "2005"}

(c) Column Store: Each storage block has data only from only one column (Cassandra, HBase etc.)

(d) NoJOINS: NoSQL databases do not support JOINS (compensated by allowing embedded documents as in MongoDB)
(e) No standard SQL but there are query languages (e.g., MongoDB, query language)

NoSQL DBMS Vendors

The NoSQL DBMS vendors are: (a) Amazon offering Dynamo DB (also deployed by LinkedIn and Mozilla), (b) Facebook offering Cassandra (also deployed by Netflix, Twitter, eBay) and (c) Google offering Big Table (also deployed by Adobe Photoshop).

New SQL

We also have 'New SQL' coming up with NoSQL DBMS capabilities of Big Data, along with the standard SQL-based ACID properties of SQL, with a higher per-node performance, with shared-nothing architecture, added with a new concurrency control mechanism (where real-time reads will not conflict with writes).

Key/Value Databases

From an API perspective, key/value stores are the simplest NoSQL data stores. From the data store for a given key, the value can be stored ('put') or retrieved ('get'). The value is a binary long object (blob).

Popular Key/Value Data Stores

'Riak', Redis (or Data structure server) 'Memcached', 'Berkely DB', 'Upscale DB' (for embedded use), Amazon's 'Dynamic DB' (not open source), project 'Voldemart' and 'Couchbase'. Differences do exist: While 'Memcached' data is not permanent, 'Riak' has persistent storage of data. If a node goes down, all the data 'memcached' will be lost and can be only refreshed getting from resource system not so is Riak, but we need to update the data also. Therefore, the specific application requirement dictates as to which particular key/value DBMS is to be chosen.

Document Databases

In Document Databases, the documents (in XML, JSON, BSON, etc.) can be stored and retrieved. Maps, collections and scalar values can be stored as these documents. They need not be exactly same (unlike other RDBMSs); they can be only similar. Document databases store documents in the value part of the key/value store, where value can be examined. MongoDB, for example, offers the users a strong query language and features as indexes similar to regular database. This enables easy transition to document database from the conventional RDBMS.

Popular Document Databases

MongoDB, Couch DB, Terra Store, Orient DB and Raven DB are popular. Column family databases on data stores store data in column families. They store data as rows with many columns. A row key is also there. Thus, column families are groups of related data which are often accessed together. If we compare with RDBMS environment, a column family is equivalent or analogous to a container of rows in a table in RDBMS. The only difference is that various rows used need not be exactly identical, unlike in RDBMS. A new column can be added to any row at any time without adding the same to the other rows in the table. If a column consists of a map of columns, then it is a supercolumn (it contains a name and a value which is a map of columns).

Thus, a supercolumn is a container of columns. Cassandra [2] has such super-columns. Cassandra is a very popular column family database system. HBase, Hyper-Table and DynamicDB (by Amazon) are all column database systems. Cassandra is fast and scalable. Cassandra 'write' commands are spread across the cluster, with any node in the cluster equally capable of executing any 'write' command.

Graph Databases

Graph databases enable storage of entities and the relationships between them. In the graph database, entities are represented as nodes which have properties. Properties are represented as edges of the nodes with directionality being considered. They enable us to find out interesting patterns between nodes. Thus, graph databases enable us to not only store the data once but also allow us to interpret the same, in various ways, depending on the relationships (indicated by edges). This feature is a unique strength of graph databases. If we store graph data (e.g., a relationship 'manager' for an 'employee') in a RDBMS, if there is any addition or deletion or modification, the scheme of the database is required to be change. But in a graph database, we need not do so.

In RDBMS, the graph traversal path is also preplanned and implemented, regarding the data to change every time, if the path is changed. In graph databases, this is avoided, as multiple nodes are permitted in the beginning itself.

Also since there can be unlimited number and types of relationships, a node can have all relationships (such as secondary relationships). Can be represented in the same graph Quad trees for partial spatial indexing and linked lists for sorted access can be implemented in graph database.

The main strength of the graph database is the ability to store relationships which may have a type, a start node and an end node also properties. Properties can be used to add intelligence to the relationship.

Therefore, they are used to carefully model the relationships in an application environment. Graph databases can be queued in the entity properties and their relationships. New relationships can be easily added. Also changing their relationships is similar to data migration as the changes are required to be executed on each node and each relationship.

Example Graph Databases

Neo 4 J, Infinite graph, Orient DB, Flack DB (single depth relationships) are some of the graph databases

How to Choose?

Application characteristics decide the database to be selected: If the application calls for querying the data or to store relationships between entities or to use multiple keys, we can avoid a key/value database but instead use graph databases. On the other hand, for content management systems, blogging, web analytics, real-time analytics, e-commerce, we may deploy document databases. Key/value databases will be useful for storing session-related information, user profiles, shopping cart data, etc. If complex transactions spanning multiple operations or queries against varying aggregates are used, we may avoid document databases.

Column family databases are useful in content management systems, heavy write volume long aggregation.

Schemaless NoSQL Databases and Migration

Migration in schemaless NoSQL databases is more complex than migration in conventional schema-based databases as RDBMSs. Incremental data migration techniques to update data need to be deployed.

NoSQL is not NO SQL

NoSQL database does not mean the death of RDBMS or SQL. It only opens up a new perspective with polyglot database fractionality with SQL, NoSQL, Schemaless, column databases, graph databases, etc., that is all!

I. **MONGODB**

MongoDB is a cross-platform, open-source, non-relational, distributed NoSQL, document-oriented data store.

The traditional RDBMSs have been challenged by issues like being able to handle large volumes of data, handling variety of data—semi-structured or particularly unstructured data and scaling up to Big Data scale needs of the enterprise data. There is a need felt for a database which can scale out or scale horizontally to meet sealing requirements, has flexibility with respect to schema, is fault tolerant, is consistent and partition tolerant and is easily distributed over a multitude of nodes in a cluster. It also has indexing, JSON (Java Script Object Notation), which has rich query language and features as high availability. JSON is extremely expressive. In MongoDB, BSON (Binary JSON) is used. It is open standard. It is used to store complex data structures. As against the fixed formats in CSV, in JSON we can express variable number of field occurrences (e.g., we can express and store multiple randomly variable number of phone numbers for an individual). JSON being highly expressive provides the much needed facility to store and retrieve documents in their original real form. The binary form of JSON is BSON which is an open standard, and it is more space efficient than JSON. It also becomes easily amenable to be converted into programming language format. There are available a number of drivers of MongoDB for programming languages such as C, C++, Ruby, PHP, Python and C#, and each works slight differently. By using the basic binary format (as BSON), it will be easily possible to enable the native data structures to be built quickly for each language, without the need for first processing JSON.

Each JSON document has a unique identity: its 'id Key', similar to primary key in relational database. This facilitates search for documents based on unique identifier. An index also is built automatically on the unique identifier. The unique values can either be provided by the user or MongoDB will generate the same by itself.

Database

A database in MongoDB is a collection of collections or a container of collections. It gets created first time when the first collection makes a reference to it. This can also be created on demand. Each database gets its own set of files as the file system. A single MongoDB server can browse several databases.

Collection

A collection is similar to a table of RDBMS. A collection is created on demand. It gets created the first time when we attempt to save a document that references it. A collection exists within a single database. A collection holds several MongoDB documents. A collection does not enforce a schema. Therefore, documents within a collection can have different fields or can be a different order of the same fields.

Document

A document is analogous to a row/record/table in an RDBMS table. A document has a dynamic schema. Therefore, a document in a collection need not have the same set of fields/key/value pairs.

Dynamic Queries

MongoDB supports dynamic queries on dynamic data (in RDBMS, we have static data and dynamic queries). Couch DB, on the contrary, has support for dynamic data and static queries.

Storing Binary Data

MongoDB provides grid FS to support storage of binary data. It can store up to 4 MB of data. This is adequate for photographs or small audio clips. If more data is required to be stored, MongoDB provides a different notation: It stores the metadata (data about data, along with content information) in a collection called 'file'. It then breaks the data into small pieces called 'chunks' and stores it in the 'chunks' collection. This process takes care of scalability.

Duplication

High availability is provided by redundancy, by replication. When hardware failures and service interruptions occur, recovery is made possible by this mechanism. In MongoDB, the replica set has a single primary and multiple secondaries. Each write request from the client is directed to the primary. The primary logs all write requests into its Oplog (operational log). This Oplog is then used by secondary replica members to synchronize their data. This way strict adherence to consistency is achieved. Multiple secondary replicas will be maintained. The clients usually read from the primary only. However, the client can also specify a read preference that will then direct the read operation to the secondary.

Sharding

Sharding is similar to horizontal scaling. It means that a large dataset is divided and distributed over multiple servers or shards. Each shard is an independent database and collectively, they constitute a logical database.

Advantages of Sharding Are

Sharding reduces the amount of data that each shard needs to store and manage (as a fraction of the original large dataset). Increasing number of servers (nodes) in the cluster automatically decreases the size of each shard.

1. Sharding reduces the number of operations each shard handles. When a new insertion is performed, the application needs to access only that shard column houses that data.

Updating Information in Place

MongoDB updates the data wherever it is available. It does not allocate separate space, and the indexes remain unaltered. MongoDB performs lazy writes. It writes to the disk once every record. Otherwise data is there only in main memory, thereby enabling fast access and speed performance and fast retrieval (however, there is no guarantee that the data will be safely stored in the disk).

Equivalent/Analogous Terms

RDBMS	Versus	MONGODB
Database	–	Database
Table	–	Collection
Record	–	Document
Columns	–	Fields/Key-value pairs
Index	–	Index
Primary Key	–	Primary Key (-id is a identify)

Data Types in MongoDB

String (UTF8 valid) integer (32 or 64 bit), Boolean, Double (Flouting point), Min/Max keys, Arrays, Time stamp, Null, Data, object ID (Document Id), Binary Data (images, etc.) code (to start Java Script Code into document) regular expression (to store regular expressions).

MongoDB Commands

Create: use DATABASE, Name'—will create a database.

db. drop database(); will drop a database 'show dbs' will get a list of all databases.

'db. Students. insert ({-id:1,
Roll No.1 001', Name – 'Ram'};
will insert a new record.

db.Students.find (<Student Name : <regex: "a$")})'pretty '()';
will produce all student names ending with 'a'.

db.students.find ({Hobbies: {$ kin: ['chess', 'tennis')}} .pretty ();
will identify student documents with hobbies neither 'chess' nor 'tennis'.

db.students.find ({grade:{$ he: 'x')}. Pretty ();
will food those documents where grade is not set to 'x' arrays.

db.find.update ({-id; 4}, {set: {'fruits.1='Mango'}});
will set the 1st index partition of 'fruits' array with Mango'.

db.find.update ({-id; 5}, {$add to set {fruits:'Orange'}})
will update the document with -id:5 by adding an element 'orange' to the list of elements is the array 'fruits'.

II. **CASSANDRA** [2]

Cassandra was originally developed at Facebook and from 2008 onward, it became open-source Apache Cassandra.

Cassandra is built upon Amazons' DynamoDB and Google's Big Table.

Cassandra never goes down—it is always available. As it is not master–slave architecture, there is no single-point failure possible. Non-stop business critical applications can run on Cassandra.

Cassandra can distribute gigantic amounts of data, i.e., Big Data, on commodity servers. It is highly scalable, high-performance distributed database system.

Features of Cassandra—An Overview

Cassandra is highly scalable, open-source, distributed and decentralized (server symmetry), peer-to-peer architecture. Cassandra is column-oriented DBMS (as against row-oriented RDBMS).

Cassandra is not based on ACID properties. Cassandra is BASE or Basically Available Soft State with Eventual Consistency.

Cassandra is a state-of-the-art product, successfully deployed in clients such as Twitter, Netflix, Cisco, Adobe, eBay and Rackspace.

Peer to Peer

As any other NoSQL database system, Cassandra is designed to distribute and manage large data loads across multiple nodes in an environment of commodity hardware cluster. There is no possibility of single point of failure, as there is no master–slave architecture. A node in Cassandra is structurally identical to any other node in a cluster.

If a node fails, it only impacts the performance but will not bring the cluster to a standstill. The problem of failure is overcome by employing a peer-to-peer distributed system across homogeneous nodes. It ensures that data is distributed across all nodes in the cluster; each node exchanges information across the cluster every second.

In case of a write in Cassandra, each write is written to the commit log sequentially. A write is taken to be successful, if only it is written to the commit log. Data is then indexed and pushed to an in-memory structure called 'Memtable'. When this memory structure, 'Memtable', is full, its contents are flushed to 'SS table' (Sorted String) data filed on the disk. The SS table is immutable and is append only. It is stored on disk sequentially and maintained for each Cassandra table. The partitioning and replications of all writes are performed automatically across the cluster.

Failure detection by gossip protocol (intraring communication) is used for intraring communication. It is a peer-to-peer communication protocol which eases and enables the discovery and sharing of location and state information with other nodes in the cluster.

A node has only to send out communication to a subset of other nodes. For repairing unread data, an antientropy version of gossip protocol is used in Cassandra.

A partitioner takes a call on how to distribute data on the various nodes of a cluster. It also determines the node on which to place the very first copy of the data. Basically, a partitioner is a hash function to compute the token of the partition key. This partition key helps to identify a row uniquely.

Replication and Replication Factor

The replication factor determines the number of replicas (number of copies) that will be stored across the nodes in a cluster. If only one copy of row is stored on each node, the replication factor is one and of two copies of each row are to be stored on each node the replication factor is two. Ideally, the replication factor is more than one but less than the number of nodes in the cluster. To determine which nodes to place the data on, two possible strategies are available for replication: (a) simple strategy and (b) network topology strategy.

The network topology strategy is simple, and it supports easy expansion to multiple data centers, should there be a need.

Replication of Data

To achieve fault tolerance, a given piece of data is replicated in one or more nodes. A client can connect to any node in the cluster to read data. How many nodes will be read before responding to the client is based on the consistency level specified by the client. If the client specified consistency is not met, then the read operation is blocked. There is also a possibility that a few of the nodes may respond with an out-of-date value. In such a case, Cassandra will initiate a read repair operation to bring the replica with old or stale values to up-to-date level. For repairing unread data, Cassandra uses any 'antientropy' version of the 'gossip' protocol. 'Antientropy' implies comparing all the replicas of each piece of data and updating each replica to the newest version. This read repair operation is performed either before or after returning the value to the client according to the consistency level prespecified by the client.

Writes

When a client initiates a write request, it is just written on to the commit log (only if it is written to commit log the write is taken as successful). The next step is put the write to a memory resident data structure called Memtable. A threshold value is defined in the Memtable. When the number of objects stored in the Memtable reaches a threshold, the contents of the Memtable are flushed to the disk in a file called SS Table (Sorted String Table). Flushing is a non-blocking operation. It is permeable to have multiple Memtables for a single column family. One out of them is current, and the rest are waiting to be flushed.

High Availability—Hinted Handoffs

How Cassandra achiever high availability (non-stop availability)? If we consider a cluster of three nodes, A, B and C, that node C is down for some time. With a replication factor '2', two copies of each row will be stored on two different nodes. The client makes a write request to Node A. Node A is the coordinator and serves as a proxy between the client and the nodes on which the replica is placed. The client writes Row K to Node A. Now Node A writes Row K to Node B and stores a hint for Node C (which is down).

The hint contains: (1) location of the node on which the replica is placed, (2) version Metadata and (3) the actual data.

As and when Node C recovers and comes up to normalcy, Node A reacts to the hint by forwarding the data to Node C.

Tunable Consistency

A database system can go for strong consistency or eventual consistency. Cassandra can provide either of the above two flavors of consistency, depending on the application requirement.

Looking at a scenario where, in a distributed system, among several servers in the systems, some servers are in one data center and others are in other data centers. In that case, we can have either strong consistency or eventual consistency.

1. Strong consistency implies that each update propagates to all the locations where that piece of data resides. In a single data center setup, strong consistency will ensure that all the servers that shall have a copy of the data will have it, before the client is acknowledged with a success. This may also mean additional time (a few will records) to write to all the servers.
2. Eventual consistency implies that the client is informed of success as soon as a part of the cluster acknowledges the write, i.e., as soon as the first server has executed a write (all other servers may only be eventually getting the write executed).
3. The choice between the above two types of consistency is left to the user. In case of very quick or urgent requirement of performance, this eventual consistency will be taken up, while an application that necessarily requires the complete ACID properties for each transaction execution (not withstanding the performance overload), the strong consistency option will be selected.

Read Consistency

Read consistency means how many replicas must respond before sending out the result to the client application. Multiple possibilities can be identified as below:

ONE	Returns response from the closest node (replica) holding the data concerned
QUORUM	Returns a result from a quorum of servers with the most recent time stamp for the data
Local Quorum	Returns a result from a quorum of servers with the most recent time stamp for the data in the local data center
Each Quorum	Returns a result from a quorum of servers with the most recent time stamp in all data centers
ALL	This provides the highest level of consistency of all levels and the lowest level of availability of all levels. It responds to a read request from a client only after all the replica nodes have responded completely

Write Consistency

Write consistency means or how many replicas 'Write' must succeed before rending art the acknowledgment to the client application. There are several write consistencies as below:

ALL	This is the highest level of consistency of all levels, as it necessitates that a write must be written on the commit log and Memtable on all replica nodes in the cluster. If the cluster located on multiple data centers a write should be written on the commit log and Memtable on quorum of replica nodes
EACH QUORUM	A write must be written on the commit log and Memtable on all replica nodes in the cluster. If the cluster located on multiple data centers, a write should be written on the commit log and Memtable on quorum of replica nodes
QUORUM	A write must be written to the commit log and Memtable on a quorum of replica nodes
LOCAL QUORUM	A write must by written on the commit log or Memtable on a quorum of replica nodes in the same data center as the coordinator node

We can reduce interdata center communication delays or latency. We may ensure that a 'write' must be written to commit log and Memtable of at least one (or two or three) replica nodes (for local-one a write should be sent to and successfully acknowledged by at least one replica node in data center).

CQL Data Types

The built-in data types for column in CQ Lane:

1	Int.	32-bit integer
2	Big int.	64-bit signed integer
3	Double	64-bit IEEE-754 floating point
4	Float	32-bit IEEE 754 floating point
5	Boolean	True or False
6	Blob	Arbitrary bytes (in hexadecimal)
7	Counter	Distributed counter-value
8	Decimal	Variable—precision integer
9	List	A collection of one or more ordered elements
10	Map	A JSON style array of elements.
11	Set	A collection of one or more elements
12	Time stamp	Date plus time
13	Var char	UTF-8 encoded string
14	Var Int.	Arbitrary precision integers
15	Text	UTF-8 encoded string

CQLSH

CQLSH has an objective as to get help with CQL for each input the output/result of executing the input statement will be provided.

Key Spaces

A key space is a container to hold application data. When we create a key space, it is required to specify a strategy class: 'simple strategy' class or 'network topology strategy' class. While 'simple strategy' class is used for evaluation/testing purpose, the 'network typology strategy' class is used for production usage.

The command 'CREATE KEYSPACE students WITH APPLICATION = { 'class' 'simple strategy', replication factor' :1} will create Keyspace 'students'

The command DESCRIBE KEY SPACES will be describing all Keyspaces.

The Command SELECT * FROM systemschema-Keyspaces will provide details of all Keyspaces From system schema—Keyspaces

The Command Use student will connect the client session to the specified key space (student)

The command CREATE TABLE students info (roll No. Int primary key name text JOIN date time stamp, percentage double); will create a table 'student info' in the Keyspace 'students'.

The command DESCRIBE TABLES will produce an outcome to look up all tables in the current key space or in all key spaces (if there is no current key space).

The command DESCRIBE TABLE student_info; produces an output which is a list of CQL commands with whose help the table Student_info can be recreated.

To insert data into column family Student_info, we need to use INSERT command which will write one or more columns; it is a record in Cassandra table atomically.

A bunch of Insert commands between BEGIN BATCH' and

APPLY BATCH will insert a bundle of records into table.

Viewing data from table is possible by SELECT FROM student_info where (candidates); (similar to SQL)

The command CREATE INDEX ON student_info will create an index on student name column of 'student_info' column family.

To execute a query using index defined on 'Student name' column use, we execute the command

SELECT FROM Student-Info
Where student name = 'Ram'.

To update the value held in 'student name' column of the 'student_info' column family to 'Sekhar' for the record where roll number column value = 2, we execute the command

UPDATE student_info SET student name = 'sekhar'
WHERE roll_number = 2;

To update primary key/value (to 6 from 3), we execute

UPDATE Student_info SET Roll number = 6
WHERE Roll number = 3

We can also update more than are column.

Similarly, we can perform all transactions. We have many additional features such as FILTERING ORDER BY in the SELECT events.

Collections

Set Collection

A column of type Set consists of unordered unique values. When the column is queried, it returns the values in sorted order (e.g., text values in alphabetical by sorted order).

List collection

When the order of elements matters, we should use list collection.

Map Collection

Mapping means mapping one element to another. A map is a pair of typed values. It is used to store time stamp-related information. Each element of the map is stored as a Cassandra column. Each element can be individually queried, modified and deleted. ALTER TABLE command is used to alter schema for the table (to add a column Hobbies as below)

ALTER TABLE student information ADD hobbies set <text>.

To update the table ‘student information’ to provide the values for ‘hobbies’ for the student with Roll No. = 1:

UPDATE student_information
SET hobbies = hobbies + [‘Char, table tennis’]
WHERE Roll No. = 1,

To confirm the value in hobbies column (for student Roll No. = 5)

SELECT FROM Student_information
WHERE Roll No. = 5.

To update value of list ‘language (for Roll No. = 5)

UPDATE student information
SET language = language + [‘Spanish’, Telugu’]
WHERE Roll No = 5;

COUNTER

A special column which is changed in increments is called a counter.

SET counter-value = counter-value +1
WHERE book_name = ‘introduction to data analytics’
AND
Name = ‘Ravi’;

Time to Live (TTL)

Time to Live (TTL) is optimal expiration period for a column (which is not a counter). The time period of TTL is specified in seconds.

CREATE TABLE user login (userid int primary key. Password text).
INSERT INFO user login (userid, password)
VALUES (1, 'infy') USING TTL 20;
SELECT TTL (password)

FROM user login
WHERE userid = 1

ALTER and DROP

ALTER is used to bring a change in the structure of the table/column family or charge the data type of an element.

ALTER TABLE example
ALTER example-id TYPE int (from primary type text)

We can drop a column to delete a column (except primary key)

ALTER TABLE example

DROP example—name;
We can DROP a table
The command DROP column family example;

will drop a column family 'example'.

We can DROP a database (keyspace)

DROP key space students;

EXPORT

To export the contents of the table/column family 'clearing lists' present in 'Students' database to (in steps below) to CSV file (d:ll clearing lists. C&V)

SELECT *

from clearing Lists;
copy clearing lists (id, P1, P2)
to id clearing lists. CSV);

IMPORT

SELECT FROM clearing text;

4.11 Conclusion

In this chapter, we have presented HDFS, MapReduce, Pig, Flume, Sqoop, Mahout, Ganglia, Kafka, Spark and NoSQL databases, namely MongoDB and Cassandra.

References

1. Cloudera documentation on Hadoop, HDFS, MapReduce
2. Cassandra, The Definitive guides by Eben Hewitt published by O' Reilly Media

Chapter 5
Predictive Modeling for Unstructured Data

5.1 Introduction

Huge influx of data is experienced everywhere on the Internet and also in enterprises. Understanding and analyzing the content present in this data can provide myriad applications. Earlier, correlations and patterns in the data are understood with descriptive analytics by compacting data into useful bytes of information. Now, it is no more considered effective to use descriptive analytics as they are responsive. We need to be proactive, and predictive analytics is the next possible solution. It utilizes various methods that are borrowed from statistics and computer science theory to model data with machine learning and data mining approaches. Thereby, it allows to study new and old data to make forecasts. Also, providing proactive analytics with predictive modeling helps to convert new and old data into valuable information.

Numerous applications are found in businesses, where predictive models use company-specific data to identify risks and opportunities. For example, predictive models are used in decision making for candidate transactions by capturing relationships among different features to identify risks associated with certain states. The usage of predictive modeling is also observed in analyzing customer relationships. They help to enable a company to evaluate customer-specific data by analyzing patterns that is used to predict customer behavior. Predictive modeling has also found its applicability to several tasks or domains like text and multimedia mining, selection of players for sport teams, criminal activity prediction, auctions and box office returns of movies.

Other examples can be found about companies that offer products at various levels. Predictive modeling can help a firm to analyze their expenditure patterns to make efficient sales by identifying right customers. This directly leads to higher profit per customer and better customer relationships. It is non-trivial to predict new outcomes by seeing the old observations. Various steps are involved in transforming raw data into more insightful information.

C. S. R. Prabhu et al., *Big Data Analytics: Systems, Algorithms, Applications*,
https://doi.org/10.1007/978-981-15-0094-7_5

Raw data contains many patterns, and lot of constituents defines a pattern. For example, data generated in a supermarket contains purchase patterns of the customers. A pattern observed here is the ‘combination of items’ bought together. This information provides great insights for market basket analysis [1]. Product Network Analysis (PNA) is another such application which leverages linked or relational data found in social networks in the category management domain. It automatically identifies important products that belong to the same category by creating category loyalty. It also identifies products that are most likely to acquire cross-category sales. Similar examples can be found in time series data used for stock market price analysis and streaming data obtained from sensor networks.

Identifying patterns require robust preprocessing techniques. Most of the raw data is noisy and cannot be directly used for predictive modeling. Raw data needs to be preprocessed to eliminate unwanted information. Depending on the nature of data, various preprocessing techniques are employed to extract relevant content. For example in text mining, a noisy text generated from social media forums like Twitter contains slangs, misspellings, junk characters, etc. Cleansing social media text provides improved natural language processing applications. This preprocessing can also be considered as the activity of standardization of the data, in a way changing the reference of data to new standards (OWL, HTML, RDF, etc.). Also, preprocessing or data cleaning can help to remove irrelevant, inaccurate and inactive information from database or other structured records.

Preprocessed data is then transformed into a feature space used for building predictive models. Feature engineering is considered as an important step in data analysis as it assists to present the data in a different format. Representation of data is an important task to learn solution to a problem. Key to success for predictive models is heavily influenced by feature engineering. Generally, finding right features to a problem is considered as an art [2]. There are varied subproblems involved in feature engineering. They can broadly classify as feature extraction, feature selection and feature learning.

Once the features are derived, they are used to build models using different machine learning and data mining approaches. Models form the basis for predictive analytics. Final set of features obtained after feature engineering are learned by fitting a model on training data using certain evaluation measures. Models that attain less error not only on training data but also on the testing data are considered to be robust and are scalable.

Predictive models vary based on the problems and type of data. Most of the sources generate data in the form of streams, graphs, multimedia and sequences. Models require fine-tuning by varying parameters to attain good accuracies on each of these data sources which produce images, audio, video, text or tabular data with lot of features. Generalizing or building a model which can work on every type of data is challenging.

In this chapter, we concentrate on predictive modeling approaches used for various artificial intelligence (AI) tasks pertaining unstructured data such as text and multimedia. Initially in Sect. 5.2, we motivate the reader by discussing various possible applications of predictive modeling used for AI tasks like natural language processing (NLP), automatic speech recognition, computer vision. Most of these predictive modeling applications are heavily dependent on feature engineering. In Sect. 5.3, we present feature engineering subproblems with different examples. Once data represented as features, it is provided as input to various predictive modeling approaches listed in Sect. 5.4 based on supervised, unsupervised and semi-supervised learning for clustering, classification or regression. We conclude the chapter with Sect. 5.5.

5.2 Applications of Predictive Modeling

Most of the predictive modeling approaches are used to build or develop cognitive or artificial intelligent systems with unstructured data. In this section, we show few predictive modeling approaches that are built for certain cognitive tasks like NLP, computer vision, speech processing and information retrieval.

5.2.1 Natural Language Processing

Goal of the predictive modeling approaches in NLP is to design models that imitate humans on a given linguistic task. This is achieved by making predictions of NLP systems much closer to the true value by minimizing the difference. Predictive modeling also helps to find unknown or recurring patterns in the large datasets. There are several applications that are built based on the various tasks in NLP. Most of these tasks try to infer knowledge from the unstructured data. Content present in the unstructured data can be observed from many perspectives as illustrated in Fig. 5.1.

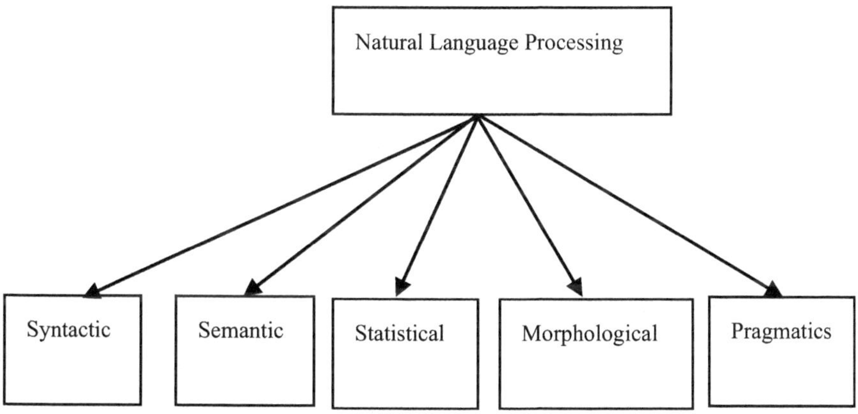

Fig. 5.1 Natural language text processing

Various tasks in NLP adapt these perspectives to describe information present in the unstructured data. Part-of-speech (POS) tagging, semantic role labeling (SRL), shallow parsing, word-sense disambiguation (WSD), named entity recognition (NER) and anaphora resolution are core tasks in NLP where predictive modeling approaches are applied. Other applications that build predictive models are based on meta-data found around product reviews like ratings for sentiment analysis etc. In the following sections, we describe each of the core tasks separately by highlighting the role of predictive modeling.

Part-of-Speech (POS) Tagging

Aim of the POS tagging systems is to label each word in the given text with a tag that indicates its syntactic role (noun, verb, adjective and so on). Below, we can see a sample sentence tagged with POS tags with NLTK2.

[(0 predictive0;0 JJ0); (0modeling0;0NN0); (0is0;0VBZ0); (0an0;0 DT0); (0art0;0NN0)]

Here, we see that 'predictive' is JJ, an adjective; 'modeling' and 'art' are NN, or noun; 'an' is DT, a determinant and so on. POS tagging is seen as a predictive modeling task where learning of POS tags is performed on a training data, while predictions on testing data. Several approaches are developed to perform predictions, and some of the methods use text windows along with bidirectional decoding algorithms for inference [3]. Furthermore, bi-directional inference is combined with SVM and MEMM classifier to perform classification.

Currently, state-of-the-art POS tagging system performance is about 97.33% on a standard benchmark by using one of the earlier mentioned approaches, but still there is a possibility of improvement up to 100% [4]. The prospects of improvement are possible only by combination predictive modeling with descriptive linguistics. Also, POS tagging predictions have been extended to different languages.

Shallow Parsing

The aim of shallow parsing is to label segments (phrases) of a text with syntactic elements (noun or verb phrases (NP or VP)). Each word in irregular length pieces of text (i.e., phrases or sentences) is assigned with a tag that represents beginning (B*) and inside (I*) of the parsed chunks. Following an example illustrated in Daelemans et al. [5], shallow parsing on an example sentence is given below.

Predictive modeling is an art which will be tagged as follows:

Predictive [NP modeling] NP is [VP an] NP art] NP:O

In general, predictive model is trained with a support vector machine (SVM) classifier in a pair-wise fashion with word n-grams and their POS tags as features. Also, several other predictive models like a second-order random fields or a maximum voting classifier are also used with words and POS tags as features. Similar to POS tagging, shallow parsing is also extended to multiple languages and had been used for tasks like machine translation, etc.

Word-Sense Disambiguation (WSD)

Identifying the meaning of words in context is an important problem to solve. Semantic disambiguation is a non-trivial task in understanding natural language. Earlier approaches attached polysemy to a word for performing disambiguation for each ambiguous word in a sentence. An example of different senses of the words according to Schutze et al. [6] can be seen in the Table 5.1. However, these senses are not definite and are corpus dependent.

Aim of the predictive model for WSD is to observe meaning of the word in context, and it has found applications in machine translation and information retrieval systems.

First machine translation system built with WSD predictive model improved the accuracy of word translation from 37 to 45%. Other important problems where predictive models of WSD have found its application are to understand the global context of text by clustering words. By grouping, different words with their real-valued vector representation show their co-occurrence in a corpus.

In general, disambiguation of an ambiguous word is carried out by identifying the nearest centroid of the nearby clusters. But with predictive models used for WSD, always a major sense of the word is captured from the training data. So the prediction of semantic disambiguation of a word is highly biased toward training data.

Semantic Role Labeling

It is as an interesting task in NLP where the main goal of predictive models built for semantic role labeling (SRL) is to identify the verbal arguments in text to categorize them into semantic roles (agent/recipient). A predefined list of semantic roles is

Table 5.1 Example of senses

Word	Sense	Description
Plant	1	Industry or a company
	2	Put an object in a place
	3	Herb
Ruling	1	Govern something or somebody
	2	Lines across paper
	3	Influence
Space	1	Area which is available and unoccupied
	2	Outer space
	3	Time interval
Capital	1	Investment (money, assets)
	2	Administrative city of a country
	3	Upper-case letter
Interest	1	A feeling of wanting to know something
	2	Dividends
	3	Benefit of a person or a group

used by SRL to match the semantic relations between predicates and its associates. Example below shows the similar semantic relations illustrated in Marquez et al. [7].

[Ramu on a bike]Agent
[called]Pred
[the girl beside him]Recipient

Different parts or roles in SRL are agents, recipients and position about the entities that engage in an action or an event. Sometimes, temporal classification of an event or about participant relations is also considered. Thus, labeling with semantic roles provides a basic semantic structure in the natural language text. Roles also indicate the action properties and possible relations between entities that are present in a sentence or a phrase. This identification of an action or event frames also benefits many other NLP applications, like machine translation, information extraction, information retrieval and also question answering.

In general, predictive models for SRL are learned using large annotated datasets. Initially for a given predicate, appropriate entities are identified by analyzing discrete or continuous sequence of words. Arguments define the semantic properties (roles) in a sentence and scoring of the arguments with certain confidence determines role labels. If there is a no argument in a sentence, it is labeled as 'no-argument' to evaluate candidate argument exclusion.

Features are considered to be an important factor for predictive models. Most of the times, identification of an argument and predictive modeling is jointly analyzed for attaining confidence scores between 'argument' and 'no-argument' label information. Identification of an argument is more about syntactic, while predictive modeling is more semantic. Once the features are well represented, SRL sequence tagging is achieved with classifiers and conditional random fields (CRFs) on tree structures approaches.

Named Entity Recognition

Identifying or predicting an entity in the text requires named entity recognition (NER) [8]. Aim of a NER is to categorize the word in a document into defined categories (i.e., person, company, time). NER also aims to do predictive modeling based on information extraction by identifying only specific information from the documents. Since entities cover major content in a document, NER is an important step toward building better information extraction systems.

NER performs parsing at the surface level and delimits the length of extracted tokens. NER is considered as a backbone for identifying relationship between two named entities that inherently help to build a semantic network between entities. Predictive models used to build NER can be built in a simple way with reasonable performance, but can be improved to handle ambiguous cases as well to attain human-level performance.

NER can be applied on noisy text like Twitter messages which are noisy and informal, but sometimes informative. NER is also part of a NLP pipeline in which POS tagging and shallow parsing are applied. Table 5.2 shows the common and Tweet specific named entities as illustrated in Ritter et al. [9].

Table 5.2 Example named entities (Twitter use case)

Type	Description
Band	Music bands
Facility	Facilities
Person	People
Geo-loc	Geographic location
TV-show	TV shows
Company	Name of the companies
Product	Products of companies
Sports team	Sports teams

Anaphora Resolution

It is considered as a very difficult predictive modeling problem in NLP. Anaphora resolution requires integrated knowledge of syntactic, semantic and pragmatic information. Various methods had been proposed for resolving anaphora. Consider a simple example referred from Carbonell et al. [10] to show why we need anaphora resolution.

Ramu took the sandwich from the fridge and ate it.
Ramu took the sandwich from the fridge and washed it.

The motor pushed the cardboard toward the transporter belt.
But, it misjudged and pushed it on its wrong way there.

Example illustrates sentences ("Ramu took…") with same noun or subject. In the second sentence, 'Ramu' can be replaced with semantics provided by pronoun (He). But in the example where 'Motor' is specified, anaphoric resolutions like 'its' and 'there' is used referring to two different forerunners. Here, subjects have difficulty in determining which denotation was used for what anaphora. This indicates the need of sophisticated semantics.

If we analyze, it is understood that 'it' in 'it misjudged and pushed' might have referred to 'motor' rather than the 'cardboard' or the 'transporter belt.' This is conveyed as the 'cardboard' cannot take action, but 'motor' can misjudge things, but it can be about a 'transporter belt' also. This is something very subtle to understand. The fine-grain semantics needs to be involved to make former one greater than the latter. This kind of difficulty in an anaphoric referent makes this a complex predictive modeling problem. Also, anaphora resolution is domain dependent. Nevertheless in a less ambitious setting, problem can be manageable and can have a major practical importance. Textual context or dialog segments can also be used to attain information to drive discourse analysis.

Another variation of anaphora resolution is co-reference resolution. A small difference between anaphora and co-reference resolution is that former concentrates on identifying the antecedent of an anaphora, while co-reference resolution aims to identify all co-reference classes or categories.

5.2.2 Computer Vision

Aim of the predictive models in computer vision is to replicate the abilities of a human vision by understanding what is present in an image. It is also concerned with the artificial intelligence systems which aim to recognize or extract information from an image present in many forms such as videos, CCTV footages and scanned medical data. There are many subdomains in computer vision like motion estimation, image restoration, object tracking and recognition, video indexing where predictive models have lot of applications to the real-world scenarios. Predictive modeling also finds applications in robotics-driven computer vision systems. Tasks such as automatic vehicle navigation, automatic inspection of industrial applications and robotics arms have shown significant progress due to predictive models. Due to the vast spread of the field, we discuss only certain tasks by analyzing the possible applications.

Object Detection

The widely used application of predictive modeling in computer vision systems for identifying or recognizing objects in images is object detection. Predictive models of object detection are used for detecting instances of semantic classes like locations, humans, monuments in digital images and videos. Figure 5.2 illustrates sample objects detected in digital images obtained from PASCAL Visual Object Classes (VOC) Challenge [11].

Object detection includes subdomains like face and pedestrian detection (discussed later) and also find applications to image retrieval and video surveillance based on CCTV footages. A more common approach for object detection is training feature descriptors that operate on image patches and then scale in an exhaustive manner across all locations of an image. A caveat in this kind of approaches is that an exhaustive search through all possible locations and scales poses a computational challenge and becomes harder as the number of classes grows.

Face Detection

Detection of faces is an important and necessary step to build face recognition systems. Predictive models for face detection identify the location and extract the face region from the image background. Figure 5.3 shows the sample images with faces detected.

Identification of faces propels several other applications from multiple areas like video retrieval and encoding, image retrieval, crowd activity surveillance and many other HCI devices. Due to the dynamic and changing object nature of the human face and irregularity in its look, building predictive models for face detection is a tough problem in the field of computer vision. Many approaches have been proposed to build predictive models, ranging from simple techniques like edge or cornet detection-based algorithms to sophisticated techniques utilizing advanced feature descriptor-based methods. These algorithms can be further classified into feature-specific or image-specific approaches.

Fig. 5.2 An example of object detection. Image reproduced with permission from Erhan et al. [29]

Fig. 5.3 Sample images with detected faces. Image reproduced with permission from Li et al. [30]

Human Pose Estimation

It is the predictive modeling process of estimating the layout of the body (i.e., pose) from a monocular image. It is one of the core issues in computer vision that has been studied for many years now. There are many applications that benefit from human pose estimation (HPE) [12] like high-level reasoning in the context of HCI and activity recognition. It is also one of the basic building blocks for motion capture technology which has further applications in character animation and clinical analysis. Figure 5.4 shows the sample images with the detected human poses.

Though predictive models of HPE find many applications to various issues, it still faces several challenges. Some of them are listed below.

Fig. 5.4 Estimated human poses. Image reproduced with permission from Belagiannis et al. [31]

- Hard to estimate the human visual appearance in the images due to heavy dependency of HPE on image quality.
- Human physique variability for each individual.
- Contrast and lighting conditions used to capture the pose.
- Silhouette dependency and its variability.
- High dimensionality of the pose makes it hard to capture.
- Image projection from 3D to 2D losses lot of information.

Gesture Recognition

Up to now, we observed scenarios where predictive models are built for the data in the digital format. Gesture recognition leverages predictive modeling to a different level by recognizing meaningful expressions from human motion. This is different from other forms of input obtained from haptics when using touch screens, mouse or stylus.

Gestures are considered as expressive motions of the body. They involve many physical movements of the body parts with an intent to convey useful information or interacting with the environment or surroundings. It is considered that gestures constitute a small subspace of possible human locomotives. Figure 5.5 shows sample data

Fig. 5.5 Different data modalities provided for gesture recognition. Image adopted from Escalera et al. [32]

modalities provided for gesture recognition from Multimodal Gesture Recognition Challenge captured with kinect.

According to Mitra et al. [13], gesture is comprehended by the surroundings as a capability of communicating compressed information somewhere and eventually reconstructed by the recipient. Predictive models that are built for gesture recognition find many applications and few of them are listed below.

- Can be used as an aid for hearing impaired.
- Can be used to design or recognize a sign language.
- Can be used to monitor patient's stress or emotional levels.
- Helpful in gaming devices or virtual environments for navigation.
- Can be used for tele-medicine or distance learning assistance.
- Gestures recognition can be used for monitoring automobile driver's drowsiness levels.

5.2.3 *Information Retrieval*

Aim of information retrieval (IR) is to find unstructured data within large collections [14] that satisfies an information need. Earlier most of the information retrieval approaches ranked relevant documents based on the standard scoring functions. Introduction of predictive modeling in IR with language modeling was the first step toward an alternative. In general, predictive models look at word interactions and offer few improvements. Below, we present some approaches that find applications in IR.

Language Modeling for IR

Predictive models generally use probabilistic models for document retrieval. But, predictive models based on language modeling estimate the relevant documents for user queries with relevance likelihood. This is visualized as the generation of language with low-entropy process doing the analysis of probability distribution over a corpus. Predictive models also concentrate on estimating the probability of relevance to achieve effective retrieval.

Language model depicts queries that are most likely to appear in the documents of interest. In a way, it helps users to opt query words that differentiate these documents from others in a corpus or collection. Inferences for indexing and retrieval are same as query generation probability.

Machine Learning for IR

Instead of using the term and document weighting functions, different sources of information are used as features in a learning problem. A predictive model is built in the form a text classifier that identifies relevant and non-relevant documents for each set of queries. Classification approach is not necessarily best, but it can be used for ordering of documents according to the confidence of a two-class classifier and also helpful to identify relevance. Main bottleneck here is to collect training examples that are assessed by annotators.

Predictive models can also be generalized to the functions with more than two variables to rank documents. There are various indicators of relevance for a document to a query such as document age, contributions, document length. These measures are obtained from the training document collection with relevance judgments. For example, a model which is learned to do binary classification with SVM or any other classifier can be used to rank documents.

Machine Learning-Based Ranking in IR

Recently, predictive models for ad hoc retrieval are built using machine learning techniques. It is thought as an ordinal regression problem. Here, the aim is to re-rank a set of documents for a given query. This formation gives additional ability. Now, the documents are evaluated relative to other prospective documents for a given query, rather than mapped to a global scale, thus weakening the problem assumptions.

Creating a rank for documents is a major problem in learning to rank approaches. It is usually formalized as a supervised learning task. In general, learning to rank has two definitions. In a broad sense, learning to rank refers to machine learning techniques applied for ranking. While in a narrow sense, learning to rank refers to machine learning techniques for building ranking models for rank creation and aggregation.

Feature creation and rank aggregation are done on training, testing and evaluation dataset. Many methods have been proposed for rank creation that can be broadly categorized into point-wise, pair-wise and list-wise approaches based on loss functions they adopt. They are also categorized based on techniques they employ, such as SVM, boosting, or neural network-based approaches.

5.2.4 *Speech Recognition*

Main goal of the predictive models used for automatic speech recognition (ASR) is to convert speech signal into textual representation. ASR is considered as a very challenging task due to high variability in features like different speaking styles of speakers, background environmental noises. There are several types of predictive models built for different variations of speeches. Some of these variations are speeches with pauses between letters and words, continuous speeches without any pauses between words. More dynamic speeches like spontaneous speeches used in a dialog and highly conversational speeches spoken during meetings.

ASR systems generally map speech signals into sequences of words or phonetic symbols. Predictive modeling in any advanced ASR system works in two different modes. First mode is about training a model, and the second mode decodes the speech. In model training process of an ASR system, models are created based on speech acoustics. Also, training text data or grammatical structure of a sentence is used along with the recognition lexicon containing identifiable markers with one or more phonetic transcriptions.

Acoustic modeling represents signals acquired from audio classes containing speech elements (i.e., monophones, allophones, pentaphones, triphones and syllables) which are context independent and context dependent. Context-independent speech elements are monophones and syllables, while pentaphones, allophones, triphones are context dependent. There are various real-world applications in which speech recognition has been successfully applied like Siri voice assistance, Google voice search, call steering and automated speaker identification.

5.3 Feature Engineering

Building applications mentioned in Sect. 5.2 requires numerous ways of generating features by preprocessing input data. It is an important stage in building a predictive model. Accuracy of predictive models is fairly dependent on the feature generation. In this section, we discuss the possible feature engineering steps to build an efficient predictive model for NLP and computer vision systems.

5.3.1 Feature Extraction and Weighing

Feature extraction and weighing is an important preprocessing step for several predictive modeling tasks especially using machine learning techniques. Various NLP tasks have many common steps like reading a corpus and obtaining n-grams from it or testing for morpho-syntactic or semantic constraints between words. Below, we provide different feature extraction and weighing approaches used for several tasks.

Text Processing

In text processing, features can be categorized into syntactic, semantic, link-based and stylistic features. Unigrams and bigrams are usual features that are extracted from textual information for several predictive modeling tasks.

Syntactic Feature Extraction and Weighing

Syntactic features can refer to word n-grams and part-of-speech (POS) tags. But for noisy text like Tweets, features such as decomposed hashtags are also used.

Stylistic Feature Extraction and Weighing

Structural and lexical style markers can be considered as stylistic features which have shown good results in web discourse.

Semantic Feature Extraction and Weighing

Currently, semantic feature extraction and weighing is driven by finding distributed representations of the textual units in an unsupervised fashion. It actually learns feature hierarchies to provide semantics. Learning distributed representations also eliminate the need for handcrafted features which can be time consuming and can be inappropriate as they can hider predictive model accuracies. Also, current approaches for distributed representations reduce often over-specified and incomplete feature dependencies. Sometimes, semantics features are learned without being task or domain specific.

Earlier, majority of the rule-based and statistical NLP approaches treated words as atomic units. For example, 'one-hot vector' representation of words 'Buy' and 'Purchase' as shown in Fig. 5.6 does not overlap due to this assumption, even though they are synonyms of one another. But changing this representation by understanding a word by its neighbors can provide semantics and is also much useful to reduce the dimensionality of a vector. Figure 5.7 shows the changed vector representation by considering semantics of its location in a sentence or a document.

This kind of representation is also referred as word embeddings [15] (henceforth). Generally, word embeddings are vectors with high-dimensional space. These vectors have some really nice properties. As shown in the previous example, it identifies words with similar meanings and makes them closer in high-dimensional space. Some other interesting things that were observed are that embeddings have semantic sense or meaning. Further, difference between word vectors seems to capture analogies. For example, the difference between 'woman' and 'man' word vector space is similar to the difference between 'queen' and 'king.' Vector ("woman")–vector ("man") = vector ("queen")–vector ("king").

Enhancement of word embeddings to larger chunks of text can be done with paragraph vectors [16] that build on word embeddings or bag-of-words representation. Word embedding vectors are used to solve natural language processing tasks that involve a word, while paragraph vectors are used to predict the words that are present in a paragraph. Concretely, a low-dimensional representation of the word

Buy [0 0 0 0 0 0 0 0 0 1 0] and Purchase [0 0 0 0 0 0 0 0 1 0 0] =0

Fig. 5.6 One-hot vector

Buy [-0.134 0.311 0.211] and Purchase [0.211 -0.112 0.567]

Fig. 5.7 Distributed representations

occurrence statistics for different paragraphs is learned as features. In the hidden layer of a neural network, vectors represent similar paragraphs together.

Image Processing

Over the years, various approaches have been developed to understand the features present in an image which form the backbone for many computer vision applications mentioned in Sect. 5.2. Feature extraction for images aims at obtaining abstractions of an image by making decisions at every image point or pixel to assert certain image features. There are several ways features can be extracted from an image. Below, we list out and discuss distinctive (feature detection) and region around features (feature description) separately.

Feature Detection

Aim of the feature detection is to take pixel coordinates of significant areas in the image that are distinctive and looks different from its neighbors. Below, we describe different image feature detectors separately.

Edge Detection

In an image, edges are those bunch of points where there is a boundary between two different image regions. Generally, an edge is of arbitrary shape and may include intersections. During implementation, a set of points in an image which contains intense gradient magnitude are edges. Also, gradient points can be linked together to form more complete description of an edge. Attributes of an edge are restricted with several constraints such as smoothness, shape and gradients. Structure of the edges is considered to be one-dimensional, and edge is considered as a contrast with strong intensity. Figure 5.8 shows an example image with detected edges.

Fig. 5.8 Edge detection with different filters **a** Left: Sobel **b** Right: Canny

Corner Detection

Point-like features in an image can be considered as the corners and have a local two-dimensional structure (e.g., contours). Corners refer to the rapid change of edges in a specific direction. Corners are helpful where explicit edge detection is no longer required like looking for high levels of curvature in the image gradient. Corner detection finds its applications in many computer vision tasks like motion detection, restoration of images, tracking videos, 3D modeling and object recognition/identification. Figure 5.9 shows an image with corners detected with Harris corner algorithm [17].

Blob Detection

Blobs describe image structures in terms of regions, as opposed to the edges and corners that are more interested in points. Sometimes, blob descriptors may also contain a preferred or interested point. But, blob detectors are mostly concentrated on regions and can detect areas in an image which are too smooth to be detected by a corner detector.

Regions detected as blob descriptors are helpful to identify objects in an image, thus showing its applicability to object recognition and tracking. In other related areas like histogram analysis, hike is identified using blob descriptors with an application to image segmentation. Blob descriptors are also generally used as main feature for texture recognition.

Fig. 5.9 Corner detection

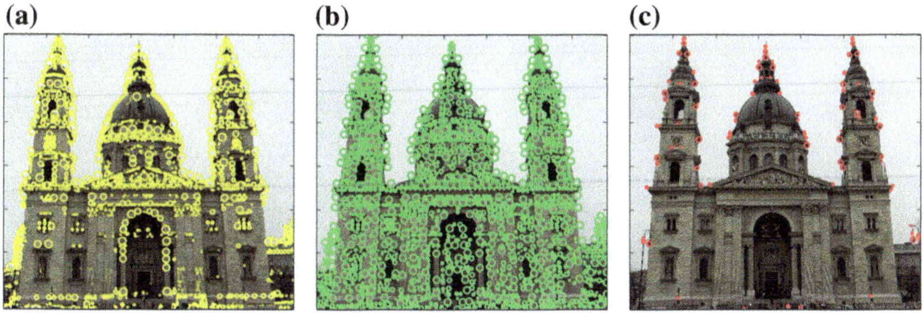

Fig. 5.10 Blob detection with different approaches **a** Left: Laplacian of Gaussian **b** Center: Difference of Gaussian **c** Right: Determinant of Gaussian

Blob descriptors have many versions, and they change according to scales. Figure 5.10 shows an image with blobs detected [18].

Ridge Detection

A ridge is considered as a long and narrow edge or angle. In general, ridge features are useful for computational tasks on gray-level medical or aerial images. Ridge is understood as a one-dimensional curve that represents an axis of symmetry with an attribute of local ridge width associated with each ridge point present in an image. Ridge features are harder to extract from gray-level images when compared to edge, corner or blob features.

Ridge features play a crucial role in identifying roads in aerial images, blood vessels in medical images, fingerprint identification in biometrics, etc. Figure 5.11 shows an image with ridges detected with differential geometry [19].

Feature Description

Aim of a feature descriptor is to output feature descriptors/feature vectors of an image. These feature vectors would usually include the location of the feature as well as other information. Below, we describe different image feature descriptors separately.

SIFT

Scale-invariant feature transform (SIFT) is a standard feature descriptor used for many computer vision applications. SIFT extracts features from scale-invariant key points in an image and computes its descriptors. SIFT algorithm [20] design is based on four major steps.

- In the first step, local extrema over space and scale are found with scale-space extrema detection by performing difference of Gaussian (DoG). For example, each pixel in an image is juxtaposed with its 8 surrounding neighbors and 9 other pixels in the next and previous scales. After comparison, if it is found as local extrema, then it is a prospective key point. It certainly means that key point is better illustrated in that scale.

Fig. 5.11 Ridge detection

- In the second step, once the prospective key-point locations are identified, points are corrected to get more precise results. To achieve it, Taylor series expansion of the scale space is used to obtain more precise location of the extrema satisfying certain threshold. By doing this, it eradicates existing low contrast and some edge key points.
- In the third step, an angle is attributed to each key point to achieve stability toward image rotation.
- While in the last step, key-point descriptors are created. Image pixels with (16×16) dimensions around proximity of a key point are acquired to divide them into another 16 subblocks with each (4×4) size. For each subblock, 8-bin histograms are created with 128 cumulative bin values. Figure 5.12 shows an image with SIFT key points.

SIFT achieved good results for 3D object recognition, location recognition, although it fails at certain cases like illumination changes etc.

SURF

Speeded up robust features (SURF) [21] is local feature descriptor used for many computer vision tasks. It is considered to be the fast version of SIFT. SURF is also

Fig. 5.12 Key points (SIFT)

rotation and scale invariant. It outperforms on previously proposed approaches on parameters such as uniqueness and robustness.

In general, SURF relies on image convolutions. It builds on existing approaches by utilizing Hessian matrix measure for the detector and a distribution-based for the descriptor. SURF depends on responses of wavelets (horizontal and vertical) direction to describe features. Image pixels with (20×20) dimensions around each key point and is acquired and further divided into (4×4) subregions. For each of these subregions, wavelet responses are captured to build a vector with 64 dimensions.

Most of the times, lower dimensions improve computation and matching speed, but cannot provide uniqueness to features. To add more uniqueness, SURF feature descriptor can be extended to 128 dimensions.

For the speedup over SIFT, Hessian matrix trace is used for central interest point. This does not add more computational cost as it was already computed while detection. To distinguish bright blobs on the dark backgrounds, the sign of the Laplacian is used. This minute details allow fast matching and do not reduce the descriptor's performance. Figure 5.13 shows an image with SURF key points.

SURF is good at dealing with image blur and rotation, but not so good at dealing viewpoint and illumination change.

HOG

Histogram of oriented gradients (HOG) [22] is also a feature descriptor used in most of the computer vision and image processing tasks. In this approach, the appearance of gradient position in localized portions of an image is counted. HOG finds its similarities with edge orientation histograms and SIFT descriptors. But, it also differs in many aspects. HOG is computed on a concentrated mesh with uniform distribution of cells. It also uses the overlay of local contrast normalization to improve precision.

Fig. 5.13 Key points (SURF)

In general, HOG creates histograms of edge orientations from certain patches in images. A patch may belong to an object, a person, or anything else and is merely a way to describe an area using edge information. HOG descriptors have been proved effective for pedestrian detection [22]. Figure 5.14 shows an example of the pedestrian detected with HOG features.

To improve results to many applications, pyramid scheme of HOG is used. Instead of extracting a single HOG vector from an image, patches inside images are extracted into several subimages along with their individual HOG vectors. The process can be repeated, and in the end, a final descriptor is obtained by concatenating all of the HOG vectors into a single vector.

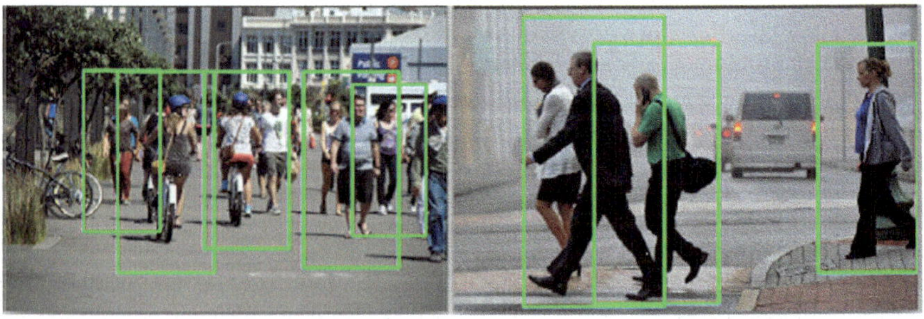

Fig. 5.14 Pedestrian detection with HOG features

5.3.2 *Feature Selection*

Feature selection methods help in removing the features which may not be useful for categorization or classification. For this, feature selection techniques select subset features. But, it is important to reduce dimensionality of feature vectors without compromising on accuracy of classifier.

Text Processing

Many approaches are proposed for feature reduction in text processing such as information gain (IG), correlation feature selection (CFS), chi-squared (c2), odds ratio, categorical proportional difference (CPD) and entropy-based category coverage difference (ECCD) [23]. For example, Mogadala et al. [24] used ECCD feature selection process for subjective classification, while Largeron et al. [23] used it for INEX XML documents' mining.

Image Processing

Several feature selection approaches are also used for processing images in different computer vision tasks. Most of the feature selection approaches like principal component analysis (PCA), factor analysis, canonical correlation analysis (CCA), independent component analysis (ICA), linear discriminate analysis (LDA), non-negative matrix factorization (NMF), t-distributed stochastic neighbor embedding (t-SNE) are used to perform dimensionality reduction.

5.4 Pattern Mining for Predictive Modeling

There are many pattern mining approaches used for predictive modeling. In this section, we cover two different learning approaches that are currently shown to perform well on large and unstructured data. These approaches are either used for clustering, classification, regression or ranking.

5.4.1 *Probabilistic Graphical Models*

A probabilistic graphical model (PGM) is a probabilistic model where conditional dependence between random variables is expressed within a graph structure. In general, PGM is considered as a union between probability and graph theory. PGMs are considered as a natural model to handle uncertainty and complexity. The core idea of any graphical model is its flexibility. Complex systems based on PGM's efficient general-purpose algorithms are constructed by combining simple parts driven by probability and graph theory.

Methods developed based on PGM have shown its usability in enormous range of application domains like search, image processing, bioinformatics, speech recognition, natural language processing, cryptography, robotics and many more areas. In general, there are two important branches of graphical representations of distributions that are commonly used.

Bayesian Networks

A directed acyclic graph representation is conventionally used for bayesian networks. The PGM model built with bayesian networks referred as directed graphical model and is represented as factorization of the joint probability of all random variables. Some special cases of a PGM-based bayesian network model are hidden Markov models and some neural networks.

Markov Networks

A PGM model built with a Markov network or Markov random field (MRF) is a model over an undirected graph. A noticeable variation of a MRF is conditional random field (CRF), and a set of global observations influence the random variables by certain conditions.

5.4.2 Deep Learning

Set of some machine learning algorithms based on multilayer neural network used for predictive modeling. Deep learning improves upon shallow machine learning approaches that involves a lot of effort to express things that a deep architecture could have done more densely. This means a deep architecture can reuse previous computations. Deep learning also reduces the chance to be stuck in local minima and improves training performance.

We see that deep learning can be divided into standard supervised/unsupervised learning approaches and for rapid implementation of them, several toolkits were developed such as cuda-convnet, Caffe, PyTorch, TensorFlow and MXNet. Deep learning is currently applied to various tasks and domains in NLP, computer vision, robotics and speech processing in the areas like building autonomous car, neural language modeling, face detection, handwriting recognition, image restoration, object recognition and sentiment analysis.

Supervised Learning

Some of the deep feature learning approaches are supervised and have shown significant improvements for language processing, speech and vision recognition tasks. Below, we see two of these approaches.

5.4.3 Convolutional Neural Networks (CNN)

CNN is an extension of the traditional multilayer perceptron [25] based on shared weights and spatial or temporal subsampling. It has many layers, and each of these layers is connected based on the combination of convolutions, nonlinearity and subsampling information. Initially, CNN was used for digit classification [26] with a convolution and subsampling layer followed by a densely connected output layer which will feed into the softmax regression and cross-entropy objective. Generally, mean or maximum pooling is used for subsampling layer. Learning is achieved with a back-propagation algorithm which calculates a gradient w.r.t. the parameters of model. Stochastic gradient descent (SGD) and momentum are the most commonly used gradient approaches. Figure 5.15 shows the LeNet-5 [26] used for digits classification.

But lately, CNN also found its applications to other visual recognition tasks like region-specific object detection, scene parsing, 3D scene understanding, action recognition. Figure 5.16 shows deep neural network architecture [27] used for classifying 1000 ImageNet7 classes.

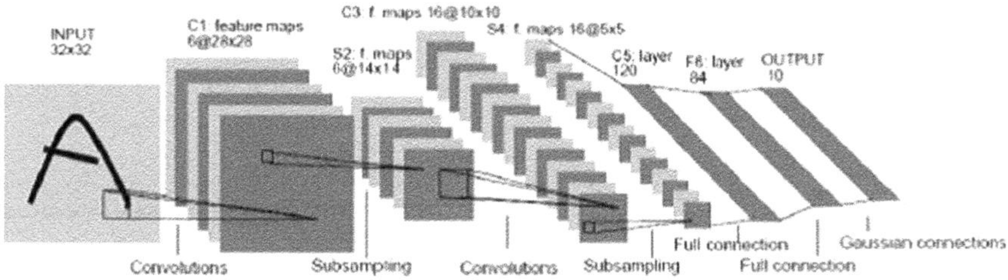

Fig. 5.15 CNN digit classifier. Image adopted from LeNet-5 [26]

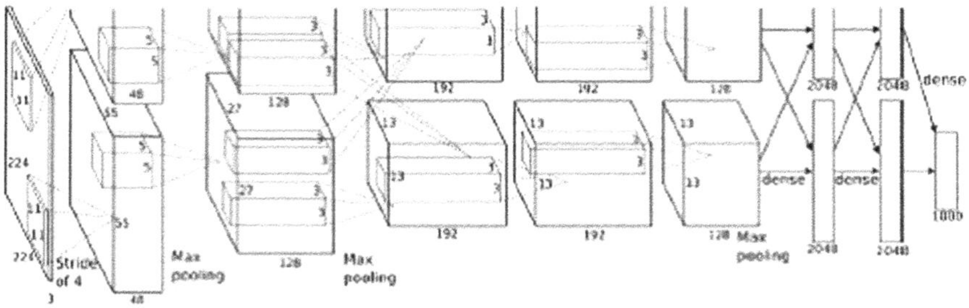

Fig. 5.16 Deep CNN ImageNet classifier. Image adopted from AlexNet [27]

5.4.4 *Recurrent Neural Networks (RNNs)*

There are two important types of neural networks, feed-forward and recurrent. In the feed-forward networks, activation is in a pipeline through the network from input to output units. A recurrent neural network (RNN) is a network consisting of neurons with feedback connections. A simple example of a RNN can be observed in Fig. 5.17. It learns many operational or sequence processing tasks which are considered to be difficult for the other traditional machine learning methods. With the help of RNNs, computers now can learn mapping sequences with or without supervision. They are computationally powerful and are more compelling than other existing approaches like hidden Markov models (HMMs), feed-forward networks and SVMs.

Recently, RNNs have shown interesting solutions to many applications like speech recognition, machine translation, music composition, stock market predictions and many other related problems. Long short-term memory (LSTM) is a variant of RNN. It is a feedback network that overcomes the primary problems existing with traditional RNNs. It also effectively learns to solve many complex tasks. LSTM networks consist of many connected cells and are very efficient in managing time.

Unsupervised Learning

Most of the deep feature learning approaches are unsupervised and have shown significant improvements for artificial intelligence tasks. Below, we see few of these approaches.

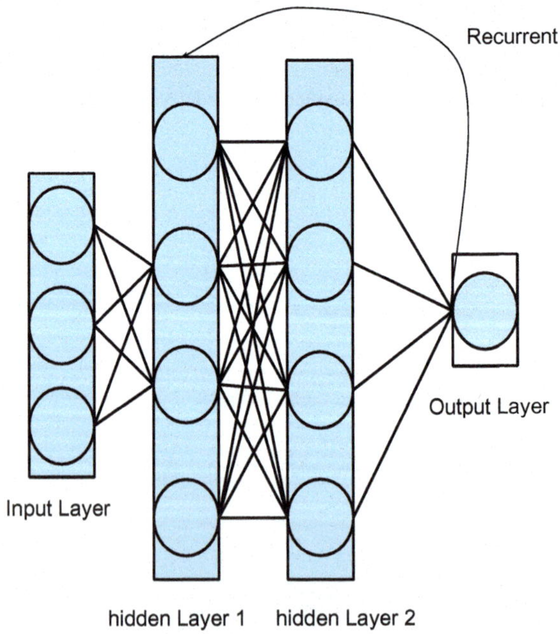

Fig. 5.17 Recurrent neural network

5.4.5 Deep Boltzmann Machines (DBM)

A deep Boltzmann machine (DBM) [28] is considered as a Boltzmann machine that contains many layers with hidden variables. It can be used for learning a generative model of data that consists of multiple and different input modalities. This means that a model can be used to learn a unified representation which combines different modalities. Figure 5.18 shows the pre-training of a stack of modified restricted Boltzmann machines as illustrated in Ruslan et al. [28] that are combined to construct a DBM. Like a deep belief network (DBN), DBMs also have ability to learn complex internal presentations thus allowing DBM's to solve object and speech recognition problems effectively. In DBMs, representations are built from a largely available unlabeled inputs (i.e., sensory units) and very limited labeled data to fine-tune a model for the specific task. Also, approximate inferencing procedure is used to incorporate top-down feedback along with bottom-up pass. This allows DBMs to propagate uncertainty and deals with more robust and unclear inputs. Generally inside DBMs, expectations that depend on the data are guessed with variational approximation procedure which only focuses on a single mode. But expectations that are data are estimated with Markov chains.

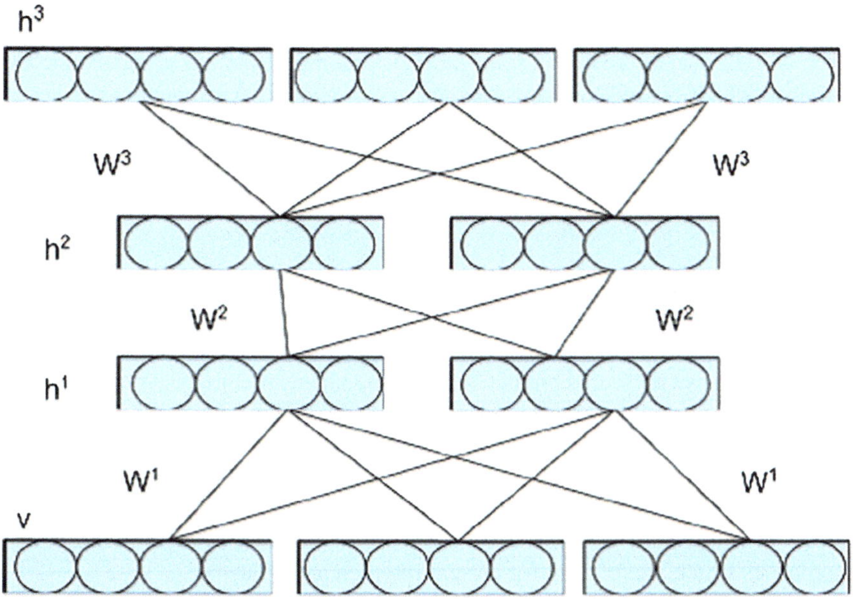

Fig. 5.18 Deep Boltzmann machine

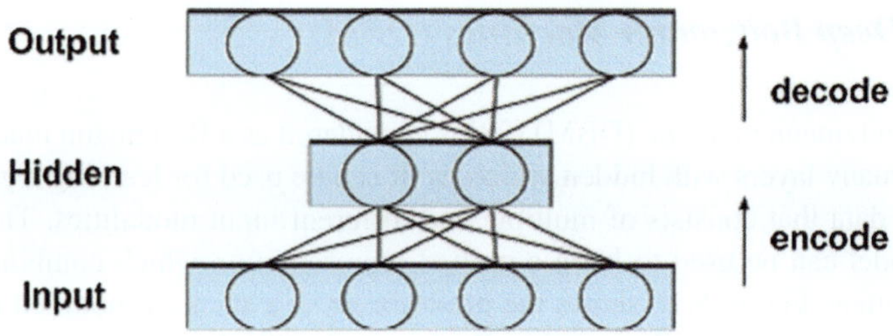

Fig. 5.19 Autoencoder

5.4.6 *Autoencoders*

Autoencoders are considered to play an important role in unsupervised learning and in deep architectures for the various tasks. Autoencoders are simple learning circuits which aim to transform inputs into outputs with the least possible amount of deformation.

Autoencoder is a variant of neural network which is feed-forward and non-recurrent. It resembles a multilayer perceptron (MLP) consisting of one input and output layers with one or more connecting hidden layers. Biggest difference with MLP is that output layer of an autoencoder contains equal nodes as that of input layer. Also instead of predicting some labels y for any given inputs x, an autoencoder is trained which is used to reconstruct the given inputs x. A simple example of an autoencoder can be observed in Fig. 5.19.

There are different variations of autoencoders such as stacked autoencoders, denoising autoencoder, stacked denoising autoencoder and recursive autoencoders.

5.5 Conclusion

In this chapter, we presented predictive modeling on unstructured data by evaluating different stages in building artificial intelligence (AI) systems. Initially, we identified different applications in which predictive models are used to support AI systems. As most of the current predictive models are heavily dependent on feature engineering, we identified state-of-the-art feature extraction and selection approaches. Predictive models which are built using these features are discussed later by highlighting certain approaches which are relevant for current research and industry.

5.6 Review Questions

1. Explain a scenario where proactive analytics with predictive modeling is useful in converting data into valuable information.
2. What are the techniques available to process raw data, e.g., Twitter data to eliminate unwanted information.
3. What is feature engineering and what are the subproblems involved in feature engineering.
4. Provide details about various forms of unstructured data available on the web.
5. Explain various natural language processing perspectives.
6. Explain differences among tasks in NLP, for example, POS tagging and NER.
7. What kind of data computer vision tasks deal with. Give some examples.
8. Provide details about human pose estimation for predictive modeling.
9. What are the applications of face recognition to human–computer interaction (HCI).
10. Explain several methods used for information retrieval.
11. Describe various feature extraction and weighing methods.
12. Provide details about different image processing techniques.
13. How is feature extraction and selection different.

References

1. M.J. Berry, G. Linoff, *Data Mining Techniques: For Marketing, Sales, and Customer Support* (Wiley, 1997)
2. P. Domingos, A few useful things to know about machine learning. Commun. ACM, 78–87 (2012)
3. Y. Tsuruoka, J.I. Tsujii, Bidirectional inference with the easiest-first strategy for tagging sequence data, in *Proceedings of the Conference on Human Language Technology and Empirical Methods in Natural Language Processing* (Association for Computational Linguistics, 2005), pp. 467–474
4. C.D. Manning, *Part-of-Speech Tagging from 97% to 100%: Is It Time for Some Linguistics* (2011)
5. W. Daelemans, S. Buchholz, J. Veenstra, *Memory-Based Shallow Parsing* (1999). arXiv preprint arXiv:cs/9906005
6. H. Schutze, Dimensions of meaning, in *Supercomputing* (IEEE, 1992), pp. 787–796
7. L. Marquez, X. Carreras, K.C. Litkowski, S. Stevenson, Semantic role labeling: an introduction to the special issue. Comput. Linguist. **34**(2), 145–159 (2008)
8. G. Zhou, J. Su, Named entity recognition using an HMM-based chunk tagger, in *Proceedings of the 40th Annual Meeting on Association for Computational Linguistics* (Association for Computational Linguistics, 2002), pp. 473–480
9. A. Ritter, S. Clark, O. Etzioni, Named entity recognition in tweets: an experimental study, in *Proceedings of the Conference on Empirical Methods in Natural Language Processing* (Association for Computational Linguistics, 2011), pp. 1524–1534
10. J.G. Carbonell, R.D. Brown, Anaphora resolution: a multi-strategy approach, in *Proceedings of the 12th Conference on Computational linguistics*, vol. 1 (Association for Computational Linguistics, 1988), pp. 96–101

11. M. Everingham, L. Van Gool, C.K. Williams, J. Winn, A. Zisserman, The pascal visual object classes (VOC) challenge. Int. J. Comput. Vis. **88**(2), 303–338 (2010)
12. L. Sigal, Human pose estimation, *Computer Vision* (Springer, US, 2014), pp. 362–370
13. S. Mitra, T. Acharya, Gesture recognition: a survey. IEEE Trans. Syst. Man Cybern. Part C Appl. Rev. **37**(3), 311–324 (2007)
14. C.D. Manning, P. Raghavan, H. Schutze, *Introduction to Information Retrieval* (Cambridge University Press, Cambridge, 2008)
15. T. Mikolov, K. Chen, G. Corrado, J. Dean, *Efficient Estimation of Word Representations in Vector Space* (2013). arXiv preprint arXiv:1301.3781
16. Q.V. Le, T. Mikolov, *Distributed Representations of Sentences and Documents* (2014). arXiv preprint arXiv:1405.4053
17. C. Harris, M. Stephens, A combined corner and edge detector, in *Alvey Vision Conference*, vol. 15 (1988)
18. R.T. Collins, Mean-shift blob tracking through scale space, in *Computer Vision and Pattern Recognition (CVPR)* (IEEE, 2003)
19. T.H.T. Thi, A. Lux, A method for ridge extraction, in *6th Asian Conference on Computer Vision*, vol. 2 (2004)
20. D.G. Lowe, Object recognition from local scale-invariant features, in *Proceedings of the Seventh IEEE International Conference on Computer Vision* (IEEE, 1999), pp. 1150–1157
21. H. Bay, A. Ess, T. Tuytelaars, L. Van Gool, Speeded-up robust features (SURF). Comput. Vis. Image Underst. **110**(3), 346–359 (2008)
22. N. Dalal, B. Triggs, Histograms of oriented gradients for human detection, in *Computer Vision and Pattern Recognition (CVPR)* (IEEE, 2005), pp. 886–893
23. C. Largeron, C. Moulin, M. Géry, Entropy based feature selection for text categorization, in *Proceedings of the ACM Symposium on Applied Computing* (2011)
24. A. Mogadala, V. Varma, Language independent sentence-level subjectivity analysis with feature selection, in *26th Pacific Asia Conference on Language, Information and Computation (PACLIC 26)* (2012)
25. E.B. Baum, On the capabilities of multilayer perceptrons. J. Complex. **4**(3), 193–215 (1988)
26. Y. LeCun, L.D. Jackel, L. Bottou, A. Brunot, C. Cortes, J.S. Denker, V. Vapnik, Comparison of learning algorithms for handwritten digit recognition, in *International Conference on Artificial Neural Networks* (1995), pp. 53–60
27. A. Krizhevsky, I. Sutskever, G.E. Hinton, ImageNet classification with deep convolutional neural networks, in *Advances in Neural Information Processing Systems* (2012), pp. 1097–1105
28. R. Salakhutdinov, G.E. Hinton, Deep boltzmann machines, in *International Conference on Artificial Intelligence and Statistics* (2009), pp. 448–455
29. D. Erhan, C. Szegedy, A. Toshev, D. Anguelov, Scalable object detection using deep neural networks, in *Computer Vision and Pattern Recognition (CVPR)* (IEEE, 2014), pp. 2155–2162
30. H. Li, Z. Lin, X. Shen, J. Brandt, G. Hua, A convolutional neural network cascade for face detection, in *Proceedings of the IEEE Conference on Computer Vision and Pattern Recognition* (2015), pp. 5325–5334
31. V. Belagiannis, A. Zisserman, *Recurrent Human Pose Estimation* (2016). arXiv preprint arXiv: 1605.02914
32. S. Escalera, J. Gonzàlez, X. Baro, M. Reyes, O. Lopes, I. Guyon, V. Athitsos, H. Escalante, Multi-modal gesture recognition challenge 2013: dataset and results, in *Proceedings of the 15th ACM on International Conference on Multimodal Interaction* (ACM, 2013), pp. 445–452

Chapter 6
Machine Learning Algorithms for Big Data

6.1 Introduction

Growth of data provided from varied sources has created enormous amount of resources. However, utilizing those resources for any useful task requires deep understanding about characteristics of the data. Goal of machine learning algorithms is to learn these characteristics and use them for future predictions. However, in the context of big data, applying machine learning algorithms rely on the effective processing techniques of the data such as using data parallelism by working with huge chunks of data. Hence, machine learning methodologies are increasingly becoming statistical and less rule-based to handle such scale of data.

Statistical machine learning approaches are dependent on three important sequences of tasks.

- Large collection of data showing how humans would perform a task.
- Designing of a model to learn from the collected data.
- Learn parameters of the model by leveraging data.

However, machine learning is seen beyond the only design of the algorithms. It is observed from a bigger perspective, where understanding of data for making provision for data analysis, preparation and decision making is important. Also, ideas from basic mathematics, statistics and domain expertise are leveraged in machine learning for supporting the big data challenges. It informs us that machine learning, in general, combine or leverage information from varied areas. Pattern recognition, data mining and knowledge.

Discoveries are few such areas where machine learning has an overlap. The goal of these former areas is to extract insights from the data. However, the core goal of machine learning is to predict, thus going beyond the only discovery of patterns in data.

When designing machine learning algorithms, assumption about underlying distributions about the data is made prior. These assumptions remain same for both seen

C. S. R. Prabhu et al., *Big Data Analytics: Systems, Algorithms, Applications*,
https://doi.org/10.1007/978-981-15-0094-7_6

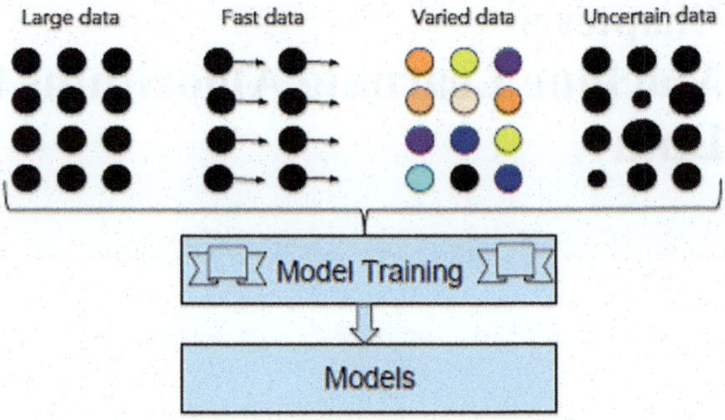

Fig. 6.1 Machine learning models learned from different big data types: large data (volume), fast data (velocity), varied data (variety) and uncertain data (veracity)

and the unseen data. Machine learning also faces several challenges for big data as stated by Singh et al. [1]:

- Learning for large scale of data.
- Learning for varied types of data.
- Learning for streaming data.
- Learning for incomplete data.

Given such challenges, machine learning models learn from the data by effectively processing the data to provide new insights. In Fig. 6.1, different types of data are leveraged to build models for varied tasks.

6.2 Generative Versus Discriminative Algorithms

Machine learning algorithms can be broadly classified into generative and discriminative algorithms. The generative algorithms follow Bayesian approach by using prior information about data distributions while learning a model. While, the discriminative algorithms leverage frequenters approach where prior data distribution is ignored when learning a model.

Usually, generative methods model class-conditional probability density functions (pdfs) and prior probabilities with an ability to generate synthetic data points. However, discriminative models directly estimate posterior probabilities and do not consider the underlying probability distributions. They only concentrate on the given task for achieving better performance.

To emulate generative models behavior, we understand it with an example. Using Bayes' rule, where posterior probability $p(y/x)$ is given as a joint probability distribution $p(x; y)$ of training sample x, its label y and the underlying distribution of x, y as $p(xjy)p(y)/p(x)$. Hence, the equation which generate x for any given y is given as: $f(x) = yp(x/y)p(y)$. However, for the discriminative approaches, to predict the label y from the training sample x, we only use $f(x) = yp(y/x)$. That is just choosing what is the most likely class y when given x. To be specific, discriminative models learn a boundary between classes while generative approaches model the distribution

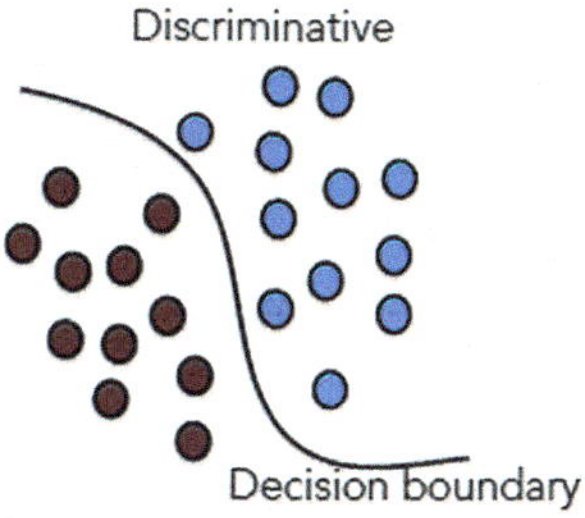

Fig. 6.2 Discriminative versus generative models: visualization with data points

of individual classes. Figure 6.2 shows the visualization of the difference between generative vs discrimination approaches.

Generative or discriminative algorithms are used for different applications in machine learning. For example, classification is a task of categorizing objects. It can be done with both generative and discriminative approaches. We comprehend it with a real-world example. Suppose there are two classes of fruits [apples ($y = 1$), oranges ($y = 0$)] and x is the feature of fruits. Given the training data, a discriminative approach will find a decision boundary that separates the apples and oranges. Furthermore, to classify a new fruit as either an apple or an orange, it verifies which side of the decision boundary it falls, and makes a prediction. However, a different approach is followed by a generative model. First, looking at apples, we build a model of how apples look like. Then, looking at oranges, we can build a separate model of what oranges look like. Finally, to classify a new fruit, we match the new fruit against both apple and orange model, to see whether the new fruit looks more like apples or oranges that was previously observed in the training data.

In practice, according to Ng et al. [2], the generative model has a higher asymptotic error than the discriminative model. However, it is observed that the generative model approaches its asymptotic error faster than the discriminative model possibly with training samples that is only logarithmic, rather than three linear, in the number of parameters. Usually, it can be assumed that discriminative model can be correct even when the generative model is incorrect, but not vice versa.

There are several generative models that exist such as Naïve Bayes, mixtures of experts, Bayesian networks, Hidden Markov models (HMM), Markov random fields (MRF), etc. Similarly, there are many discriminative models such as logistic regression, SVMs, conditional random field (CRF), neural networks, etc. Generally, discriminative and generative algorithms form pairs to solve a particular task. For example, Naïve Bayes and logistic regression are a corresponding pair for classification, while HMM and CRF are a corresponding pair for learning sequential data. From the perspective of explanatory power, generative models have an upper hand over the discriminative models.

Both generative and discriminative approaches emerged due to limitations in each of them. In the case of discriminative approaches,

- They lack the elegance observed in generative approaches such as modeling of priors, structure and uncertainty.

- Alternative notions of penalty functions, regularization and kernel functions.
- Mostly black-box, where relationship between variables is not explicit.

However, there are attempts [3] to combine best of generative and discriminative approaches, where a general framework is used for discriminative estimation based on the maximum entropy principle. Computations are made with distributions over parameters rather than specific settings and reduce to relative entropy projections.

6.3 Supervised Learning for Big Data

Machine learning algorithms for the big data is divided into several areas. One of those areas which is applied extensively is the supervised machine learning. In general, supervised learning works with two different variables mainly input training data (x) and the output labels (y), where the learning algorithm learns a mapping function from the input (x) to the output (y), i.e., $y = f(x)$. Goal of this mapping function is to predict the output label y for the new input data ($\times 0$). This approach is referred as supervised learning as the process of learning an algorithm from the training data can be thought of as the supervision of learning process. As the true labels are known prior, algorithm predictions are corrected by comparing against the ground truth. Learning is stopped when the algorithm achieves a satisfactory performance.

Supervised learning algorithms from a broader perspective can be categorized into either classification or regression methods. The aim of the classification algorithms is to classify the input data (x) into discrete classes (y), while regression methods model independent variables (x) to predict real-valued dependent variable (y). Although they perform different tasks, the input training data used in both scenarios is provided by human annotators. Figure 6.3 summarizes the supervised classification approach.

Many supervised learning methods have been proposed in the literatures that leverage big data. In the following, some of them are discussed in detail.

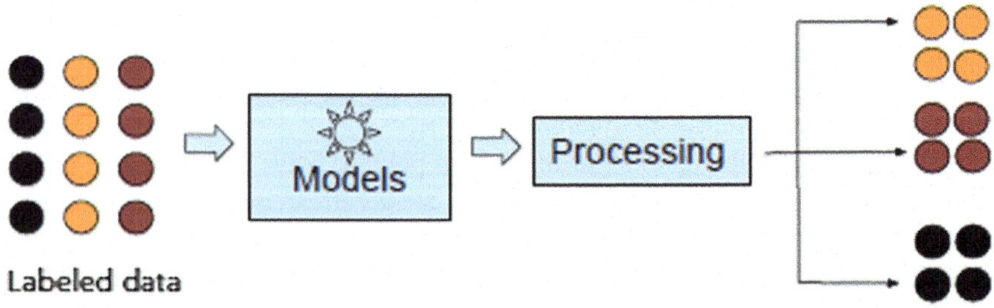

Fig. 6.3 Supervised machine learning for classification

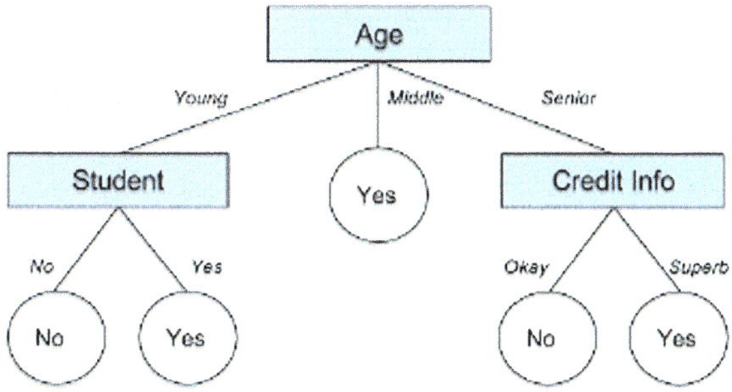

Fig. 6.4 Example representation of a decision tree for articulating a concept: who will buy laptop? with its attributes (age and credit information)

6.3.1 *Decision Trees*

A decision tree is based on inductive inference where many observations are made to understand a pattern to infer an explanation. The learning is achieved by approximation of discrete-valued, real-valued or missing attributes to build a tree.

Representation

As the name suggests that the decision trees are structured in the form of a tree. Hence, the decision trees classify a sample instance by evaluating them down the tree, i.e., from root to the leaf nodes. Figure 6.4 illustrates it with a similar example presented in Han et al. [4]. The aim of this decision tree is to classify a person based on his/her willingness to "buy a laptop" depending on their attributes such as age and financial status.

In general, decision trees are build based on constraints arising across attributes either through conjunctions or disjunctions. However, different variations of decision trees exist, which enhance over such constraints and will be discussed in the following section.

Variations

Over the years, various versions of the decision tree algorithms were proposed. Of them, we briefly introduce three highly used such as Iterative Dichotomiser 3 (ID3) [5], C4.5 [6] and CART [7].

6.3.2 *Logistic Regression*

The logistic regression is a predictive algorithm whose aim is to build a model that finds connection between dependent and the independent variables. In general, the dependent variable is binary (0 or 1) and independent variables can be nominal, ordinal, etc. While, the coefficients generated in the logistic regression are estimated with an equation that includes transformation of probabilities about features acquired from the training data.

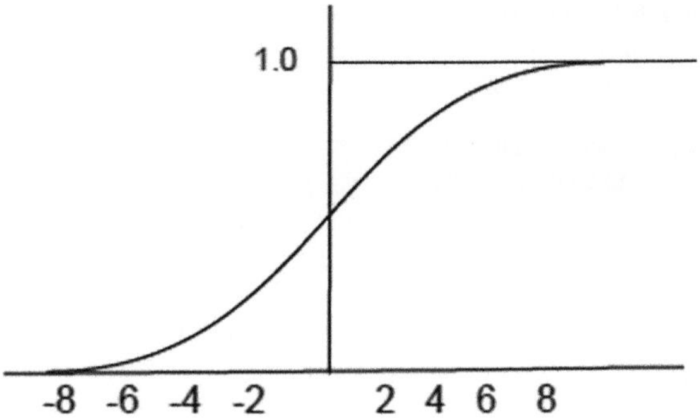

Fig. 6.5 Logistic regression with x-axis constituting integers and y-axis show the resulting values of Eq. 6.1

Representation

The representation of logistic regression can be simply achieved with Eq. 6.1 constituting _ and _ as the coefficients. While, visualization can be shown with a graph plotted between dependent (y) and independent variable (x) as presented in Fig. 6.5.

$$y = \frac{1}{1 = e^{-(\alpha + \beta x)}} \tag{6.1}$$

6.3.3 Regression and Forecasting

The goal of forecasting is to make prediction about the future values of the data and regression is one such approach to achieve it. The regression analysis is a statistical approach to examine the quantitative data for estimating the parameters of a model to make future predictions. Also, the regression analysis calculates the statistical correlation among variables and do not find the causal relationship among them.

6.3.4 Supervised Neural Networks

Before diving into supervised neural networks, we first understand what are neural networks. In general, neural networks are termed as artificial neural networks (ANN). They are loosely inspired from human brain and exist with many names such as parallel distributed processing (PDP), neuro-computing, connectionism, etc. ANNs inherently support multiple processor architectures without much alteration and suitable for large-scale processing. To be concise, ANN is considered as architecture that is trained to predict an output when it identifies a given input pattern.

The ANNs are made of simple artificial neurons, which form the basic computational element often denoted as a unit. A single unit receives input from other units along with the associated weight w. The unit then computes a function f of the weighted sum of its inputs given by Eq. 6.2.

$$f\left(\sum_{j} w_{ij} y_j\right) \tag{6.2}$$

The function f acts as an activation to squeeze the amplitude of output to a bounded range. In most cases, the behavior of the function will describe the characteristics of a neuron. To leverage the power of neurons, they are combined into a network by representing them into single or multiple layers.

The difference between single and multiple layers network is that the multilayer network can be seen as the cascading of single-layer networks. Once the networks are built, to learn weights w we need to adopt a learning process. The training set is used to assist the learning process by adjusting the network to satisfy the constraints. Here, the approach to learning is with supervised training, thus making the neural networks supervised.

For the supervised neural networks, both inputs and outputs are provided such that the inputs are processed to generate outputs for further comparison against the desired outputs. Error is then calculated, making the system to adapt weights. This process is repeated many times such that the weights are continually adjusted.

6.3.5 *Support Vector Machines*

The support vector machines (SVM) are well-explored supervised learning algorithms mainly used as a discriminative classifier with a separating optimal hyperplane. The SVM builds on the core principles of geometry. Figure 6.6 shows separating hyperplanes, i.e., margin discriminating two different classes.

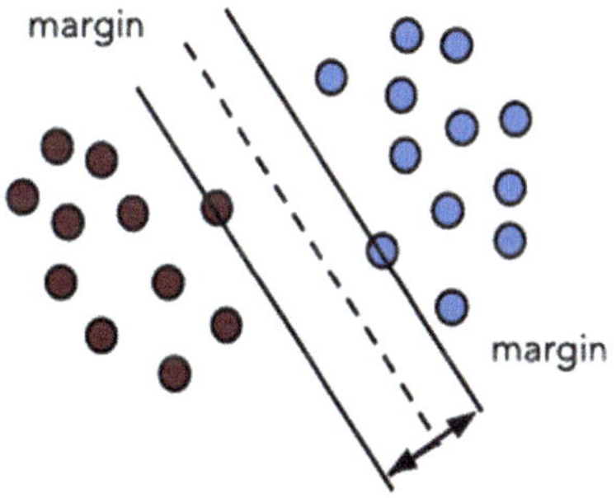

Fig. 6.6 SVM classifier used to separate two different classes

6.4 Unsupervised Learning for Big Data

The machine learning approach which learns from the data without any labels or supervision usually referred as unsupervised learning, i.e., we have only input data (x) and no corresponding output labels (y). The goal for unsupervised learning is to model the underlying structure or distribution in the data in order to learn more about the data.

In unsupervised learning, unlike supervised learning there are no correct answers and there are no ground truth output labels. Unsupervised learning algorithms discover and present the interesting structure observed in the data. Similar to supervised learning algorithms, unsupervised learning algorithms from a broader perspective can be categorized into either clustering or association methods. The aim of the clustering algorithms is to discover the inherent groupings in the input data (x), while the association rules learning approaches aim to discover rules that describe large portions of input data (x).

Unsupervised learning has the ability to leverage large amounts of the data as no output labels are required to be created by the human annotators. This makes it a viable option to deal big data. Figure 6.7 summarizes the supervised classification approach.

Many unsupervised learning methods have been proposed in the literature that leverage big data. In the following, some of them are discussed in detail.

6.4.1 Spectral Clustering

A well-known technique to group data for identifying similar behavior is 'Clustering'. Several traditional algorithms exist for clustering such as k-means and k-nearest neighbors. However, a relatively better approach known 'spectral clustering' has shown to outperform many traditional approaches. It is also shown to be effectively solved using standard linear algebra methods.

Idea of spectral clustering is built on graph Laplacian matrices. It is always assumed that G is an undirected, weighted graph with weight matrix W, where $w_{ij} =$

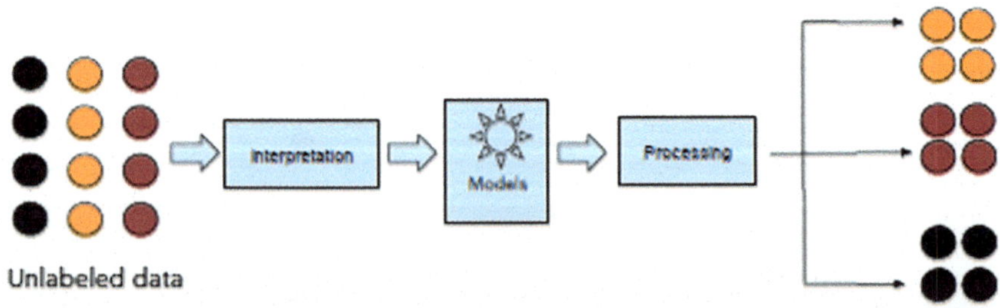

Fig. 6.7 Unsupervised machine learning for clustering

$w_{ji} \geq 0$. When using eigenvectors of a matrix, it is not assumed that they are normalized. For example, the constant vector and a multiple for some will be considered as the same eigenvectors. Eigenvalues will always be ordered increasingly, respecting multiplicities.

To perform a spectral clustering there are three main steps:

- A similarity graph between N objects to cluster is created.
- The first k eigenvectors of its Laplacian matrix to define a feature vector for each object is computed.
- k-means on these features is ran to separate objects into k classes.

6.4.2 Principal Component Analysis (PCA)

Aim of principal component analysis (PCA) is to provide dimensionality reduction by identifying most import components from the vector space using fundamentals of matrix factorization. It also a technique used to emphasize variation and get out patterns in a dataset. From the mathematical perspective, it is seen as a precompiled that transforms correlated variables into a smaller number of uncorrelated variables called principal components. The first principal component accounts for as much of the variability in the data as possible, and each succeeding component accounts for as much of the remaining variability as possible. Principal components analysis is similar to another multivariate procedure called factor analysis [8].

The Factor analysis is a procedure for identifying interrelationships that exist among variables, i.e., to identify how suites of variables are related. It can be used for exploratory or confirmatory purposes. In PCA, which is based on factor model where factors are based on the total variance, unities are used in the diagonal of the correlation matrix computationally implying that all the variance is common or shared. Figure 6.8 shows the representation of PCA with translated coordinates.

Although PCA is a linear model, it can leverage kernel trick to involve nonlinear modeling [9] with other variations such as iterative kernel PCA, etc. In the space of

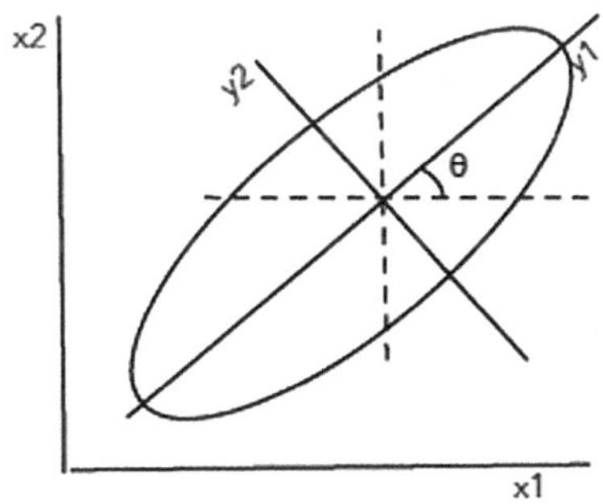

Fig. 6.8 PCA with principal axes $y1$ and $y2$

PCA, nearest neighbor search in the low-dimensional heuristics for selecting large inner products is possible.

6.4.3 *Latent Dirichlet Allocation (LDA)*

Proposed by Blei et al. [10], the goal of latent Dirichlet allocation (LDA) is to build a generative probabilistic model for modeling discrete collections (e.g., textual corpora.). A vanilla version of the LDA is a three-level hierarchical Bayesian model, where each item in the collection is modeled as a finite mixture over an underlying set of topics. Depending on the type of discrete data collection chosen, topic probabilities provide an explicit representation for an item in the collection. However, as in standard graphical models, exact inference is intractable and efficient approximate inference techniques are adopted for Bayes parameter estimation.

LDA which was initially created to model discrete data such as textual collections and has an underlying assumption of changeability. This is interpreted with random variables as: conditionally independent and identically distributed. However, similar assumption may not hold for other discrete form of data collection. To comprehend the vanilla version of LDA, we leverage discrete textual data collection, although it can be leveraged for other domains such as cross-modal retrieval and bioinformatics.

LDA for textual documents is represented as random mixtures over latent topics, where each topic, i.e., latent multinomial variables is characterized by a distribution over words. In the following, LDA generative process for each document is briefly described:

- A word is assumed as the basic unit of the data with indexed vocabulary.
- A document contains sequence of words.
- A collection contains documents.

Now, a generative process for each document in a collection (M) is given as:

- Choose a Poisson distribution for a sequence of N words.
- Identify a parameter _ which follows Dirichlet distribution over _.
- For each of N words: (i) choose a topic (z) which follows multinomial distribution over _. (ii) choose a word based on conditionally dependent on topic and parameterized matrix.

Figure 6.9 presents the plate notation of the LDA model. However, there exists many variations of LDA model for the textual data.

Few examples include:

- Correlated topic model [11]: models same type of data as LDA and only differs in the first step of the generative process. Instead of drawing _ from a Dirichlet distribution it assumes that _ is drawn from a logistic-normal distribution.
- Dynamic topic model [12]: use state-space models on the natural parameter space of the underlying topic multinomials, as well as on the natural parameters for

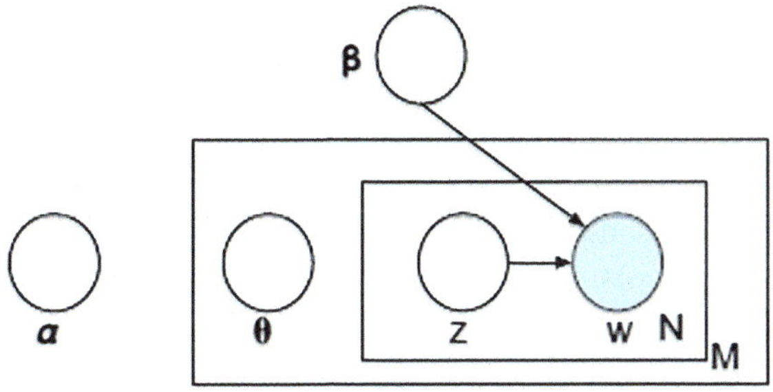

Fig. 6.9 LDA plate model

the logistic-normal distributions used for modeling the document-specific topic proportions. Also, the sequential structure between models is again captured with a simple dynamic model.

- Continuous-time dynamic topic models [13] leverage a sequential collection of documents. Brownian motion is used to model the latent topics. When compared against the discrete-time dynamic topic model which requires that time be discretized, continuous-time is better in better in terms of perplexity.

Going beyond textual documents, LDA is applied for visual content as well for object recognition [14] and also for generative decomposition of visual categories for unsupervised learning and supervised detection [15].

6.4.4 Matrix Factorization

Goal of matrix factorization is to fit a matrix X with a low-rank matrix Y by minimizing the sum squared error, which is usually achieved explicitly in terms of the singular value decomposition of X [16]. However, few problems persist such as with loss functions. Other than the squared-error loss function, many functions yield non-convex optimization problems with multiple local minima. This is mainly attributed to the low-rank approximations, which add constraints such as limiting the dimensionality of factorization, sparsity and non-negativity [17].

Nevertheless, low-rank approximations have shown effective results for applications such as collaborative filtering and many more. Usage of non-negativity is found to be a useful constraint for matrix factorization, which can learn representation of the data in parts. Formally, non-negative matrix factorization (NMF) is derived by finding non-negative matrix factors W and H of a non-negative matrix X, i.e., X_WH.

Earlier; NMF was successfully applied to the statistical analysis of multivariate data [18]. To comprehend the working of NMF, let's consider a standard example, where multivariate data vectors belonging to n dimensions are placed in the columns of an $n\ \_\ p$ matrix X, where p is the number of examples in the dataset. Factorization of X approximately leads into an $n\ \_\ r$ matrix W and an $r\ \_\ p$ matrix H. Generally, r is chosen to be smaller than n or p. Here, W can be regarded as containing a basis that is optimized for the linear approximation of the data in X. It is considered a good approximation only when the basis vectors discover structure that is latent in the data.

Furthermore, techniques of factorization are extended to deal with large/big data by leveraging distributed algorithms [19]. For this, a variant of stochastic gradient descent is used for an iterative optimization in a sequential setting. Sparse matrix factorization is also explored in certain scenarios.

Recommendation systems have been heavily benefited with matrix factorization techniques [20]. It is shown in one of the recommendation engine challenge that the matrix factorization models are superior to classic techniques for product recommendations. Models allowed the incorporation of additional information such as implicit feedback, temporal effects and confidence levels.

6.4.5 *Manifold Learning*

Goal is to achieve nonlinear dimensionality reduction. Algorithms developed for the manifold learning assume that data points in the dataset contain big feature vectors, which may be reduced or described with only a few underlying parameters. Hence, it is assumed that the data points are actually examples from a low-dimensional manifold that is embedded in a high-dimensional space [21]. Idea of using low-dimensional manifold is driven by motivation to reduce the irrelevant or misleading features, which reduce the difficulty of the optimization algorithms to find the global optima. Hence, the aim of manifold learning is to uncover the manifold structure in a dataset, where the low-dimensional space reflects the underlying parameters and high-dimensional space is the feature space.

Usually, a simple manifold can be depicted with a one-dimensional curve. Also in other words, manifold is expected to lie in a high-dimensional space, but will be homeomorphic (e.g., any continuous function whose inverse is also considered to be a continuous function) with a low-dimensional space. In general, manifold is denoted as an embedding of certain dimensions and is transformed into another form by identifying the only subset of its dimensions.

Reviving the discussion on learning, we already understand that the aim of manifold learning is to learn a manifold from a set of points. Several algorithms are developed to perform it such as ISOMAP [22] and local linear embeddings (LLE) [23]. ISOMAP a shorter form for isometric feature mapping is seen as an extension to multidimensional scaling, a method for embedding dissimilarity information into Euclidean space. While, LLE had a different idea and was based on visualizing a manifold as a collection of overlapping coordinate patches.

Furthermore, Laplacian Eigenmaps [24] is also introduced based on spectral graph theory where the Laplacian is exploited to capture local information about the manifold. There also other approaches such as local tangent space alignment [13], which is based on the geometric intuitions as LLE, i.e., dataset is sampled from a smooth manifold, then the neighbors of each point remain nearby and similarly co-located in the low-dimensional space.

Manifold learning has been applied heavily for processing images [25] and also natural language processing applications [26].

6.5 Semi-supervised Learning for Big Data

Sometimes it is hard to annotate entire input data (x) with output labels (y) due to their large scale. However, partially labeled data is still possible. But, similar to supervised learning approaches, learning from partially labeled and the unlabeled data can still be achieved. Hence, when there is a large amount of input data (x) and only some of the data has output labels (y) learning approaches that leverage such a data can be devised as the semi-supervised problems.

These approaches combine effectiveness of supervised and unsupervised learning methods. Demand for such approaches is high, as most of the real-world datasets are unlabeled and it is expensive or time consuming to label all data using domain experts. Whereas, unlabeled data is readily available. Approaches which perform semi-supervision use unsupervised learning techniques to discover and explore structure in the input variables, while supervised learning learn how to make predictions from the labeled data.

Furthermore, semi-supervision feed the data back into the supervised learning algorithm as training data and use the model to make predictions on new unseen data. As mentioned earlier that the goal of semi-supervised learning is to leverage both labeled and unlabeled data. This is, however, motivated from its practical value in learning while achieving results in faster, better and cheaper manner [27]. Figure 6.10 shows the visualization of the goal of semi-supervised approaches.

Many semi-supervised learning methods have been proposed in the literature that leverage big data. In the following, some of them are discussed in detail.

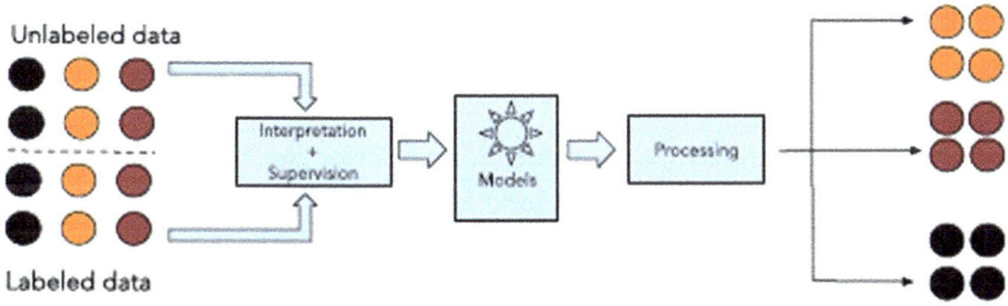

Fig. 6.10 Semi-supervised learning for the classification leveraging both labeled and unlabeled data

6.5.1 Co-training

Idea of co-training for semi-supervised learning is proposed by Blum et al. [28] inspired from Yarowsky et al. [29], who identify meaning for words in different instances to perform word-sense disambiguation. This idea was transferred to semi-supervised learning.

In co-training, there are certain assumptions:

- Extracted features are split into two sets.
- Each subset of features is sufficient to train a classifier.
- The two sets of features are conditionally independent given the class.

Sticking to the aforementioned assumptions, two separate classifiers are trained with the labeled data, on the two feature subsets, respectively. Each classifier then classifies the unlabeled data and teaches the other classifier with the sample unlabeled examples that are predicted. Classifiers are retrained with the additional training examples. This process is repeated until there is decent accuracy acquired. This improves the agreement between two classifiers on the unlabeled data along with the labeled data.

Several other approaches also emerged overtime such as co-EM [30] which performs experiments to compare co-training with generative mixture models and EM. It is also combined with other methods such as canonical correlation analysis, which reduced the requirement for the labeled data [31]. Theoretical bounds are also understood by providing a PAC-style analysis [32].

6.5.2 Label Propagation

The label propagation aims at designing a simple iterative algorithm, which can propagate labels through the dataset along high-density areas defined by unlabeled data. Earlier approaches such as k-nearest-neighbor (kNN) used in traditional supervised learning can be one possible solution. However, efficient algorithms are required to propagate labels. This is achieved by propagating labels to all nodes according to their proximity using minimum spanning tree heuristic and entropy minimization-based parameter optimization [33].

However, the procedure described earlier does not guarantee an equal distribution of classes, as the algorithm will not have any control over the final proportions of the classes. When the classes are not well separated and labeled data is scarce, incorporating constraints on class proportions can improve classification.

6.5.3 *Multiview Learning*

Removing assumptions in co-training is desirable. In multiview learning, different hypotheses are trained from the same labeled dataset, and are expected to find similar predictions on any given unlabeled instance. Several related works have been explored in the context of regression [34] and large structured outputs [35].

6.6 Reinforcement Learning Basics for Big Data

The goal of reinforcement learning is to make machines learn by understanding their progress. It is very close to unsupervised learning, but differs mainly in giving control to agents for determining the ideal behavior within a context. This connection is established to maximize the performance of machine.

Feedback is used to inform the machine about its progress in a task such that appropriate behavior is learned.

Reinforcement learning is considered as a complex approach, and requires plethora of different algorithms. In reinforcement learning, an unknown environment is present, and an agent interacts with this environment to see what happens, i.e., an agent decides the best action based on the current state in the environment. The reinforcement learning (RL) is also seen as a mix between supervised and unsupervised learning as some form of learning scheme is decided based on the observed signals. It is also seen as supervised learning in an environment of sparse feedback. Figure 6.11 shows the overall procedure adopted for the reinforcement learning.

Many reinforcement learning methods have been proposed in the literature that are used for big data. In the following, some of them are discussed in detail.

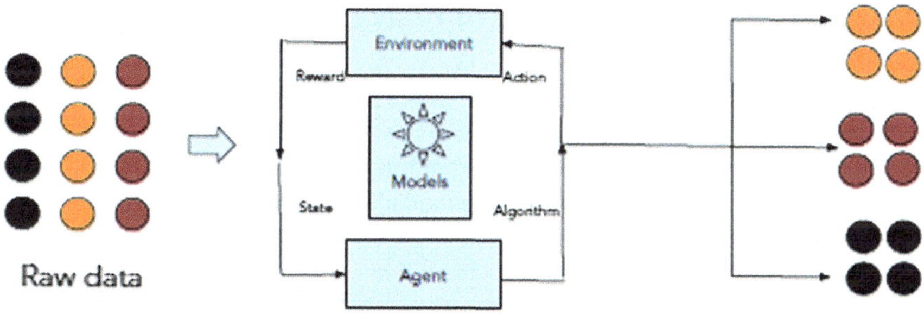

Fig. 6.11 Reinforcement learning for the classification of raw input data

6.6.1 Markov Decision Process

The Markov decision process (MDP) [36] is a framework which is helpful to make decisions in an environment that is stochastic. Goal in MDP is to find a policy which is a map that gives all optimal actions on each state in the environment. MDP is usually considered powerful than the planning approaches as the optimal policy allows to performing optimal actions even if something has wrong along the way.

Important notations usually used in MDP are:

- States: A set of possible states.
- Action: Things that can be performed on a particular state s.
- Reward: Value that is given for been on a state.
- Model: Probability of the state s going to the new state $s0$, when an action a is performed.
- Policy: Is a map that tells the optimal action for every state.
- Optimal policy: Is a policy that maximizes expected reward.

6.6.2 Planning

Mostly deals with principled and efficient way for cooperative multiagent dynamic systems. Methods explored are coordination and communication between the agents which is derived directly from the system dynamics and function approximation architecture [37]. Entire multiagent system is viewed as a single, large MDP, which is represented in a factored way using a dynamic Bayesian network (DBN).

6.6.3 Reinforcement Learning in Practice

Applications of reinforcement learning are where information about the environment is limited. For example, RL is applied to robot control and simple board games. Recently, combining it with deep learning simplified the states and made the computer program to won against the second-best human player in the world [38]. Also, it has shown its applicability to other areas such as robotic control (where robots perform several tasks similar to humans), online advertising, dialogue systems, etc.

6.7 Online Learning for Big Data

A variant of big data is the streaming data. It is sometimes referred as the fast data [39] which captures the notion that streams of data are generated at very high rates.

In order to leverage this type of data for actionable intelligence, it requires quick analysis satisfying its properties such as time-variance and unpredictability.

Fast data dealing cannot be done with the traditional machine learning algorithms as they allow only batch learning and are bounded by the following constraints:

- They assume entire training data is available during learning.
- There are no time constraints, learning algorithm waits until all data is loaded for training.
- Usually a single model is trained and no further updates are made to the trained models.

However, earlier mentioned constraints are considered to be rigid when we deal with the fast data. Expectation from the learning algorithms which use fast data is that: (i) model needs to be updated whenever a new sample arrives. (ii) data cannot be accessed in a single go and is available in streaming. To handle such scenarios, online machine learning [40] is the best strategy.

The idea of online machine learning is presented in an overview [41] explaining its goal, i.e., to learn a concept incrementally by processing labeled training examples one at a time. After each data instance has arrived, model is updated and the instance is discarded. This also has other consequences. As only a small amount of the data can be kept for analysis. It allows certain uncertainties such as suboptimal solutions on parts of the data that is already used and difficult to unwind. Nevertheless, for the fast data incremental or online learning is the best solution.

The traditional machine learning approaches use batch learners for building non-dynamic models from the finite data. In contrast, online learning models face different challenges from multiple perspectives. First challenge is from the data, as data is generated from a continuous, non-stationary flow in a random order. Models generated from online learning are affected by concept deviations. Second is from the distribution, this occurs when the data volume is large as it can effectively reduce the time for computation.

Online learning usually tries to address above-mentioned challenges by designing statistical tests to choose better hypotheses [42]. There is also possibility of providing theoretical guarantees by adding constraints on the distribution of the data. Also, as discussed earlier that the fast data streams change over time, thus online learning algorithms are expected to handle such concept drifts. Also, there is a possibility of leveraging streaming computational models. This means having training adapted to the concept drift by overwriting parameters of the model [41]. However, it will be adhered to the task, the data and the choice of the approach.

The solutions developed for the online learning keep into consideration above challenges arising from the fast data to build models. This means that suboptimal decisions made in an earlier stage cannot be undone as the data is no longer available for the algorithm to reconsider it for training. Combining both batch and online learning approaches is also possible. They have high impact in practice, as models trained in batches can be used as the precompiled sample to be further used for prediction of samples arising from the fast data [43].

Closely related area of the online learning is having its distributed version. It is usually achieved with either data or model parallelism. Data parallelism partitions the training data and builds separate models using the subset of the data to be merged eventually. However, model parallelism [44] partition across attributes in the same training instance of data. This is in contrast to partitioning the training data coming from the fast data stream where each data instance is split into its constituting attributes, and each attribute is sent to a different processing element. For instance, models that use gradient descent for optimization, coefficients can be stored and updated by accessing a distributed storage. Nevertheless, a setback in this approach is that to achieve good performance, there must be sufficient inherent parallelism in the modeling algorithm.

A well-known example of model parallelism is parameter server [45]. Goal is to achieve distributed online learning applications by allowing access to shared parameters as key–value pairs. Typically tasks that are parallel are asynchronous, but the algorithm design allows flexibility in choosing an appropriate model. Several variants of the parameter server exist [46, 47] with different configurations. For example, Smola et al. [45] use distributed key–value stores such as memcached to store the parameters. Other advanced solutions [48] use specific parameter access interfaces such as a general platform for synchronous operations.

Parameter server makes online learning serve other learning approaches, i.e., supervised and unsupervised. For example, neural networks for classification [49], multiple additive regression trees [50], etc. Key modifications are done to standard algorithms such as gradient descent into asynchronous distributed version help it to scale for large-scale systems. Nevertheless, parameter server approach also fits to standard batch models. However, it serves better for online learning methods and other approaches such as reinforcement learning [51].

Sometimes, online learning is designed from the task perspective, e.g., classification, regression, etc. Thus, it will introduce the task-specific constraints in the algorithm design. For example, iteration over the entire data multiple times is ruled out. Also, training instances are assumed to be independent, i.i.d. and generated from a stationary distribution. Hence, different prediction models have to be applied and evaluated at different times [52].

For online learning new or modified evaluation measures are developed, as the existing standard cross-validation and other strategies which are used for batch learning are not applicable. It is usually evaluated in the literature with overall loss as a sum of losses experienced by individual training examples [53].

Other approaches include area under curve (AUC) modified for online learning [54].

Several popular supervised, unsupervised and reinforcement learning approaches which are discussed in earlier sections are modified to support online learning. Few examples include online kernel PCA [55], online matrix factorization [20], online LDA [47], reinforcement learning fitting fast data and non-stationary environments handling both data and actions [56], online gradient descent [57] and online SVM [58].

6.8 Conclusion

In this chapter, we presented machine learning basics on big data by exploring different techniques. Initially, we divided the machine learning approaches into different categories such as supervised, unsupervised, semi-supervised and reinforcement techniques for learning in the big data setting. Furthermore, we saw how some techniques can be leveraged in online setting as apposed to batch learning. We also presented some of the practical applications of these techniques that are applied in the industry settings for solving multiple challenges.

6.9 Review Questions

(1) What are the different types of learning algorithms and how do they differentiate between each other?
(2) How is supervised learning different for regression and classification. Give examples to support your claim?
(3) What is graph Laplacian and how is it used in spectral clustering?
(4) What are advantages and disadvantages of online learning when compared against batch learning approaches?
(5) Provide an algorithm to create online version of the stochastic gradient descent?
(6) Which type of batch learning algorithms can be converted into online variants?
(7) Does semi-supervised learning can be leveraged for building models that can work for classification and regression?
(8) What type learning algorithms can work better for large volume of data?
(9) What are the applications of reinforcement learning?

References

1. S.P. Singh, U.C. Jaiswal, Machine learning for big data: a new perspective. Int. J. Appl. Eng. Res. **13**(5), 2753–2762 (2018)
2. A.Y. Ng, M.I. Jordan, On discriminative vs. generative classifiers: a comparison of logistic regression and naive bayes, in *Advances in Neural Information Processing Systems* (2002), pp. 841–848
3. T. Jaakkola, M. Meila, T. Jebara, Maximum entropy discrimination, in *Advances in Neural Information Processing Systems* (2000), pp. 470–476
4. J. Han, J. Pei, M. Kamber, *Data Mining: Concepts and Techniques* (Elsevier, 2011)
5. P.E. Utgoff, Incremental induction of decision trees. Mach. Learn. **4**(2), 161–186 (1989)
6. J.R. Quinlan, *C4. 5: Programs for Machine Learning* (Elsevier, 2014)
7. L. Breiman, J. Friedman, C.J. Stone, R.A. Olshen, *Classification and Regression Trees* (CRC Press, 1984)
8. S. Wold, K. Esbensen, P. Geladi, Principal component analysis. Chemometr. Intell. Lab. Syst. **2**(1–3), 37–52 (1987)
9. B. Schölkopf, A. Smola, K.-R. Muller, Nonlinear Component Analysis as a Kernel Eigenvalue Problem. Neural Comput. **10**(5), 1299–1319 (1998)
10. D.M. Blei, A.Y. Ng, M.I. Jordan, Latent dirichlet allocation. J. Mach. Learn. Res. **3**, 993–1022 (2003)

11. D.M. Blei, J. Lafferty, Correlated Topic Models, in *Advances in Neural Information Processing Systems* (2005)
12. D.M. Blei, J.D. Lafferty, Dynamic topic models, in *Proceedings of the 23rd International Conference on MACHINE Learning*, (2006), pp. 113–120
13. J. Wang, Local tangent space alignment, in *Geometric Structure of High-Dimensional Data and Dimensionality Reduction* (2012), pp. 221–234
14. J. Sivic, B.C. Russell, A.A. Efros, A. Zisserman, W.T. Freeman, Discovering objects and their location in images, in *ICCV* (2005), pp. 370–377
15. M. Fritz, B. Schiele, Decomposition, Discovery and Detection of Visual Categories Using Topic Models, in *CVPR* (2008)
16. N. Srebro, J. Rennie, T.S. Jaakkola, Maximum-margin matrix factorization, in *Advances in Neural Information Processing Systems* (2005), pp. 1329–1336
17. D.D. Lee, H. Sebastian Seung, Learning the parts of objects by non-negative matrix factorization. Nature **401**(6755), 788 (1999)
18. D.D. Lee, H. Sebastian Seung, Algorithms for non-negative matrix factorization, in *Advances in Neural Information Processing Systems* (2001), pp. 556–562
19. R. Gemulla, E. Nijkamp, P.J. Haas, Y. Sismanis, Large-scale matrix factorization with distributed stochastic gradient descent, in *Proceedings of the 17th ACM SIGKDD International Conference on Knowledge Discovery and Data Mining* (2011), pp. 69–77
20. Y. Koren, R. Bell, C. Volinsky, Matrix factorization techniques for recommender systems. Computer **8**, 30–37 (2009)
21. L Cayton, *Algorithms for Manifold Learning*. University of California at San Diego Tech. Rep 12, no. 1–17: 1 (2005)
22. J.B. Tenenbaum, V. De Silva, J.C. Langford, A global geometric framework for nonlinear dimensionality reduction. Science **290**(5500), 2319–2323 (2000)
23. L.K. Saul, S.T. Roweis, Think globally, fit locally: unsupervised learning of low dimensional manifolds. J. Mach. Learn. Res. **4**, 119–155 (2003)
24. M. Belkin, P. Niyogi, Laplacian eigenmaps and spectral techniques for embedding and clustering, in *Advances in Neural Information Processing Systems* (2002), pp. 585–591
25. R. Pless, R. Souvenir, A survey of manifold learning for images. IPSJ Trans. Comput. Vis. Appl. **1**, 83–94 (2009)
26. I. Labutov, H. Lipson, Re-embedding words, in *Proceedings of the 51st Annual Meeting of the Association for Computational Linguistics (Volume 2: Short Papers)*, vol. 2, (2013), pp. 489–493
27. X. Zhu, Semi-supervised learning, in Encyclopedia of Machine Learning (2011), pp. 892–897
28. A. Blum, T. Mitchell, Combining labeled and unlabeled data with co-training, in *Proceedings of the Eleventh Annual Conference on Computational Learning Theory* (1998), pp. 92–100
29. D. Yarowsky, Word-sense disambiguation using statistical models of Roget's categories trained on large corpora, in *Proceedings of the 14th Conference on Computational Linguistics*, vol. 2 (1992), pp. 454–460
30. K. Nigam, A.K. McCallum, S. Thrun, T. Mitchell, Text classification from labeled and unlabeled documents using EM. Mach. Learn. **39**(2–3), 103–134 (2000)
31. X. Zhu, T. Rogers, R. Qian, C. Kalish, Humans perform semi-supervised classification too. AAAI **2007**, 864–870 (2007)
32. S. Dasgupta, M.L. Littman, D.A. McAllester, PAC generalization bounds for co-training, in *Advances in Neural Information Processing Systems* (2001), pp. 375–382
33. X. Zhu, Z. Ghahramani, Learning from labeled and unlabeled data with label propagation (2002)
34. V. Sindhwani, P. Niyogi, M. Belkin, A co-regularization approach to semi-supervised learning with multiple views, in Proceedings of ICML Workshop on Learning with Multiple Views (2005), pp. 74–79
35. U. Brefeld, C. Bscher, T. Scheffer, Multi-view discriminative sequential learning, in *European Conference on Machine Learning* (2005),pp. 60–71

36. W.S. Lovejoy, A survey of algorithmic methods for partially observed Markov decision processes. Ann. Oper. Res. **28**(1), 47–65 (1991)
37. C. Guestrin, D. Koller, R. Parr, Multiagent planning with factored MDPs, in *Advances in Neural Information Processing Systems* (2002), pp. 1523–1530
38. D. Silver, A. Huang, C.J. Maddison, A. Guez, L. Sifre, G. Van Den Driessche, J. Schrittwieser et al., Mastering the game of Go with deep neural networks and tree search. Nature **529**(7587), 484 (2016)
39. W. Lam, S.T.S. Lu Liu, A.R. Prasad, Z. Vacheri, A. Doan, Muppet: MapReduce-style processing of fast data. Proc. VLDB Endow. **5**(12), 1814–1825 (2012)
40. Ó. Fontenla-Romero, B. Guijarro-Berdiñas, D. Martinez-Rego, B. Pérez-Sánchez, D Peteiro-Barral, Online machine learning, in *Efficiency and Scalability Methods for Computational Intellect* (2013), pp. 27–54
41. G. Widmer, M. Kubat, Learning in the presence of concept drift and hidden contexts. Mach. Learn. **23**(1), 69–101 (1996)
42. P. Domingos, G. Hulten, Mining high-speed data streams, in *Proceedings of the Sixth ACM SIGKDD International Conference on Knowledge Discovery and Data Mining* (2000), pp. 71–80
43. D. Crankshaw, X. Wang, G. Zhou, M.J. Franklin, J.E. Gonzalez, I. Stoica, Clipper: a low-latency online prediction serving system, in *NSDI* (2017), pp. 613–627
44. M. Li, D.G. Andersen, J.W. Park, A.J. Smola, A. Ahmed, V. Josifovski, J. Long, E.J. Shekita, S. Bor-Yiing, Scaling distributed machine learning with the parameter server. OSDI **14**, 583–598 (2014)
45. A. Smola, S. Narayanamurthy, An architecture for parallel topic models. Proc. VLDB Endow. **3**(1–2), 703–710 (2010)
46. B. Fitzpatrick, Distributed caching with memcached. Linux J. **124**, 5 (2004)
47. Q. Ho, J. Cipar, H. Cui, S. Lee, J.K. Kim, P.B. Gibbons, G.A. Gibson, G. Ganger, E.P. Xing, More effective distributed ml via a stale synchronous parallel parameter server, in *Advances in Neural Information Processing systems* (2013), pp. 1223–1231
48. M. Li, L. Zhou, Z. Yang, A. Li, F. Xia, D.G. Andersen, A. Smola, Parameter server for distributed machine learning. Big Learn. NIPS Works. **6**, 2 (2013)
49. J. Dean, G. Corrado, R. Monga, K. Chen, M. Devin, M. Mao, A. Senior et al., Large scale distributed deep networks, in *Advances in Neural Information Processing Systems* (2012), pp. 1223–1231
50. J. Zhou, Q. Cui, X. Li, P. Zhao, S. Qu, J. Huang, PSMART: parameter server based multiple additive regression trees system, in *Proceedings of the 26th International Conference on World Wide Web Companion* (2017), pp. 879–880
51. A. Nair, P. Srinivasan, S. Blackwell, C. Alcicek, R. Fearon, A. De Maria, V. Panneershelvam et al., Massively parallel methods for deep reinforcement learning (2015). arXiv preprint arXiv: 1507.04296
52. A.A. Benczúr, L. Kocsis, R. Pálovics, Online machine learning in big data streams (2018). arXiv preprint arXiv:1802.05872
53. A.P. Dawid, Present position and potential developments: some personal views: statistical theory: the prequential approach. J. Royal Stat. Soc. Series A (General), 278–292 (1984)
54. P. Zhao, S.C.H. Hoi, R. Jin, T. Yang, Online AUC maximization, in *ICML* (2011)
55. S. Agarwal, V. Vijaya Saradhi, H. Karnick, Kernel-based online machine learning and support vector reduction. Neurocomputing **71**(7–9), 1230–1237 (2008)
56. R.S. Sutton, A.G. Barto, F. Bach, *Reinforcement Learning: An Introduction* (MIT Press, 1998)
57. J. Langford, L. Li, T. Zhang, Sparse online learning via truncated gradient. J. Mach. Learn. Res. **10**, 777–801 (2009)
58. A. Bordes, L. Bottou, The huller: a simple and efficient online SVM, in *European Conference on Machine Learning* (2005), pp. 505–512

36. W.S. Lovejoy, A survey of algorithmic methods for partially observed Markov decision processes. Ann. Oper. Res. 28(1), 47–65 (1991)
37. C. Guestrin, D. Koller, R. Parr, Multiagent planning with factored MDPs, in Advances in Neural Information Processing Systems (2002), pp. 1523–1530
38. D. Silver, A. Huang, C.J. Maddison, A. Guez, L. Sifre, G. Van Den Driessche, J. Schrittwieser et al., Mastering the game of Go with deep neural networks and tree search. Nature 529(7587), 484 (2016)
39. W. Fan, [illegible], A.R. Prasad, Z. Vachel, A. Lyon, [illegible], [illegible] of [illegible]-data, Proc. VLDB Endow. 8(12), [illegible] (2015)
40. C. [illegible]-Romero, [illegible], [illegible], [illegible], [illegible], Online machine learning on big streams and scalability methods for [illegible] (2013), pp. [illegible]
41. G. Widmer, M. Kubat, Learning in the presence of concept drift and hidden contexts. Mach. Learn. 23(1), 69–101 (1996)
42. P. Domingos, G. Hulten, Mining high-speed data streams, in Proceedings of the Sixth ACM SIGKDD International Conference on Knowledge Discovery and Data Mining (2000)

Chapter 7
Social Semantic Web Mining and Big Data Analytics

7.1 Introduction

In the context of Big Data Analytics and Social Networking, Semantic Web Mining is an amalgamation of three scientific areas of research: (1) Social Networking (2) Semantic Web and (3) Big Data Analytics, including Web Mining and Data Mining.

Semantic Web is defined and used to provide semantics or meaning to the data on the web, while Big Data Analytics and data mining are aiming to identify and extract interesting patterns from homogeneous and less complex data in or outside the web.

In view of the huge growth of the sizes of data on the web and social networks, we have the Big Data scenario coming up and in view of the growth of the complexity of meaning of data in the web we have the Semantic Web coming up. Due to the rapid increase in the amount of Semantic Data and Knowledge in various areas such as biomedical, genetic, video, audio and also social networking explosion, there could be a transformation of the entire data scenario into a perfect target for Big Data Analytics leading to both in the terms Semantic Web Mining and Social Semantic Web Mining. The Petabyte scale of huge data sizes which are required to be mined on the web which can be either a normal (syntactic) web- or knowledge-based Semantic Web; we have research possibilities of Social Semantic Web Mining Techniques. Semantic Web is changing the way in which data is collected, deposited and analyzed [1] on the web.

7.2 What Is Semantic Web?

We define Semantic Web as a Web which provides meaning to its contents as against the present 'Syntactic Web' which carries no meaning but only display. It is about providing meaning to the data from different kinds of web resources so as to enable machine to interpret or 'understand' these enriched data to precisely answer and satisfy the requests from the users of the Web [2–4]. Semantic Web is new generation

C. S. R. Prabhu et al., *Big Data Analytics: Systems, Algorithms, Applications*,
https://doi.org/10.1007/978-981-15-0094-7_7

Web 2.0 or beyond. Semantic Web is an extension of the present (Syntactic) Web (WWW) to enable users to represent meaning of their data and also share it with others.

Why Semantic Web?

Semantic Web was initiated for working on specific problems [5]: (1) to overcome the limitation of data access in the web as: instead of retrieving all kinds of documents from the web for a given query but focus better with more knowledge-based best fit searches and (2) the delegation of task problems (such as integrating information) by supporting access to data at web scale and enabling the delegation of certain classes of tasks.

Finally, it is the objective of Semantic Web to represent knowledge on the web instead of simple data where meaning is not machine understandable. This calls for techniques of knowledge representation (KR) being taken from artificial intelligence (AI).

Ontology in Semantic Web represents the knowledge repositories.

7.3 Knowledge Representation Techniques and Platforms in Semantic Web

XML

Extended markup language (XML) is a mechanism to represent all types of data. XML can be read by computers for interoperability between applications on the web. XML pages generated by applications based on schema can be read by humans also and XML can be interpreted by computers. XML provided data interoperability but not meaning—the data in XML could not be described for its meaning—meaning/semantics could not be described. It could be used only for defining syntax. Semantics integration and interoperability (e.g., British units and Metric units) calls for explicitly semantic descriptions to facilitate integration. Prior to XML, one had to code by hand the modules to retrieve the data from the data sources and also construct a message to send to other applications. XML facilitated building systems that integrate distributed heterogeneous data. XML allows flexible coding and display of data by using metadata to describe the structure of the data (using DTD on XSD). The first step in achieving data integration is to take raw data such as text, tables or spreadsheets and convert them to well-formed XML documents. The next step is to create DTD on XSD for each data source to analyze/document its structure. XML is incorporated in web services which can interoperate among themselves by using SOAP for invocation.

RDF
Resource description framework (RDF) can be read by machines, i.e., computers. Resources (web sites) are linked to form the web—to give meaning to resources and links a few new standards and languages are being developed—to allow precisely describe the resources and their links. RDF and OWL are standards that enable the web to be able to share documents and data RDF is semantic definition of the details of the description of a web site.

7.4 Web Ontology Language (OWL)

Web ontology language (OWL) is more complex language with better machine interpretability than RDF.

OWL primarily defines the nature of resources and their inter-relationships [6]. To represent the information in Semantic Web, the OWL uses ontology.

7.5 Object Knowledge Model (OKM) [7]

The RDF and RDFS provide the mechanism for knowledge representation using triplets (subject-predicate-object). This is only to the sentence level of Semantics which is being captured. More deeper semantics can be captured if the Object Knowledge Model (OKM) is deployed which gives greater or word level semantics in terms of root, suffix indicating the case, gender, person, etc., depending on whether the word is a noun or verb, etc. (as identified after part of speed (POS) tagging). OKM represents complete natural language semantics of a sentence in a web page. Thus, instead of user identifying the metadata in terms of RDF entities, etc., it will be jointly to analyze the sentences in the web page themselves to identify not only the metadata in terms of entities and their properties and relationships with other entities, but also details of such relationships including case, number, gender and person.

Ontology
Ontologies are similar to taxonomies but use richer semantic relationships among terms and attributes and also strict rules on how to specify terms and relationships (Ontologies were originally from artificial intelligence (AI) used for inferencing but are also recently being applied to semantic web area).

An ontology is a shared conceptualization of the world. Ontologies contain definitional aspects such as high-level schemas, associated aspects such as entities, attributes, inter-relationships between entities, domain vocabulary and factual knowledge—all connected in a semantic manner [8–11].

Ontologies provide a common understanding of a particular domain.

The domain can be communicated between people, organizations and application systems using ontologies. Ontologies provide specific tools to organize and provide useful descriptions of heterogeneous content.

The major uses of ontologies

1. To assist in communication between human beings.
2. To achieve interoperability among software systems.
3. To improve the design and quality of software systems.

Technically, an ontology model is likely an ordinary object model in object-oriented programming. It contains classes, inheritance and properties.

In many cases, ontologies are considered as knowledge repositories.

7.6 Architecture of Semantic Web and the Semantic Web Road Map

Figure 7.1 indicates the architecture and the road map of the Semantic Web as indicated by W3C (Copyright).

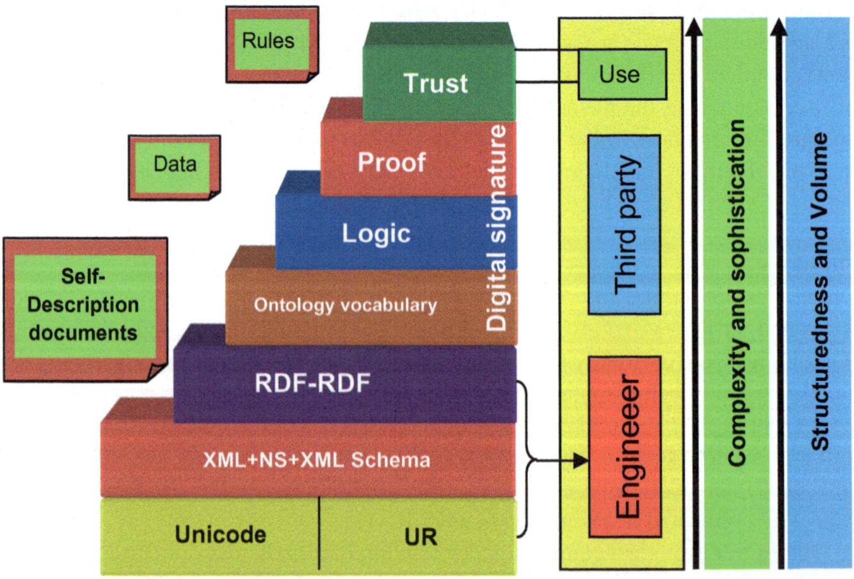

Fig. 7.1 Semantic web layer architecture [4]

7.7 Social Semantic Web Mining

The World Wide Web includes the social web or social networking-based web in addition to the normal conventional web. It represents the vast repository where millions of people on thousands of sites across hundreds of diverse cultures and languages and leave trail or footprint about various aspects of their lives, both professional and personal. Social systems are understood to be purely systems of communication and only communication, but therefore leave trails by analyzing what we can analyze in the social life itself: the political trends, cultural trends, media responses, trends of business and financial transactions and also Science.

There are several automated methods and techniques dealing with various aspects of information organization and retrieval that can be applied and used in the social science research based on Social Networks. Especially in the context of elections in the USA, India and other countries, political attitudes of voters become very important and it becomes very critical to analyze the trends and predict the results of the elections. The social web is also known to have influenced political discourse and a lot of trails about political standpoints and political views/sentiments of the citizens can be found on all over the web the use of Big Data Analytics.

Design and development of repositories of heterogeneous content with central themes is possible. Studies, surveys, databases and data warehouses on central themes can be prepared. Such themes could be specific problems such as attitudes toward government policies and projects or social trust among people, biases, prejudices, etc. Automated search engines are used to form certain hypothesis or model can be prepared by researchers and such hypothesis can be verified from the existing data repositories.

A separate section outlines the possibilities of automated multimedia processing by describing Speech Recognition and Machine (Computer) Vision techniques.

A separate section gives natural language processing (NLP) techniques overview and a separate section provides a survey of sentiment analysis techniques, and in a separate section, we describe Recommender systems and user profiling which allows us to automatically generate profiles of users based on their behaviors in a given domain and accordingly suggest what could be interesting to them. A separate section provides the concept Semantic Wiki Systems which extend usual wiki systems like Wikipedia with semantic description (e.g., metadata) and thus allow the creation of vast text and multimedia repositories that can be analyzed automatically based on AI techniques like automated reasoning. A separate section identifies the challenges and a separate section identifies the possible issues and research directions.

Data mining or knowledge discovery in large input data is a broad discipline under which web mining is a branch or subdomain. Web mining deploys techniques of data mining on web information so as to extract useful information from the web. The outcome is valuable knowledge extracted from the web.

Web mining can be classified into (a) Web content mining, (b) Web structure mining and (c) Web usage mining. In (a) Web content mining, the text and images in web pages are required to be mined, requiring to adopt natural language processing

(NLP) techniques. Web structure Mining is a technique to analyze and explain the link and structure of web. Web usage mining can be used to analyze how websites have been used much as determining the navigational behavior of users in terms of user behavior and therefore linked to social and psychological sciences.

Web is a source of invaluable social data by virtue of the ability the users to create, search and retrieve web pages containing valuable information.

Social Networking activities are a part of this mass data creation trend. They can be considered as special instance of Web Mining, i.e., social Web mining. Social Networking is understood to be a continuous networking activity to connect individual users to other individuals or groups of users on a daily, online or even real-time basis. This will produce many social relationships, including closeness and intimacy between individuals or groups. Experiences in life, ideas and thoughts are data which including video and photograph are shared among people. A variety of Social Networking Services is available today as:

Full-fledged personal Social Networking as Facebook and Twitter.

Professional Social Networking as LinkedIn
Blogging services as Word press
Video-sharing services as Vimeo
Photograph-sharing services as Flickr
e-commerce services as Quickr and many others as Amazon, Snapdeal, Flipkart, etc.

All these different Social Networking are continuously generating very large quantities of information on a nonstop, continuous basis, reflecting the instantaneous behavior of their users, reflecting their true life, relationships, interpersonal communication, thereby creating ready to use datasets, fit for analysis.

Although data is available, analyzing it is not easy. Several and different types of data exist and they are required to be analyzed accordingly, using appropriate tools technology and methodologies. Structured data in databases can be analyzed using SQL type of queries in DBMS such as Oracle. Data in data warehouses can be analyzed using OLAP tools and also data mining techniques and tools. Machine learning techniques such as Classification, Clustering and Regression can be deployed on the data, if it is adequately structured to find out trends, patterns and even forecast the possible future happenings on the data. This type of data could be from online sources as exchange rates, weather, traffic, etc.

Text data in web pages, emails or in blogs can be analyzed using NLP techniques of artificial intelligence (AI).

Semantic Web techniques can be deployed on creating metadata and Semantic Web Documents on web pages. They will be in RDF, RDFS and OWL kind of environments.

Image data, both static or still photographs and video clips, can be analyzed by image processing techniques (e.g., flash applications).

Since there is a time dimension also in the WWW, the temporal analysis of data is possible. This enables the trend analysis as in OLAP or regression. Adding geographical dimension to temporal data is of great value. Analyzing Social Networking data in time and using location information produces valuable results. For example, in a Cycle Race by the following the spectator's comments, it is possible keep track of the Race in real time. Analyzing data of social networks with all components as time and location (temporal and geographical) can lead to information which can make it easier to follow and also to keep check of the voters in an election. By gathering such information, it is possible to analyze it and find out why and how voters are thinking about the Party and the Candidate. Based on that data, it is possible to influence the voters by knowing what is important wherein a locality of population. In 2012 election in the USA and in 2014 election in India, a detailed analysis of social networking activity of the voters was extensively used to reach the voters and conveying them information which influenced their voting behavior and decide voting patterns to win the election finally. Those candidates who used such techniques benefited immensely.

Context in data is critical. Unless the context is understood, the data cannot make any sense or any meaning for the purpose of analysis. Sometimes, statements in Social Networks can be misleading, especially when they are sarcastic and therefore they mean the opposite of their true intention. Suddenly, they may lose their sarcasm. For example, in a satirical portal or online magazine, if the true context of the background is not understood, then the opposite meaning will be interpreted by the software or for analysis purpose. Therefore, it becomes extremely important to know the context for data creating, publication and consumption.

Social and Conceptual Analysis and Tag Clouds (II)

The new Science of Networks or Network Theory allows and enables us to study networks of any origin or subject domain. In the context of social Web analytics, we have two types of networks as relevant and of interest: social and conceptual. While social network provides the linkages (communication, friendship, interactions, trade, cooperation, etc.) between social entities (people, organization, social groups, political parties, countries, etc.), the conceptual networks provide insights into structure (homonymy, mutual context, synonymy, hyperlink, etc.) and dynamics (evolution of context) of concepts (words, ideas, phrases, symbols, web pages, etc.).

On the Social Semantic Web, both Social Networks and Conceptual Networks are ubiquitous. If two people can be connected on the social web, it can be any of the following: connection, friendship, co-opinion, cooperation, classmates, co-voter ship or they may have liked the same audio or same video on a news casting site or two people have similar political interest or attitude. When we multiply this situation for millions of individuals on the Social Semantic Networks, we can deploy social network analysis techniques as finding connected components or clustering or classification to identify networks of people who form groups of similar features such as similar political ideology or attitudes or belonging to the constituency of same political attitude.

If we observe such Social Networks over time, we might also reason about the dynamics of evolution of groups how and when, where groups or subgroups were formed. By having such data, we can develop agent-based models (A&M) that might predict future behavior based on the past behavior.

In addition, Social Networks can be analyzed so as to provide visualization of appealing, yet informative insights into a social system and find its peculiarities, if any, calling for appropriate action.

On the other hand, conceptual networks are not so directly observable or obvious: two web pages with one linked to the other or two keywords used in the same article, two tags on the same online video, two concepts extracted from the same text, etc.

Tag Clouds, Tag Maps, folksonomies, visualization and many other usable methods exist for analyzing Conceptual Networks.

Tag Clouds

Content created by humans can be tagged. Such tags are metadata on the context. What are Tags? Tags are keywords that can provide insights to readers of the contents of a given web page. Multiple tags can be created for better accuracy.

In Social Tagging, the so-called Tag Clouds are implemented which are visualization of tags provided by users. The most used tags are shown in large-sized letters and less used tags in small-sized letters. Tools such as 'Wardle' even have basic NLP capabilities and are capable of auto-summarizing the text in form of beautiful keyword clouds.

Tag or keyword clouds are not following network analysis method, but they do create and present visual representations of text and the context that provides an insight into the matter with a simple look on it.

Topic Maps

Topic Maps are a methodology for Text Visualization. Let us take a context of text, for example, in newspaper articles, we want to analyze. If we wish to see the development of certain topics keywords in time in the given context of a textual discourse, then we can do such visualization by deploying Topic Maps. If a topic is used more often in a given text, then it covers greater surface on the Topic Map. Even though such mechanisms are used in online social blogging or forms, they can be used in any series of text with spatio-temporal frame.

7.8 Conceptual Networks and Folksonomies or Folk Taxonomies of Concepts/Subconcepts

Conceptual networks comprise of clusters to give well-connected components. Complex conceptual network visualization may have hundreds of thousands of nodes. The visualization of complex network is an art and a science. Visualization has to be appealing, informative and concise. Numerous algorithms for complex network visualization have been proposed and developed, each for different types of networks.

Processes on Networks

There are techniques that allow in the study the dynamics and evolution of networks. Examples such as virus spreading or rumors spreading can be represented as Network

Processes. It is possible to develop agent-based models that allow us to understand the spreading of certain themes, whose behavior is to be studied. For example, in order to study the process in which the particular political attitude can spread through a social network, we might be required first to harvest the communication between people on the network (forum or blog), then we can conclude text (messages, articles, blogging, past, etc.) that deal with the name subject and also temporal data (time of publication). Then by deploying advanced analytical techniques to (a) construct the social network of people involved in the discourse, (b) identify different attitudes toward the actor (politician) concerned by using NLP techniques or sentiment analysis and finally (c) define the actual process of information spreading in the form of an agent-based model.

7.9 SNA and ABM

Another distinct method that enables analysis of a distinct set of primarily social entities and their interaction is by using the combined power of Social Network Analysis (SNA) and Agent Based Modeling (ABM) so as to enable huge amounts of data to be embedded into user friendly models.

The SNA and ABM go together. ABMs are modeling agents. They interact on the basis of a mathematical model which is also area-specific. The SNA plays a role in unveiling the structure of interaction of the modeled agents. Both models complement each other. As an example, 'Recipe World' (by Fontana and Torna) stimulates the emergence of a network out of decentralized autonomous reaction. Agents are provided with recipes or steps to be taken to achieve a goal. Agents are activated and the network emerges as an effect.

This model is based on four distinct sets: (1) The actual work populated by entities and their actual network (2) in an ABM (agent-based model) with agent which base their behavior on the orders and recipes derived from (1) above (3) represents the network generated by (2). The events are: the possibility of populating (3) and (4), respectively, by knowing (4) representing them as data on the network and using influence and a kind of reverse engineering.

Agent-based model (ABM) can be created as an outcome of reverts engineering effort by combing the previously mentioned web mining and network analysis. Data (D) can be gathered from social web mining to model a political network (C), comparing data about political parties and their interactions of their voters, leaders, NGO and also Government organizations.

In turn, this network (C) can serve as the basis for a ABM (B) have allowing the simulation of real-world entities and their actual network (A).

7.10 e-Social Science

In the context of more and more digitization of life, the term e-Social Science refers to the use of electronic data in social studies, especially social networking, where much of interaction and social life is increasingly now only in online mode and online world only. The online records of social activity offer a great opportunity for the social science researchers to observe communicative and social patterns in much larger scale than ever before. Having removed the human element, it becomes easy to sample data at arbitrary frequencies and observe the dynamics in a much more reliable manner. A large diversity of topics can be studied in e-Social Science. There are many diverse electronic data sources in the Internet for e-Social Science research. These include electronic discussion networks, private, corporate or even public groups such as Usenet and Google groups, blogs and online communities, web-based networks, including web pages and their links to other WebPages.

The expansion of the online universe also means a growing variety of social settings that can be studied online. Extraction of information from web pages also called information retrieval or information extraction can extract networks of a variety of social units such as individuals or institution. The most important technical issue or hurdle to overcome is the problem of person name disambiguation, i.e., matching of references to actual persons they represent. Person names are neither unique nor single: a person may have different ways and methods of writing or even different spellings of his own name and also several persons can have the same name. Even if Father's Name is associated with a few duplicates and ambiguities may appear. When Father Name and Address combination is attached to a single name then ambiguity or duplication may be reduced, if not eliminated. However, a unique identity number (as Social Security Number or PAN Number or Aadhar Number) can disambiguate the name. Semantic Web can aim at identifying a named object identity formation for all entities. Semantic Web sites such as 'Live Journal' that already import data in a Semantic Format do not suffer from this problem. In their case, a machine-processable description is attached to every web page, containing the same information that appear on the web page but in a way that individuals are uniquely identified. This method can also be used for merging multiple electronic data sources for analysis.

Speech Recognition and Computer Vision

The Social Semantic Web is not plain text but has substantial audio and video content which needs to be processed and understood for its knowledge content by using appropriate advanced techniques of Speech Recognition and Computer Vision.

Speech recognition comprises of human speech-to-text conversion. Various recorded speeches need to be processed to directly identify the speaker and then given the speech so it convert to text. Speech recognition technology and speech-to-text conversion techniques are very much advanced but have a serious limitation of language dependency. It requires substantial text base in that specific language and speech to text system for that specific language only. If such language base or speech to text system is not available for a new language, one has to develop from the beginning such a system and that is a long-drawn process which also a difficult task.

Computer vision is a field of computer science that aims at allowing the computer systems to analyze images and/or video data in order to recognize the objects that are present on a given image/video. Unlike speech, there is no language or depending on a language in computer vision, but there is even a worse problem that for every type of object that needs to be recognized there needs a system that recognizes that object. This is more true for person recognition or biometrics where one needs to trace the system to recognize every single person of interest (this process is called enrollment).

Even then, there do exist certain types of objects which can more or less be recognized by even single system like text or any image or in a video. Such text can be used along with the recognized speech.

Natural Language Processing (NLP)

Natural language processing, or NLP, considered a part of artificial intelligence, aims at acquiring an understanding of the contents of a given text in any particular natural language such as English or French or Hindi. This is applicable to spoken languages also, in addition to the recorded text. To quote Cohen, natural language processing (NLP) comprises hardware and/or software components which have a capability to analyze or synthesize spoken or written language, similar to the humans. The main problem for achieving this objective is the complexity of the natural language as characterized by two well-known constraints: ambiguity and context sensitivity. Further, languages always keep evolving in time.

Every natural language has a grammar that controls such evolution and restricts changes by formulating logical rules (grammar) for the formation of words, phrases and sentences which are compulsorily required to be adhered to by the user of the language if it has to be a valid language construct. In addition to logical rules of grammar, every language requires a Lexicon or Dictionary. These two (the grammar and the Lexicon) are the characteristic requirements for the language itself. Then, there are specific linguistic tools that make it easier for the algorithms to access, decompose and make sense of sentences already available for use in processing of NLP data. Such tools include:

- Tokenizers or sentence delimiters which detect sentences based on delimiters (as a fullstop).
- Stemmers and POS taggers—morphological analyzers that link a word with its root and therefore identify the part of speech (POS) as noun, verb, adjective, etc.
- Noun phrase and name recognizers—labeling works with noun phrases (e.g. adjective + noun) and recognizing names.
- Parsers and grammar.
- Recognizing well-formed phrases and sentence structures in a sentence.

Steps involved in NLP can be summarized as:

Firstly, the entire text is required to be broken down into sentences. Secondly, the sentences are required to be broken down into words and words to be tagged for part of speech (POS). Then, each word so identified is to the broken down for its root and its suffix to indicate the meaning of the word by looking into the Lexicon by using Stemmers.

Stemmers can be ineffectual (identify the tense, gender, etc.) or derivational which link up several words to the same roots (say the same verb root for several words).

History of NLP can be traced back to ancient times to Sanskrit grammar associated with Panini and Patanjali as the grammarians who successfully processed the language of Sanskrit using the human brain alone without any computer systems.

In modern times, NLP developed since the 1950s with (a) symbolic approach and (b) statistical approach of language processing.

Symbolic approach is outlined above with the grammar as the basis for processing the language.

Statistical natural language processing is based on machine learning algorithms. Machine learning algorithms learn from training data and use that learning during testing phase. This approach requires corpora of the language with handmade annotations of labels such as nouns and verbs. Using such corpora, learning is performed by learning algorithms. Such learning is utilized in testing phase. Decision trees can be used.

NLP has many applications in processing some corpora of text, auto-summarization, automated classification of documents and other similar applications.

7.11 Opinion Mining and Sentiment Analysis

This amazing technology enables us to identify the sentiment or feelings of the user (a person who had these feelings while writing some text or message). This has many applications including: (1) election time attitude and sentiment analysis, (2) product or services research for marketing purposes, (3) recommendation systems, (4) campaign monitoring, (5) detection of 'flames', (6) public health monitoring (e.g., emotional/mental status, detecting potential suicidal victims), (7) government intelligence (8) trend monitoring (social, cultural and political trends etc.).

Sentiment analysis toward a political personality is very critical to an election. This needs to be integrated with newspaper feedback also. For performing sentiment analysis using NLP, we require to maintain a database of sentiments analyzed already. Normally, the sentiments are classified as positive or negative only (with various degrees of polarity between these two extremes) and therefore complex qualitative sentiments are hard to detect and hence not implemented. Sentiment analysis systems have a huge commercial value and therefore not freely available for usage.

Recommender Systems and User Profiling
Recommender Systems recommend items available through the same systems to its user. Online shopping sites as Amazon, eBay, news portals and other portals offer the Recommendation Services in their own contexts. The factors that are to be considered are: (1) the context of each information node with descriptors generated by the use of NLP tools, (2) user profiles which are generated based on browsing data and (3) user preferences. The Sources of data for recommender systems include: data from web server access logs, proxy server logs, browser logs, registration data, user sessions or transactions, cookies, user queries, bookmark data, mouse clicks and clickstream data (or scrolls) and any other data that gets generated during the interaction of the users with system. In one approach by Seva (in 2014), user profiles were generated from user-visited documents, and newly generated content was recommended based on the underlying taxonomy-based classification. The objective is to provide individual personalized recommendations (usually user groups only arc provided recommendations) clusters of users are given recommendations by the system.

7.12 Semantic Wikis

Wikis allow users to collaboratively work on a repository of content.

Semantic Wikis include ideas from Semantic Web (whose objective is to bring in machine-readable metadata to web pages in order to allow computer systems to understand the content they are processing).

Therefore, Semantic Wikis can be used to manage large repositories of textual and multimedia (audio or video) data on which Semantic Web features such as automated reasoning, complex querying mechanism and automated categorizations of content are possible.

When collaboratively edited, metadata as keywords, categories, classification, etc., is added to Wikipedia's a social taxonomy can be generated (as happened in the case of TaOPIs). This taxonomy allows for automated summarizing of text using built-in queries. Intelligent Agents can then be built and deployed to final conceptually similar terms and descriptors of each concept described in Semantic Wikis. They provide an excellent tool to organize research data and yield the needed structural metadata for data mining. A Collaborative System can be thus built.

7.13 Research Issues and Challenges for Future

As was clear from the above sections, Social Semantic Web Mining and Big Data Analytics can provide a substantial addition to Social Science research methodology.

On any specific topic if Social Science research is required to be conducted, then it is possible to create a large Semantic Wiki, added with Semantic Social Network

Mining, audio, video, text inputs from surveys, studies and TV programs. Such a topic could be just political, for example (say on poverty and immigration).

One possible goal will be to identify the main actors (politicians), leaders, their key statements, the position of government, etc. All these can be achieved through NLP techniques and also through Social Semantic Conceptual Network Analysis and also expert opinions. Specialized user groups can be identified through user profiling. The discussions between interested groups can be analyzed using Topic Maps. Sentiments can be measured and analyzed to track the spread of attitudes and concepts and Social Concept Networks can be identified. The above was just an example. The real challenges which can be addressed to be solved in future could be outlined as:

(1) The huge 'Big' data is heterogeneous, structured, semi-structured, unstructured, text, audio and video. This has to be really integrated. They are from different file and data formats. They need to be cleaned up, integrated into a single Big Data Repository for Analysis purposes.
(2) NLP adoption is a challenge given the multiple natural languages generating huge text, audio to analyzed. Do we have multilingual NLP tools and techniques to meet these challenges? (OKM offers on approach for solving this problem).
(3) Sources of data for a specific purpose or topic or objective are required to be identified.
(4) Unique Named Object Identity to provide 'single version of truth' in the Semantic Web is still elusive with multiple dispersed, distributed sources of data which exist severally today across the web and social networks. This is required to be developed.

7.14 Review Questions

1. What is Semantic Web? What is Semantic Web Road Map?
2. What are the various knowledge representation (KR) techniques in Semantic Web? Explain each one of them.
3. What is Object Knowledge Model (OKM)? Explain its application.
4. What is Social Semantic Web Mining?
5. What are conceptual networks? Explain.
6. What are SNA and ABM? Explain.
7. Explain the evolution of natural language processing (NLP).
8. Explain various stages involved in NLP.

References

1. N. Lavrac, A. Vavpetic, L. Soldatova, I. Trajkovski, P.K. Novak, Using ontologies in semantic data mining with SEGS and G-SEGS, in *Proceedings of the 14th International Conference on Discovery Science*, Espoo, 5–7 Oct 2011, pp. 165–178
2. O. Mustapasa, A. Karahoca, D. Karahoca, H. Uzunboylu, Hello world, web mining for e-learning. Procedia Comput. Sci. **3**(2), 1381–1387 (2011). https://doi.org/10.1016/j.procs.2011.01.019
3. D. Jeon, W. Kim, Development of semantic decision tree, in *Proceedings of the 3rd International Conference on Data Mining and Intelligent Information Technology Applications*, Macau, 24–26 Oct 2011, pp. 28–34
4. V. Sugumaran, J.A. Gulla, *Applied Semantic Web Technologies* (Taylor & Francis Group, Boca Raton, 2012)
5. J. Domingue, D. Fensel, J.A. Hendler, *Handbook of Semantic Web Technologies* (Springer, Heidelberg, 2011)
6. A. Jain, I. Khan, B. Verma, Secure and intelligent decision making in semantic web mining. Int. J. Comput. Appl. **15**(7), 14–18 (2011). https://doi.org/10.5120/1962-2625
7. C.S.R. Prabhu, Extending a semantic e-governance grid with OKM, a frame based knowledge representation framework for enabling semantic—knowledge search on the contents of the web pages, in *7th International Conference on E-Government (ICEG-2010)*Bangalore, India, 22–24 Apr 2010, Paper No. 6
8. H. Liu, Towards semantic data mining, in *Proceedings of the 9th International Semantic Web Conference*, Shanghai, 7–11 Nov 2010, pp. 1–8
9. V. Nebot, R. Berlanga, Finding association rules in semantic web data. Knowl. Based Syst. **25**(1), 51–61 (2012). https://doi.org/10.1016/j.knosys.2011.05.009
10. V. Nebot, R. Berlanga, Mining association rules from semantic web data, in *Proceedings of the 23rd International Conference on Industrial Engineering and Other Applications of Applied Intelligent Systems*, Cordoba, 1–4 June 2010, pp. 504–413
11. M. Schatten, J. Seva, B. Okusa Duric, An introduction to social semantic web mining and big data analytics for political attitudes and mentalities research, AI Lab, Faculty of Organization and Informatics, University of Zagreb, Croatia

Chapter 8
Internet of Things (IOT) and Big Data Analytics

8.1 Introduction

Having surveyed in the last chapter how Big Data Analytics is applied in Social Semantic Web, in this chapter we shall delve into another very important application domain, IOT. We shall examine the interaction between IOT and Big Data Analytics.

In the evolutionary phase of Big Data on the Internet, we have IOT or Internet of Things [1] or Internet of Everything (IOE) as the recent and one of the latest trends and big time thrust of development. IOT may be described as interaction and interoperability domain of divergent areas of activity such as the telecommunications industry, software industry and hardware industry, including device manufacturing industries with a promise of great opportunities for various sectors of industry and commerce.

To provide a formal definition, 'IOT is a seamless connected network of embedded objects/devices, with identifiers, in which M2M (machine to machine) communication without any human intervention is possible using standard and interoperable communication protocols (phones, tablets and PCs are included as part of IOT)'.

Internet of Things (IOT) [2] made a beginning driven by inputs from sensor networks with trillions of sensor devices which are interfaced with smart and intelligent subsystems and in millions of information systems. The Internet of Things (IOT) shall drive unforeseen new vistas of business and customer interests which will require more and more smart and intelligent systems which automatically drive an increasingly large number of opportunities for business in IT/IOT industry and their supply chain companies.

The number of Internet-connected devices exceeding 12 billion had far surpassed the number of human beings of 7 billion and by 2020, the number of Internet-connected devices is expected to reach 30–50 billion globally [3].

C. S. R. Prabhu et al., *Big Data Analytics: Systems, Algorithms, Applications*,
https://doi.org/10.1007/978-981-15-0094-7_8

8.2 Smart Cities and IOT

To provide the better quality of living to the people living in both urban and rural areas, IOT devices, such as sensors, are being deployed in smart cities and smart villages. Globally, smart cities have been proposed, designed and are being implemented in large numbers in all countries.

The applications of IOT in smart city implementation are the following subjects:

1. Smart parking
2. Smart urban lighting
3. Intelligent transport system [4]
4. Smart waste management
5. Tele care
6. Women safety
7. Smart grids
8. Smart city maintenance
9. Digital signage
10. Water management.

Smart home [5] IOT services contribute to enhancing the personal lifestyle by making it easier and more convenient to monitor and operate home appliances such as air conditioners and refrigerators.

While interacting with each other, the IOT devices produce large amounts of data. Processing this data and integrating the IOT devices lead to establishment of a system. The development of smart city system and smart village system is based on IOT and Big Data. Such system development and implementation include data generation and collecting, aggregating, filtration, classification, preprocessing, computing and are finished at decision making.

Sectoral Applications

In addition to smart cities, smart and remotely controlled devices can help solve various problems being faced by the industry in divergent sectors such as agriculture, health, energy, security and disaster management.

This is an opportunity to telecommunication industry on the one hand and system integrators on the other, to enhance their revenues by implementing IOT applications that offer a variety of services that can be offered by this technology. In addition, the IT industry can offer direct services and also analytics-related services or other services independent of IOT.

8.3 Stages of IOT and Stakeholders

8.3.1 Stages of IOT

Four stages of IOT implementation can be identified:

Stage 1: Identify the sensors appropriate for the application.
Stage 2: Develop the application.
Stage 3: Server should receive the sensor data from the application.
Stage 4: Data Analytics software to be used for decision making process.

All the major countries have taken initiative in implementing IOT applications.

8.3.2 Stakeholders

The key stakeholders in IOT implementation are:

1. The citizens
2. The government and
3. The industry.

Each stakeholder has to show commitment to collaborate to produce the results. The participation of stakeholders at each stage or step is essential. Promotional policies are essential for developing answers on the questions: 'What data will give service to the Citizens?' IOT should clearly strategize with a single goal of 'Value Up' and 'Cost Down' models.

8.3.3 Practical Downscaling

In terms of practical realities, the downscaling of IOT from a vision of billions to a reality of thousands of IOT devices in vicinity, loosely connected with each other is to be realized. Also, many of such devices are connected or embedded with mobile devices with P2P, 2G/3G/4G and Wi-Fi connectivity.

Cloud-centric connectivity providing for data collection and the appropriate Big Data Analytics which may be personalized may also be provided. Open ecosystem without vendor lock-in is essential.

8.4 Analytics

Analytics can be performed on either 'Little' data coming from sensor devices or on 'Big' Data from the cloud. Both need to be combined as per the needs.

8.4.1 Analytics from the Edge to Cloud [6]

We cannot push all data to cloud for analytics purposes. In the context of smartphones and potentially intermittent communication, we need to decide if/when to push a subset of 'Big' Data to the phone, if/when to push a subset of the 'Little' data to the cloud and how to make collaborative decisions automatically. Communication and compute capability, data privacy and availability and end-use application inform these choices.

8.4.2 Security and Privacy Issues and Challenges in Internet of Things (IOT)

The primary issues of security and privacy that pose challenges in IOT scenario are the standard issues of (a) privacy of data, (b) confidentiality of data, (c) trust, (d) authentication, (e) integrity, (f) access control, (g) identity protection and (h) privacy of user.

These issues can be addressed in middleware level and also at the other layers of IOT ecosystem such as: (a) at sensor or hardware level, (b) sensor data communication level, (c) content annotation, content discovery level, (d) content modeling, content modeling and content distribution level. The security issues are required to be examined in a wide variety of dimensions such as the types of threats in IOT, protocols and network security, data privacy, management of identity, governance, static trust, dynamic trust, fault tolerance, management of security and privacy. Let us focus on some of the key issues identified above:

(1) To achieve privacy of sensitive details of financial, technical, personal data transmission or exchange (in the IOT ecosystem comprising of radio links, etc.), the data control techniques such as anonymization, encryption, aggregation, integration and synchronization may be deployed to hide the critical information.
(2) Authentication of identity and access control: This comprises steps to prevent intrusion and permit only authentic access to legitimate and authenticated persons into restricted areas. This may be applicable to identification of individuals, vehicles, measurement of humidity and temperature tracking of products, surveillance in sensitive areas.

(3) Trust: Trust comprises transparency and reliable guarantee of security between any two intelligent beings—persons or technological objects. Trust will be leading to a global system of reliable system of information exchange and communication between specific source and specific destination. Trust will depend upon (a) the ability of self-protection despite hostile environment and (b) the verifiability of trustworthiness of an object or node by interrogating it.

(4) Reliability: In any process, the reliability comprises of reliable information collected and the results reported being reliable. This implies that all the steps involved in the process such as sensing, sensing devices, metrology, calibration, processing of signal or data, diagnosing and diagnostics finding out exceptions anomalies, support and maintenance, etc. To insure reliability, automatic, feedback control systems are also required to be established.

(5) How to insure privacy in IOT systems? In IOT application systems existing pervasively, sensitive or confidential data may be stored in a distributed way. Hence, we have to set up adequate controls in managing and disclosing data to third parties according to the respective levels sensitivity. Data usage at processing is done at various levels such as collection transmission and storage of data. In each stage, the appropriate technologies have to be adopted and mechanisms are required to be set up to insure and guarantee data confidentiality, data integrity and data security. Possible methodologies and techniques that can be adopted for these purposes can include: anonymization, ciphers (block and stream), hashing functions, random number generator functions and public key (lightweight) infrastructure.

    Privacy in access concerns itself with the manner in which people can access the private and personal information. We need to implement effective policies to insure access privacy.

8.5 Access

We need to insure that we bring the network close to the sensors. The extensive proliferation of tens and hundreds of thousands of sensors will lead to constraints in power and communication bandwidth. Instead of relying on possible large-scale deployment of custom sensor networks, it may be of greater value to piggyback on the existing standards. Peer-to-Peer data may be utilized for last-mile connectivity purposes. We can also integrate highly functional gateways and clouds. Asymmetric architecture may be preferable. The penetration of technology has not been uniform across different countries or even across different regions in the same country, reflecting the inequalities and disparities in infrastructure, access, cost, telecom networks, services, policies, etc. Therefore, affordable sensor devices, affordable networking cost and affordable analytics solutions are to be offered accordingly.

8.6 Cost Reduction

Reuse of deployed infrastructure shall be the inevitable requirement for the successful implementation of IOT. Reusable devices and sensors to be deployed for novel usage ways are preferable in developing or poor countries as the model adopted in advanced nations with advanced infrastructure will not give relevant or cost-effective solutions for the developing or poor countries.

8.7 Opportunities and Business Model

The data that flows in IOT devices and networks shall open up endless and immense opportunities for analytics. The effective interplay between the technology of sensor devices, telecommunication infrastructure and the data flowing through them results in new business models.

Technology model will fail or succeed based on the business models that generate revenues. For Google on Internet, the boom of revenues comes from advertisements using personal data of Google users in return for free information search and retrieval services to end user. With Internet of Things, the information captured will be large, minutely microscopic and fast. Its use, reuse sharing and payment for all these activities are important. Data brokering services can be offered where open data can be utilized for sharing with businesses. This may produce rewards such as greater common good, in addition to revenue that encourages open data sharing by users with businesses in return for clear rewards, be they monitory, peer recognition or the greater good of open data policies.

8.8 Content and Semantics

Being human-centric, content and semantics form the core of data for decision making. Content-based mining and analytics become more meaningful and relevant to the content. Therefore, there has to be a mechanism to capture semantics which can describe system behavior and social behavior.

Sometimes, it is possible that intelligent agents or software agents may replace human users and therefore may become aware of personal information. For example, we have Siri and Cortana which are software agents. They will interact with other agents of service provider, utility companies, and vendor companies.

8.9 Data-Based Business Models Coming Out of IOT

Semantic content can extend the data content in M2M data interchange. Business models based on data will be evolving out. The new vistas of technology and applications that spring out of Internet of Things are characterized by low power, cheaper devices, more computation based on robust connectivity. This technology provides an opportunity of a window to look at our life and environment at microscopic level of detail, as almost everything can be monitored.

IOT business models can be either horizontal or vertical which can be integrated with the supply chain which can become a value proposition for the end user. Such value proposition span is divergent and large in number with multitude of user requirements or aspirations to be fulfilled.

The first step is to develop cost-effective sensors and actuators which make low cost and microlevel observations. The next step is to build a business model which can perform collection, storage, brokering and curation of data coming out. The third business model will be targeting at the analytics portfolio on the data. End-to-end system integration is the fourth business model.

To summarize, the development of IOT-based business models will require concerted efforts to support the modularization of the architecture, to provide access to capabilities through service-based models for solving real customer problems.

8.10 Future of IOT

8.10.1 Technology Drivers

The future of technology drivers includes low cost and accurate sensor and actuator development along with their networking technology along with computer and memory capabilities. Therefore, we can state that understanding this technology of IOT comes out of understanding of devices, their networking and their computing profiles. The devices and sensors are getting miniaturized continuously and their costs are going down too. The 'Smart Dust' technology pushed by the defense in the 90s was based on devices using MEMS and electronic textiles and fabrics or wearable computers which are all drivers for IOT-based devices and sensors.

Smart fabrics and their commercial applications in sports or medicine have become conference topics.

8.10.2 Future Possibilities

By 2020, it is expected that most hitherto manual processes could be replaced with automated processes and products and supply chains could have been embedded with sensor and active devices. Wearable devices or clothes integrated with devices could help in sports and art (dance) training, as also for patient monitoring in medical settings.

8.10.3 Challenges and Concerns

What about privacy concerns in IOT-based sensor networks? While it may be a good idea to have the sunglasses to recognize a man in a room, will it be correct to monitor the self through a cloud-based monitoring application? Numerous commercial opportunities spring up for many new initiatives and application ideas but the questions of ethics and security and privacy concerns also come to the fore. IOT has innumerable and endless industrial and commercial applications: transportations and truck fleet tracking, courier and mail tracking, environmental sensing and monitoring that engage special devices and sensors integrated into smartphones.

In sports, balls and players can be attached with sensors to obtain better accuracy of goal making. Tracking and locating goods and packages in airports, in warehouses and during transport. Wandering robots with sensors can reduce the stocking quantities in warehouses. Tracking temperatures in data center, monitoring insects, bees, migratory birds, gesture recognition, video games, virtual reality (VR), etc. are possible applications. Urban and rural consumer applications such as smart home, smart farms and smart dairy, where parameters such as temperature, moisture, etc. will be monitored online to enhance feedback-based efficient power or water usage are possible. In Europe, South Korea and USA such applications including smart grids, smart cities and smart villages, all use IOT devices and sensors for monitoring and feedback-based action steps. Oil and gas industry had already deployed such devices for long time.

Applications which do not get affected by privacy concerns can be: monitoring traffic, load on bridges, smart aeroplane wings that adjust according to airflow currents and temperatures, etc.

IOT can improve quality of life of citizens and tourists when compared with manual city guidance centers, indicating best routes, guided tours, car-sharing automation, automated self-driving car, wildlife applications as animal monitoring, fish monitoring, bird monitoring are all possible.

The global scope of IOT with sensors in all kinds of different settings provides opportunities to facilitate our daily lives, saving environment, better monitoring and implement law and order (police applications). The list is endless.

Critical factors are the ability to embed sensors and devices in important and relevant locations to monitor all kinds of phenomena and connect them to networks to monitor the data being collected. What determines investments in IOT? 'Microservices' for citizens, tourists, sportspersons, industrial houses, transportation systems, police and security requirements, etc. all provide revenues. Legal and government hurdles, if any, are required to be cleared before implementation and roll out.

8.11 Big Data Analytics and IOT

The significant and substantial increase in the connected devices that are going to happen in Internet of Things (IOT) will lead to an exponential surge in the size of data that an enterprise is expected to manage, analyze and act upon. Therefore, IOT becomes a natural partner match for Big Data simply because it provides the requisite data volumes for Big Data Analytics.

As Howard Baldwin says 'we will have data spewing from all directions – from appliances, from machinery, from train tracks, from shipping containers, from power stations etc.'. All this real-time data needs to be handled, analyzed to sense actionable signals.

IOT is still in its infancy. Soon, the data will start flowing from the sensors and devices. The actionable insights can be identifying purchasing habits of customers or efficiency of machine performance. For example, LexisNexis has open-source HPCC Big Data platform, a solution for data integration, processing and delivery of the results by integrating, machine learning and BI integration.

8.11.1 Infrastructure for Integration of Big Data with IOT

The integration of Big Data with IOT is dependent on the infrastructure environment. This includes storage and cloud infrastructure. Many organizations are attempting to move to PaaS (platform as a service) cloud for hosting and analyzing the huge IOT data since maintaining own storage for this purpose will be very expensive. PaaS cloud will be expected to provide scalability, flexibility compliance and effective sophisticated architecture to store cloud data arriving from IOT devices. If the data is sensitive, then private cloud architecture can be deployed. Otherwise, public cloud services such as AWS (Amazon) or Azure (Microsoft) can be deployed. Many cloud services provide and offer their own IOT platform for implementing IOT Applications.

8.12 Fog Computing

Data is continuously generated by IOT devices. Such data has characteristics of Big Data such as volume, variety, velocity and veracity. Latency becomes a critical requirement in IOT applications which are real time in nature, as they expect real-time response. Cloud computing cannot deliver real-time response, as latency will be very large and significant. Hence, a new concept called 'Fog Computing' has come up of late. Fog server is located near the edge (as a kind of extension of the cloud to the edge). Analyzing IOT data close to the collection point minimizes latency. It offloads network traffic from the core network and also keeps sensitive data inside the network with at most security.

Fog applications include locking a door, changing equipment settings, applying the brakes on a train, zooming a video camera, opening a valve in response to a pressure reading, creating a bar chart or sending an alert to a technician to make a preventive repair.

Fog computing is mainly useful when the IOT devices are placed globally, data is collected from extreme edges like vehicles, ships, factory floors, roadways, railways, etc. and requirement of data analysis at the same time as data collected.

8.12.1 Fog Data Analytics

I. **Introduction**

Analytics near the edge itself will be possible if we deploy Fog server near the edge and perform analytics in the Fog server itself. This will prevent the transmission of data from the edge to the cloud and therefore avoid the network latency delays in the application execution at the edge.

II. **Fog Computing Environment and Data Analytics**

Fog computing is a new technology paradigm to reduce the complexity, scale and size of the data actually going up to the cloud. Preprocessing of raw data coming out of the sensors and IOT devices is essential and it is an efficient way to reduce the load of the big data on the cloud.

The Fog server, located very near to the edge devices, offers the possibility of preprocessing and even completing local analytics, to take fast decisions for the real-time local edge requirements. Only, the aggregate or summary data, small in size, needs to be sent to the cloud. This will lead to the benefits that accrue from Fog computing that include local, fast processing, storage for geo deductible and latency-sensitive applications, drastically reduced communication overheads over the network and the Internet, thereby having a substantially reduced volume and velocity of data that will be required to be sent to the cloud.

Applications such as augmented reality, interactive gaming and event monitoring require data stream processing, a processing scenario in contrast with a ready data bank, assumed to exist in conventional Big Data application ecosystems.

III. **Stream Data Processing**

Stream data is abundant, in RFID applications, weblogs, telecom call records, security monitoring, network monitoring, stock exchange data, traffic and credit card transactions and so on. It is characterized by high speed, transient and continuous nature of the data.

IV. **Stream Data Analytics and Fog Computing**

Stream data analytics in Fog servers can be achieved by deploying products like tensor flow and even in mobile edge devices ('Mobile Tensor flow'). In spite of such implementations, the open unsolved challenges do exist—how to perform load balancing among multiple Fog servers and edge devices, without affecting the performance? While we have APIs for Fog streaming in IoT and stream processing platforms like Kafka, APIs for differential equations and control error estimations for Fog-based real-time applications are yet to come up.

V. **Different Approaches in Fog Analytics**

In the following sections, we present a survey of the different approaches for Fog Analytics.

A. **'Smart Data'**

As a complete capsule of structured IoT data, its metadata and its VM, 'Smart Data' is a product available for Fog Analytics.

B. **Fog Engine**

Fog Engine, a product, provides data analytics on premise. It enables processing of data in cloud as well as IoT devices in a distributed manner. One unit of Fog Engine can collaborate with another. Data can be offloaded into the cloud periodically. Several scenarios can be identified for Fog Engine deployment, depending on multiple receivers, multiple or single analyzers, multiple or single transmitters. The Fog Engine deployment can partially undertake the burden of network backbone and data analytics at utilities side and reduce the dependence on the cloud. While computations are done locally, only a fraction of the data that is cleaned and analyzed by the Fog Engine is transferred to the cloud, thus drastically reducing the volume of data transferred over the network, resulting in substantially reduced network congestion and delay due to latency.

C. **Other Products**

Other products in Fog Analytics include Microsoft Azure Stack and also CardioLog Analytics by Intlock which offers on-premise data analytics. Oracle delivers Oracle Infrastructure-as-a-Service (IaaS) on-premise with capacity on demand that enables customers to deploy systems based on Oracle in their own data centers. IBM's Digital Analytics on-premise is the core Web Analytics software component of its Digital Analytics Accelerator solution.

D. **Parstream**

CISCO's Parstream is capable of offering continuous real-time data analytics functionality. It can be deployed on standard CPUs and GPUs. Parstream is well integrated with many platforms such as R Language. It can analyze large streams of data with time series analysis for historical analysis purposes. Alerts and actions are used and raised to monitor data streams, create user-friendly procedures that generate alerts, send notifications and execute actions; derives models and hypotheses from huge amounts of data by applying statistical function and analytical models, using advanced analytics.

VI. **Comparison**

All the above products have their own respective strengths and weaknesses. Although all of them offer on-premise data analytics services, they lack in providing a holistic approach based on the Fog concept which is the intermediate layer between the edge and the cloud.

VII. **Cloud Solutions for The Edge Analytics**

Solutions by cloud services providers (CSPs) such as Amazon are also available—Amazon's AWA IOT offers implementing data collection through HTTP, Web sockets, MQTT and integrates with REST APIs with device gateway in cloud. Amazon QuickSight is available for machine learning purposes.

Microsoft offers Azure IOT Hub using HTTP, AMQP, MQTT and also custom protocols for data collection; offers REST APIs integration; offers stream analytics and machine learning uses Azure IOT gateway (in-premise gateway).

IBM offers IBM Watson IOT using HTTP and MQTT for data collection, integration with REST and real-time APIs. Data analytics is offered through IBM's Bluemix Data Analytics platform.

Google offers Google IOT uses HTTP only for data collection, integrates with REST APIs and RPC. Analytics is offered through cloud data flow, Big Query Datalab and Dataproc and uses general gateway (on-premise) to cloud.

Alibaba offers Alicloud IOT, uses HTTP, integrates with REST APIs, uses own analytics solution, Max Compute and uses cloud gateway to the cloud.

In this section, a survey of the approach, techniques and products for Fog analytics is presented.

8.12.2 Fog Security and Privacy

Provisioning security and privacy in Fog computing is quite different from provisioning the same in the cloud. Wireless carriers who have control of home gateway or cellular base stations may build Fog with their existing infrastructure. Cloud service providers who want to expand the cloud up to the edge also may build the Fog infrastructure.

Authentication

The main security issue is authentication at various levels of fog nodes. Traditional PKI-based authentication is not scalable. Near-Field Coins (NFC) can be used effectively in Fog computing to simplify authentication procedures. Biometrics-based authentication (such as the Aadhar card in India) can also be effectively used in Fog computing ecosystem. Authentication becomes important at various levels of gateways or at the level of device itself. Each device such as a meter in smart grids or such as an i-pad in any Fog environment should have any authentication-biometric-based authentication or otherwise, to prevent misuse, manipulation or impersonation. For example, smart meters can encrypt the data and send to the Fog devices such as home area network (HAN) where the data can be decrypted, the results are aggregated and then pass them forward to the cloud, for example.

Privacy Issues

Privacy issues pertaining to the devices which device was used, when, for what purpose, etc. are required to be analyzed. Encryption can be used to provide encrypted result which cannot be decrypted, by the individual devices.

Man in the Middle Attack

The Fog is vulnerable and the man in the middle is an example of this vulnerability. In this situation, the gateway serving as the Fog device may be compromised and replaced with fake or malicious access paths which provide deceptive SSIDs as public, legitimate ones. Thereby, the attacker can take control of the gateways and thus the private communication will be hijacked. Man in the middle attack in Fog computing can be very stealthy. It is very difficult to protect the Fog devices from it.

8.13 Research Trends

The current Big Data tools such as Hadoop are not suitable for IOT, as they do not offer online or real-time analytics for IOT purposes and applications [6]. Also, the amount of IOT data is too huge to be processed by such tools, in real time. One alternative possibility is to keep track of only the interesting data coming out of IOT devices. For this purpose, approaches such as principle component analysis (PCA), pattern reduction, dimensionality reduction, feature selection and distributed computing methods are identified [6].

Another important direction is to provide a common platform of analytics as a cloud service, instead of providing application-specific analytics software. It is proposed [7] to offer time series analytics as service or TSaaS, using time series data analytics to perform pattern mining on large sensor data.

8.14 Conclusion

IOT is here to stay, driven by device and sensor technology advances, the opportunities created by billions of smartphones which can be supplemented by IOT devices and sensors, Internet connectivity through mobile networks, resulting in millions of applications in cost reduction through automation, reduced loses or wastages and shorter duration in supply chain processes in all aspects of human life and industrial sectors. In this chapter, we have also seen how the techniques of Big Data Analytics become essential analyzing the data originating from IOT devices. We have also seen how Fog computing becomes essential; in addition to the cloud services being available, we have examined various products of Fog analytics. We have also probed into the security and privacy issues in Fog computing. Finally, we attempted to see how future IOT will shape with new developments such as wearable devices and 'Smart Dust' are coming up.

8.15 Review Questions

1. How the IOT phenomenon is going to impact the Society?
2. How IOT is essential for smart cities?
3. What are the stages in IOT implementation?
4. Explain Analytics from the edge to the cloud.
5. What is the data-based business model coming out of IOT?
6. What is the future of IOT? What are its technology drivers?
7. What are the challenges and concerns for the future of IOT?
8. What are the dynamics of linkage between IOT and Big Data Analytics?
9. What is the infrastructure requirement for integrating IOT with Big Data?
10. Explain Fog computing, its role and its importance for future.
11. What are the research trends in integrating IOT and Big Data?
12. What is Fog computing? Explain its characteristics and benefits.
13. Explain Fog analytics and its products with comparison.
14. Explain Fog security and challenges.
15. Explain Fog privacy and challenges.

References

1. J.R. Frederich, F.W. Samuel, D. Maithaias, Introduction to internet of things and big data analytics, in Mini Track, 2015 48th Hawaii International Conference on System Science
2. A. Al Faquha, M. Guizani, M. Mohammedi, M. Aledhari, M. Ayyash, Internet of things: a survey on enabling technologies, protocols and applications. IEEE Commun. Surv. Tutor. **17**(4) (2015)
3. J. Gantz, D. Revisel, The digital universe in 2020: big data, bigger digital shadows and biggest growth in the far east, in *IDC, iView: IDC Analyze the Future*, vol. 2007, pp. 1–16, Dec 2012

4. http://www.intel.in/contain/lan/www/program/embedded/internet-of-things/blueprints/IOT-building-intelligent-transportsystem
5. N. Komninos, E. Phillipon, A. Pilsillides, Survey in smart grids and home security: issues, challenges and counter measures. IEEE Commun. Surv. **16**(4), 1933–1954 (2014)
6. C. Tsai, C. Lai, M. Chiang, L.T. Yong, Data mining for IOT: a survey, 1st part. IEEE Commun. Surv. **16**(1), 77–97 (2014)
7. X. Xu, S. Huang, Y. Chen, K. Browny, I. Halilovicy, W. Lu, TSAaaS: time services analytics as a service on IOT, in *Proceedings of IEEE ICWS*, pp. 249–256 (2014)

Chapter 9
Big Data Analytics for Financial Services and Banking

9.1 Introduction

Having surveyed the scenario of the application of techniques of Big Data Analytics in the context of Internet of things (IoT), let us now examine how the application of Big Data Analytics techniques is impacting the financial services and banking section. In a highly competitive business of financial services, we have companies vying with each other to grab their potential customers. This calls for their monitoring closely the customer opinions and feedback in all different platforms of Internet-enabled world, from mortgage applications to twitter postings—which provide unprecedented data for drawing insights. The Big Data phenomenon has resulted in expanding the range of data types that can be processed, enabling the banks and financial institutions to better digest, assimilate and respond in a better way to their physical and digital interactions with the customers.

Financial services companies look into emerging big data tools for discovering hidden customer sentiment on real-time basis [1, 2]. The big data tools enable companies to analyze far greater quantities and types of data in a short span of time. Structured, semi-structured and even unstructured data such as RSS feeds, SMS text messages and emails can be analyzed to uncover the rare insights of customer sentiments. Other sectors such as e-commerce and retail have already deployed and financial services companies need to deploy such tools for analyzing to identify customer segmentation, for product development and also for customer services. In the next few sections of this chapter, we shall delve deeper how these techniques can be deployed in the contexts of banks and financial services companies.

C. S. R. Prabhu et al., *Big Data Analytics: Systems, Algorithms, Applications*,
https://doi.org/10.1007/978-981-15-0094-7_9

9.2 Customer Insights and Marketing Analysis

In a scenario of customers having business relationships with multiple banks and financial services companies, a specific bank no longer has a clear understanding of how customer behave, the buying patterns and the spending patterns [3, 4]. In other words, a specific bank cannot monopolize on monitoring the customer behavior as many other e-commerce sites such as Amazon, Flipkart, etc., and payment gateways as Paytm and other financial players are engaged with a particular customer. At the same time, it is essential to get a comprehensive picture of customer behavior, in order to continue to achieve customer satisfaction and retention. What is the solution? The banks should be able to obtain such information on customer behavior from all possible other sources such as customer call center records, customer emails, customer postings on social networks as Facebook or Twitter and also the insurance claims of the customer. In 2012, American Express offered several offers as schemes, after studying customer purchase history and other buying patterns by partnering with a location-based platform Foursquare.

In the above scenario, it is possible only for the Big Data technologies to provide a comprehensive view of the ecosystem by being able to augment and integrate the structured transactional data with unstructured data originating either from within the same organization or from external agencies to provide a 360° view of customer behavior and psychography.

Another common problem is that all banks and finance services companies maintain their customer data in silos or islands, independent of each other, in a variety of applications such as savings or current accounts, term deposits, term loans, car loans, housing loans, etc., for the same individual customer. This prevents tracking of customers and prevents the marketing department to offer customized schemes to suit a specific individual or business. Better interest rates can also be considered to be offered by using Big Data technologies. Having single thread of comprehensive customer information is beneficial and helpful in all aspects, from customer credit monitoring, fraud detection and mitigation to offering better deals for the customers. Even the loan default calculations and risk assessment calculations will be possible only with a comprehensive single thread customer data for applying the techniques of Big Data Analytics.

Today large global banks determine at the point of sale (POS), whether to permit the ongoing purchase or not, by evaluating the legitimacy of the transaction by deploying the techniques of high-speed real-time analytics.

For a multidigital customer of today, comprehensive real-time data enables real-time offer management, relationship pricing, all much more current and valuable than the good old trend analysis based on historical transaction systems.

For achieving very effective results in fruitful marketing, it is essential and will be impactful, if the real needs of the individual customers are understood exactly correctly, in advance, in real-time so as to meet the same needs of the customer exactly at the right time. This is possible by the deployment of predictive analytics based on sentiment analysis.

To address the challenges of customer retention, financial services companies need to implement sentiment analysis and predictive analytics. Such tools provide economic value by providing the technology to tailor the financial products according to customer needs and desires as well as help understand fraud patterns, reduce credit risk in addition to building strategy according to customer expectations.

9.3 Sentiment Analysis for Consolidating Customer Feedback

Customers share their thoughts and sentiments through social networks as much as to the representatives of the banks and financial services companies. When appropriately captured and managed, this information provides valuable, unfiltered and un-tampered insights into what exactly the customers are thinking. Distancing away from the traditional customer feedback and customer sentiment analysis by using survey research and focus groups, the sentiment analysis tools can provide the banks and financial services company's innovative ways to improve their financial service products and also predict customer behavior. This will also enable analysis on a real-time basis allowing fast decision-making, including any intervention or reactions to negative sentiments and opinions of the customers that may have emerged by appearing so in social networks. For example, in 2011, when an American financial institute proposed to enhance the fee for debit card, it had to withdraw that proposal due to uproar and protests in the customer feedback against the step.

Banks can assess the possible and potential impact of their decisions by monitoring and capturing customer feedback from social media platforms and customer service interactions among other platforms. They attempt at linking up words in customer feedback in unstructured communications of customers reflecting their emotions, sensing they as key inputs for strategic decision-making.

In the context of loyalty and reward programmes for attracting and retaining the customers, the customer sentiment analysis and feedback play a major role. By examining the customer confidence indices that are given to specific data elements, banks and financial services companies attempt at judging the mood of the market and accordingly reward the customers who are loyal to the bank. While such techniques are still evolving, already matured techniques help identify the likes or dislikes and preferences for deciding financial product improvements and also service improvements, thereby attempt at gaining competitive advantage in a highly competitive banking and financial services sector.

The main applications of Big Data Analytics in financial services and banking are:

1. Fraud detection and fraud prevention is possible by finding exceptions in hidden patterns in data.

2. Segmentation: Customers can be segmented or classified into various categories based on classification techniques for launching sales or promotional marketing campaigns.
3. Support regulatory frameworks.
4. Managing Risk: By adopting efficient central risk management platform for meeting regulatory requirements also.
5. Offering personalized products: Based on customer habits and friends of behavior in expenditure.

9.4 Predictive Analytics for Capitalizing on Customer Insights

Customers have a need for performing inexpensive, fast, easy and simple transactions in both financial sectors and purchasing activity. This itself becomes a challenge, as the customer needs are increasing in diverse directions.

Predictive analytics techniques enable their users to mine large amounts of historical data to determine the likely occurrence of events in the future. In this scenario by querying, visualizing and reporting these past datasets, the financial service companies can get actionable insights of illuminating behavior and transactional patterns to move forward with decision-making on strategies for products and services to offer.

9.5 Model Building

These tools can help companies to build model based on customer spending behavior and product usage to pinpoint which particular products or services are popular and found most useful with customers and which particular ones they should focus delivering more effectively. By doing so the banks can increase their share of incomes, garner loyalty and increase their profitability.

9.6 Fraud Detection and Risk Management

Banks and other financial services companies can effectively deploy predictive analytics to help detect frauds in financial service sector. By compiling large and comprehensive customer's data it will be possible for banks and brokers to better detect fraud, earlier than what was possible by the use of conventional approaches.

'Predictive scorecards' can help determine the likelihood of customer defaulting payments also enabled by the emerging analytics tools to help in risk management and mitigation by the banks and financial institutions.

9.7 Integration of Big Data Analytics into Operations

Both sentiment analysis and predictive analytics techniques can be integrated into the operations and operating model of the banks and financial institutions.

9.8 How Banks Can Benefit from Big Data Analytics?

Internationally, banks have started deploying the techniques of Big Data Analytics to derive utility across various spheres of their functionality, ranging from sentiment analysis, product cross-selling, regulatory compliance management, reputation risk management, financial crime management and for many more possibilities. In India and other developing countries also, the banks are attempting to catch up with this trend.

In all the cases, the analysis is of primary nature while the data used is secondary data.

9.9 Best Practices of Data Analytics in Banking for Crises Redressal and Management

When a bank is in crisis in terms of lack of customer satisfaction resulting in customer migration or loss, what can be done? The following steps of redressal can be taken up:

1. Determine the root cause of drop in customer satisfaction.
2. Analyze the spending patterns of the cardholder customers.
3. Channel usage analysis—debit/credit description and payment modes ATM/Cards.
4. Customer behavior and cross-selling.

Methodology

The methodology begins with analyzing the customer satisfaction measurement data to identify the cause of drop out whether due to poor services or any other cause.

After segmenting the issue with the help of feedback analysis, it is possible to identify the reasons for the drop and accordingly suggest or recommend a remedy as an action to be taken by the bank as improvements.

Customer segmentation also can be performed by using classification techniques and accordingly financial products can be suggested or recommended to meet the requirements of different customer segments, based on their type.

Feedback Analysis

In any organizational context, the feedback analysis becomes critical to identify the exact problem and to attempt to solve it. In the context of a bank, the feedback is to be taken from customers, from there who visited the bank branches and also those who used online services. This feedback was taken in writing or online from customers who availed the services of the bank. They were asked to evaluate the banking services in a scale of 1–5 on the parameters as follows:

1. Customer rating of quality of service.
2. Customer rating of speed of service.
3. Customer rating of response to their queries.

The above data, when collected for a large number of customer (20,000 or more), we can analyze by plotting a graph of the data on service quality or speed or query response against time. Such graphical analysis can bring out all the surprising or unknown happenings in time. Drop-in service quality or speed or query response in time indicates a problem situation, while their improvements show the benefit of the effective steps taken by the bank to redress the situation. Steps that result in improvement can be identified and recommended.

Online Transactional Analysis
The Analysis of online transactions of the customers indicates the spending patterns. The causes for spending patterns (say drop) can be identified (such as recession or job losses or functional seasons or vacation periods, etc.). Accordingly, some suggestions or recommendations can be made to facilitate the customers.

Channel Usage Analysis
Customer behavior can be analyzed based on expenditure channel (ATM or card versus online transactions). If surplus funds are identified with certain customers, they can be offered investment plans according to their surplus capacity.

Consumption/Expenditure Pattern Analysis
If a customer demonstrates a certain type of consumption/expenditure pattern, specific products can be offered accordingly. For example, if a customer spends heavily at a particular time a credit scheme or card can be offered to him/her.

Security and Fraud Analysis
A potential threat to the banking system can be identified based on the historical transactions and consumption capacity of the customers. Frauds already performed also can be identified accordingly. The bank can take appropriate steps accordingly. This will result in improving active and passive security.

9.10 Bottlenecks

How banks and financial services companies can maximize the value of their customer data? What are the bottlenecks to be overcome? The following are the usual bottlenecks.

1. **Silos of Data**

Customer data resides in individual silos such as CRM, portfolio management, loan servicing, etc. Legacy systems can become impediments in data integration.

2. **Skills and Development Gap Needs Closing**

New skill sets are required to be developed in Big Data Analytics. New data management skills, new platform skills (e.g., Hadoop family) mathematical and statistical skills and their platforms (as R, Matlab, Weka, etc.). Data scientists need to be deployed.

3. **Lack of Strategic Focus**

Big Data is usually viewed as yet another IT project top management needs to recognize the radical change that Big Data Analytics brings in (and not look at it as another IT Project). To give top priority, prepare for investments to implement Big Data Analytics in the organization.

4. **Privacy Concerns**

Privacy concerns limit the adoption of Big Data Analytics on customer data. Analysis of sensitive and correlated customer information may become an objectionable issue for the customers.

9.11 Conclusion

In this chapter, we have summarized the possible methodologies of application of Big Data Analytics to banking and financial services sector.

9.12 Review Questions

1. Explain how banking and financial services sector is variously getting impacted by the Big Data phenomenon.
2. Explain how customer sentiments and marketing analysis are possible to be processed in Big Data.
3. Explain how sentiment analysis and customer feedback can be processed in banking and financial services sector.
4. Explain how prediction analytics can be applied in banking and financial services sector.
5. How model building is possible in Big Data Analytic in banking and financial service sector.
6. How banks can utilize Big Data Analytics for crisis redressal and management in banking?

7. What are the best practices for Data Analytics in banking for crisis redressal and management?
8. How feedback analysis is performed and what are its benefits? Explain with specific examples in banking and financial services sector.
9. What are the bottlenecks experienced by banking and financial services sector in implementing Big Data Analytics techniques.
10. Summarize the way in which Big Data Analytics techniques can be deployed in banking and financial services sector.

References

1. Oracle Enterprise Architecture, Improving banking and financial services performance with big data, White Paper (Feb 2016)
2. IBM Global Banker Services, Business analytics and optimization—execution report 2013. Analytics: the real world use of big data in financial services
3. U. Srivastav, S. Gopalkrishna, Impact of big data analytics on banking sector: learning for indian banks. Science Direct, Elsevier (2013)
4. P. Garel James, How financial sector uses big data analytics to predict client behavior. Computer Weekly.com (July 2016)

Chapter 10
Big Data Analytics Techniques in Capital Market Use Cases

10.1 Introduction

Having surveyed the applications of Big Data Analytics in Banking and Financial Services sector in the last chapter, we shall now provide an overview of the possible applications of Big Data Analytics in the Capital Market Use cases.

Application of Big Data Analytics Techniques to Capital Market is identified [1] to be possible in the following functional areas:

(a) Financial data management and reference data management comprising

   (i) historical trading, internal data management challenge and also
   (ii) overall reference data mining to find metadata to deconstruct/reconstruct data models.

(b) Regulation application focusing on fraud mitigation
(c) Risk analytics comprising

   (i) anti-money laundering (AML),
   (ii) know your customer (KYC),
   (iii) rogue trading and
   (iv) on-demand enterprise risk management.

(d) Trading analytics comprising:

   (i) analytics for high-frequency trading and
   (ii) predictive analytics.

(e) Pre-trade decision support analytics including:

   (i) sentiment measurement
   (ii) temporal/bitemporal analytics.

C. S. R. Prabhu et al., *Big Data Analytics: Systems, Algorithms, Applications*,
https://doi.org/10.1007/978-981-15-0094-7_10

(f) Data tagging

In enterprise-level monitoring and reporting, it is hard and difficult to match and reconcile trades from various systems built to different symbology standards, usually resulting in invalid, duplicated and mixed-up trades. Data tagging can easily identify trades and events such as corporate actions and help regulators to detect stress signs easily.

10.2 Capital Market Use Cases of Big Data Technologies [2, 3]

Previously, high-velocity market data was the focus of analytics. But due to changing trading dynamics, unstructured data is now becoming very important. Unstructured data comprises of daily stock feeds, Twitter chats, blogs, etc.

10.2.1 Algorithmic Trading [4–7]

For gaining an edge in Wall Street trading, it is essential to convert unstructured information into machine-readable data. This data is then used by algorithmic traders to produce some alpha (return from market) from the news. These untapped sources can help traders to gain a competitive edge in trading.

As against the past practices of the traders depending on slow incoming news of company specific and industry specific events having occurred over a period of time, today, in the online world, it is the instantaneous and online feedback inputs coming from online news, blogs, Twitter feeds and online stock data that help the trader in making decisions based on Algorithmic Engine. Such Source and Response system to handle responsiveness of important events in the external world influencing trading decisions is shown in Fig. 10.1.

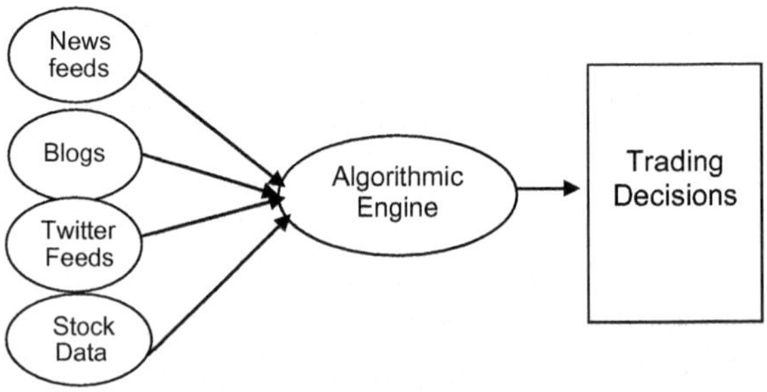

Fig. 10.1 Source and response system in trading

10.2.2 Investors' Faster Access to Securities [3–5, 7]

With today's instantaneous access to the Internet-based stock market information to the investors and the general public as well, information pouring in from quarterly reports or breaking news stories can dramatically affect the share price of a security. As the primary source of event information is provided on the Internet, the impact of such information on trading in stock markets is significant. Therefore, the methodology and technology involved in extracting and using information to support decision-making process become critical.

10.3 Prediction Algorithms

In order to predict the stock market, we have different algorithms and models from the academic community and also industry. We shall now examine and survey the developments in the prediction algorithms and models along with their comparative performance evaluation. For greater accuracy of prediction, we need to examine the impact of various global events on the stock market and the issues they raise.

10.3.1 Stock Market Prediction [3–5, 7]

Due to the huge financial gains at stake, both investors and traders have a great interest in the prediction of stock market which has been identified as an important subject in Finance sector, as also in Engineering, Computer Science and even Mathematics.

As a very huge amount of capital is traded globally every minute, the stock market is an outlet for maximum investments. Therefore, researchers have been striving hard to predict the financial market and hence, stock market prediction generates a great amount of appeal to the researches all over the globe. Inspite of all these research efforts, till today, no specific method has been developed to correctly and accurately predict the movement of the stock prices in the stock markets. Even without a full-fledged and successfully consistent prediction method, some limited successes were noted till now in the prediction process. Auto-regressive and moving average techniques are some of the famous prediction techniques which have dominated the time series prediction techniques for many decades. Inductive learning techniques have also been developed and deployed. K-nearest neighbor (KNN) algorithm and neural network techniques have been applied for prediction. However, their weakness is that they depend heavily on the structural data only while totally neglecting the impact and influence of non-quantifiable information such as news articles.

10.3.2 Efficient Market Hypothesis (EMH)

In this theory, it is stated that financial markets are already and always 'informational efficient'. This means that the current prices at a given instance of time are such that they readily (already) reflect all the known information and automatically, instantaneously change, so as to reflect, at any time, the latest and new information. Therefore, logically, accordingly to this theory, it is impossible to consistently outperform the market by depending on and using the same information which the market already is having with it (except through 'luck', if at all). 'Information' is anything which may affect the prices. There has been a lot of debate on whether financial market is predictable at all or not. Information categories have been proposed as three options: (1) weak (2) semi-strong (3) strong. In weak 'information', only historical information is incorporated and embedded into the current files. In 'semi-strong' case, both historical and current public information are incorporated and embedded into the current files. In 'strong' case the historical information, the current public information and also the current private information such as financial information are incorporated and embedded into the current share prices. The basic tenet of EMH theory is that it is impossible to outperform the market since the market reacts automatically and instantaneously to any given developments such as news or other sudden developments.

10.3.3 Random Walk Theory (RWT)

Random Walk Theory (RWT) maintains that it is impossible to outperform the market and that the prices are determined randomly, even though the historical data and even the current public information has its own impact (same as semi-strong EMH).

10.3.4 Trading Philosophies

Based on the above two theories (EMH and RWT), we have two philosophies: (a) fundamental trading philosophy and (b) technical analysis trading philosophy.

(a) **Fundamental Trading Philosophy**

This philosophy states that the stock price is determined (indirectly) by the vital economic parameters such as indices of inflation, unemployment, Return on Equity (RoE), debt level and individual Price to Earnings Ratio and also especially from the financial performance of the Company itself.

(b) **Technical Trading Philosophy**

This philosophy states that stock price is dependent on historical time series data. This school of philosophy believes that the time of investing in market is the most crucial factor, investment opportunities can be identified by carefully investigating the average price (historically) and value movement in comparison with the current prices. It also believes that the psychological factors of perception (high or low) of price barrier such as support level and the resistance level may also indicate where the opportunities may open up.

10.3.5 Simulation Techniques

Both the Fundamentalists and Technicians have developed certain methodologies and techniques for predicting the price from financial trade articles. Therefore, they adopted simulation techniques where the stock market is analyzed based on simulated (not real) stock markets with simulated (not real) traders by mimicking the real human traders. Being artificial, it is practically possible to perform dissections for identifying key parameters of information. The traders in the simulated situations are programmed to follow a rule hierarchy while trading in response to market changes, especially those based on news articles are updates. The response time of such simulated traders was made to vary, based upon the time elapsed between the point of time of receipt of information and the point of time of reaction. The results were found to be astounding when it was noted in observation that the length of reaction time is dependent on a particular trading philosophy—the traders who showed quick response formed technical strategies while those who waited long formed fundamental strategies. Therefore, we can surmise that technicians may have capitalized on the time lag by acting instantly, before all other traders. This research is able to demonstrate that there exists a week's ability to forecast only for a brief period of time.

10.4 Research Experiments to Determine Threshold Time for Determining Predictability

In a research experiment by Gidafalvi, about 5000 news articles pertaining to 12 stocks were analyzed and it was concluded that in the time interval of 20 minutes before and 20 minutes after some 13 news articles were released, there was a week possibility of predictability of the direction of market movement.

The weakness in predictability is due to the fact that the news articles concerned got repeatedly reprinted across all the news agencies and wire services.

Stronger predictability exists if the first release of the article is isolated, and by using automatic text analysis techniques, it makes possible to capitalize 20 minutes before the human traders start acting.

10.5 Experimental Analysis Using Bag of Words and Support Vector Machine (SVM) Application to News Articles

In an experiment in 2005, Schumaker picked up a large number of news articles (9211) and a very large number of stock quotes (10,259,042) from S&P 500, over a five-week period. Then, the analysis of news articles was performed and the terms which appeared more than three times in the articles were retained. Bag of Words was created with about 4000 terms from 2500 articles with about 5000 noun phrase terms and 2800 named entities, 2800 terms from 2600 articles. The above, when processed by support vector machine (SVM) algorithm derivative, using regression, three metrics M1, M2, M3 were defined for 'closeness' and 'derived accuracy'. M1, M2, M3 are the three models used. M1 uses only extracted articles terms with no baseline price; M2 uses extracted article terms and stock price when the article was released; M3 uses extracted terms at estimated 20 minutes of stock price.

SVM had to perform learning on which terms result in changes in share prices and accordingly adjust their weights according to the severity of price changes. From Closeness results and Directional Accuracy Results, it was found that model M2 gave the closest and the most accurate prediction (for +20 min stock price).

Results showed that M2 which uses the news articles and regression together performs better than pure regression. Therefore, it is essential that impact of news articles be considered for any prediction. This was the conclusion reached in this research.

10.6 Textual Representation and Analysis of News Articles

Many methods are possible to analyze the text contents in a news article. One simple way is tokenizing and using very word in the given text document. This is, however, a human-oriented technique as every word is deployed to indicate syntactic structure of the document. For machine learning algorithms, such structural markings are not required. In order to perform easy and efficient text processing, the normal approach followed is 'Bag of Words'. This is a standard approach for text processing due to its simplicity and ease of use. Over and above, certain parts of speech can be used as features. Noun phrases can be indentified through parsing and a dictionary (or lexicon) is used to identify nouns which may be aggregated, using syntax rules (as nearby words) to form noun phrases.

10.7 Named Entities

Another method called 'Named Entity Identification' is based on nouns and noun phrases. By using dictionary and also semantic lexical hierarchy, we can classify nouns and noun phrases into entities such as persons or organizations or locations. By generating a lexical profile across all noun phrases (after analyzing synonyms), it is possible to determine the semantic hierarchy of nouns or entities. Thus, the named entities capture far greater semantics than ordinary 'Bag of Words' or even just noun phrases. Even greater semantics is captured by Object Knowledge Model (OKM).

10.8 Object Knowledge Model (OKM) [8]

Objection Knowledge Model (OKM) enables much greater capture of semantics of a text article than other methods (as Bag of Words or Noun Phrases Named Entities) since it captures semantics of activities also, in addition to entities. Thus, not only Named Entities and their attributes are identified but Named Activities and their attributes are also identified in Object Knowledge Model (OKM).

10.9 Application of Machine Learning Algorithms [7]

All machine learning algorithms perform simple linear regression analysis of the old (historical) security trading data for a given time period in recent times in order to determine the price trend of a given stock. In addition, to determine the impact of textual reports or comments, a simple textual analysis technique algorithm by using 'Bag of Words' approach is performed in order to determine the keywords in the given text. Finally, all the above inputs are classified into the prediction of stock movement as (1) upwards (2) downwards or (3) unchanged.

Research has been performed on applying (1) genetic algorithms (for classification into two categories), (2) Naïve Bayesian (for classification into three categories) and (3) SVM (for classification into three categories) based on texted news articles or text postings in chat room chats. The outcome of such research indicated that apply to genetic algorithms, the chat room chats were analyzed and stock prices were classified by utilizing the postings and number of words posted on an article daily.

Research on SVM applications for the stock data and articles produced results that this technique is mildly profitable. An attempt was also made to produce an optimum profit trading engine.

10.10 Sources of Data

In all the above experiments and also later experiments, data was collected from secondary data sources such as Yahoo Finance Website which provides intraday 5-min interval data for that day (many primary servers may not provide intraday stock movement data). The news articles are taken from new agencies as Reuters and also from newspapers.

10.11 Summary and Future Work

To summarize, the process of integrating text analysis of news articles with Regression and other machine learning algorithms (as SVM, Naïve Bayesian and genetic algorithms) in an interval from −20 to +20 min is indicated in the following diagram (Fig. 10.2).

Future work can be aimed at improvements and a 'multimodel regression' and sentiment analysis on textual data can be integrated to obtain greater accuracy.

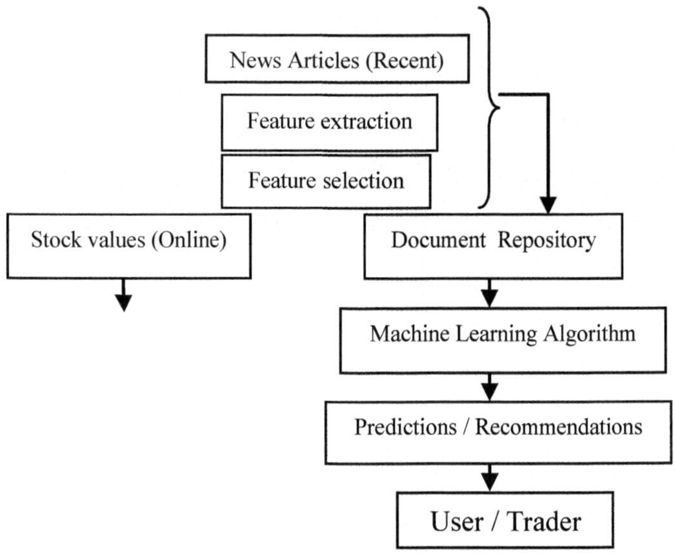

Fig. 10.2 Integrating text analysis with machine learning algorithm

10.12 Conclusion

In this chapter, we have surveyed the use cases, techniques, algorithms for predication and the research performed in determining the predictability of prices in stock markets.

10.13 Review Questions

1. List out where in Capital Market use cases the techniques of Big Data Analytics are applicable.
2. Explain Source and Response System in Trading (with diagram) in Algorithmic Trading.
3. How to provide faster investor access to securities?
4. Explain Prediction Algorithms and Stock Market Prediction.
5. Explain Efficient Market Hypothesis (EMH).
6. Explain Random Walk Theory.
7. What are the two major trading philosophies?
8. Explain simulation techniques deployed in Capital Market Analysis.
9. Explain how Experimental Analysis using Bag of Words and Support Vector Machine (SVM) are applied to new articles.
10. Explain how textual representation and analysis is performed for news articles.
11. Explain the application for machine learning algorithms in Capital Market Analysis.
12. Explain how to integrate text analysis with machine learning algorithms in news article analysis in Capital Market.

References

1. Thomson Reuters, Big data in capital markets: at the start of the journey, White Paper (2014)
2. M. Singh, Big data is capital market. Int. J. Comput. Appl. **107**(5) (2015)
3. T.H.A. Uheug, S.Y.-H. Wu, Trade data service for capital markets, Honors, School of Computer Science and Engineering UNSW (2003)
4. Trading technology survey of exchange technologies (2003). http://www.tradingtechnology.com
5. Capital Market Cooperative Research Centre (CMCRC), http://www.cmcrc.com
6. F.A. Rakhi, B. Benatallah, An integrated architecture for management of capital market system. IEEE Netw. **16**, 15–19 (2002)
7. K.-C. Li, H. Jiang, L.T. Young, Big data algorithms, analytics and applications. https://books.google.co.in/books?isbn=1482240564
8. Object knowledge model definition, Ph.D. thesis, C.S.R. Prabhu Sunrise University, 2015

10.12 Conclusion

In this chapter, we have surveyed the use of ML techniques/algorithms for prediction and the research performed in determining the predictability of prices in stock markets.

10.13 Review Questions

1. List out where in Capital Market use cases the techniques of Big Data Analytics.

2. [illegible]

Chapter 11
Big Data Analytics for Insurance

11.1 Introduction

In the last few chapters, we have seen the application of Big Data Analytics to various application domains. In this chapter, we shall examine its role in insurance.

The insurance domain has been a long-time user of conventional data processing techniques and therefore the information about customers, market trends and competition is abundantly available in legacy systems. In addition to the legacy systems, there is a huge amount of unstructured data coming from emails, social networks, messages, blogs, all put together usually referred to as "Big Data". Analyzing such variety of data will be of substantial value for insurance activities [1, 2] such as marketing and underwriting, in addition to reducing costs in operational activities which can enable better strategy formulation and risk reduction in insurance.

However, the Big Data applications call for Big Infrastructure which only the top-tier insurance companies have in place. The new type of infrastructure based on Hadoop type of environments requires technical support teams for configuring, administering and managing Hadoop clusters and the associated Hadoop family of software modules and also more recent and modern tools to handle deployment, scalability and management. Most of these tools require management of changes to configurations and other pieces by writing scripts of software. Making changes to one appliance entails a manual, time-intending process that leaves doubts in administrator's mind whether the changes have been implemented throughout the application cluster. This delay due to heterogeneous system was the reason behind the delays in Big Data deployment and implementations in organizations which proposed to implement Big Data. Alternatively, proprietary vendors as Oracle, IBM, EMC^2, Teradata who provide homogeneous software (and hardware) appliances for Big Data Applications are known to be very expensive and therefore very limited.

However, some proprietary solution (as Stack IQ) is now offered for implementing the applications of Big data environment for insurers.

In this chapter, we shall survey the methodology of deployment of big data application in Insurance sector.

C. S. R. Prabhu et al., *Big Data Analytics: Systems, Algorithms, Applications*,
https://doi.org/10.1007/978-981-15-0094-7_11

11.2 The Insurance Business Scenario

The greater longevity of customers added with the business in times of financial crises and recession competition provides challenges. Such challenges motivate the identification of new products in insurance business and also about which methodologies and techniques are required to be adopted for marketing and advertising of the insurance products, and also assessment of risk and fraud detection. Actuaries [3] have been deploying analytical techniques for pricing of insurance policies. Earlier limited data windows were available to them to analyze and plan for implementation of the schemes. However, now we have a contrasting situation of data deluge. Today, we have a new opportunity to enhance the incomes and reduce the costs and thereby be more competitive. Better quality of claims processing and assessment of risk is possible given the large data, based on the demographic and psychographic trends.

Applying analytical tools to their new and huge volumes of data requires a distinctly different infrastructure from traditional database architectures and query products. This big data infrastructure needs to be installed (Hadoop clusters, for example) and analytics tools are required to be deployed on that infrastructure to obtain analytical results and insights which are required to be trusted by the company concerned for acting upon them [3].

11.3 Big Data Deployment in Insurance [4]

Major insurance companies such as MetLife and Travelers, Bajaj Insurance and Capital One are deploying these Big Data Analytical techniques. The insurance company, MetLife, has been analyzing their Big Data for identifying patterns and how risk mitigation can be aimed; product performance can be monitored by trend analyses. Travelers is another company which uses analytics techniques on their own Big data in order to identify new products or rationalize existing products and also better understand risks globally. Other companies such as Progressive Insurance Company and Capital One have experimented on how to segment their customers by deploying classification techniques. They tailor their products accordingly and make special offers based on segmented customer profiles.

11.4 Insurance Use Cases [5]

How and what the insurance companies do in deploying Big Data Analytics? The following are some use cases:

1. **Risk Avoidance**

In contrast to the risk assessment of a human agent who sold insurance after having a firsthand knowledge of the personal life of the customer, today's virtual world requires external and internal risk assessment. This calls for building standard models based on new Big Data of customers for quantifying risk. Such application contains analysis of customer behavioral models based on the data of customers profiles over a long period of time, added with cross-reference to specific type of products. Risks inherent in specific products can be assessed.

2. **Personalized Product Formulation**

Personalized policies at appropriate premiums can be devised depending on demographic data, health record data, driving record data, etc. Therefore, the personalized policies can be evolved and offered based on personal data and specific personal needs and risks. For car insurance, sensors inside cars can help track customer behavior in tasks of driving time, brake frequency, average speed, traffic crimes, etc. When we add to this personal data the policy and customer profile data from the actuaries, we will provide the basis for how best we can rate a driver based on his performance and behavior patterns.

3. **Cross-selling and Lap Selling**

By monitoring the customer behavior through multiple channels such as social networking statements/tweets, web site click stream data and account information, it will be possible for the insurers to suggest additional products that meet and match the customer requirements and their budgets. In such applications, it may be possible to sketch customer habits to assess risks and also suggest changed customers behavior to reduce risks.

4. **Fraud Detection**

Insurance fraud can be better detected by deploying techniques such as pattern and graph analysis, in addition to social network analysis and cohort networks. The potential future and presently existing fraud can be possibly determined better by collecting data from social networks which can be analyzed for detecting normal or suspected behavior.

5. **Disaster/Catastrophe Planning**

To be able to better prepared for disasters/catastrophes which may occur anytime, statistical models can be analyzed, enhanced with direct actionable inputs from customers. This can help reducing the quantum and extent of insurance claims and also accelerate response by insurers.

11.5 Customer Needs Analysis

The insurance applications and products such as life insurance and annuity are complex and automating the discussion between prospective customers, on one hand, and their advisors, on the other hand, to improve the sale of insurance policies which will improve the efficiency.

11.6 Other Applications

Sentiment analysis, loyalty analysis, loyalty management, campaign design, campaign management and value analysis of customers are some of the other applications of Big Data Analytics in insurance.

11.7 Conclusion

In this chapter, we have summarized the application of Big Data Analytics to Insurance sector that includes use cases with applications in risk avoidance, personalized product formulation, cross-selling/lap-selling fraud detection, disaster/catastrophic planning and customer needs analysis.

11.8 Review Questions

1. Explain how Big Data Analytics Techniques are relevant to Insurance Sector Bankers.
2. What are the challenges the Insurance Sector business faces how? And how the application of analytics tools will help solve the challenges faced in Insurance business?
3. How the Big Data Analytics Techniques can be applied in Insurance?
4. What are the various use cases of Insurance Business for Big Data Analytics?
5. How personalized product formulation can be done?
6. How fraud detection can be done?
7. How disaster/catastrophe planning be performed?
8. What are the various other applications of Big Data Analytics in Insurance sector?

References

1. Mark Logic and Accord, Making sense of big data in insurance (white paper) (2013)
2. A. Woodie, How big data analytics is shaking up the insurance business? HPC, 5 Jan 2016. www.datanami.com
3. McKinsey, Unleashing the value of advanced analytics in insurance. www.mckinsey.com
4. M. Schruek, R. Shockley, Analytics: real world use of big data in insurance. IBM (m.ibm.com)
5. The TIBCO Blog, 4 ways big data is transforming the insurance industry, July 2015

Chapter 12
Big Data Analytics in Advertising

12.1 Introduction

Traditionally, advertising was nothing but communicating to a whole set of target audience. But with the advent of internet, everything changed, especially behaviorally targeted advertisements. Since 2000, the internet became the primary advertising and marketing channel for all the businesses in all sectors. But even then, the click-through rates (CTRs) flattened s after a point of time. CRTs increased 62% in 2013 and much later. Today, brands have access to a huge quantity of data in the form of reviews, tweets, followers, click, likes, etc. which offer great untapped potential. Thus, when this kind of unstructured data is combined with macro-level data from the advertising agencies, it can prove to be a valuable communication opportunity. The companies can see how they can analyze the data to gain insights on and to predict consumer behavior and also conclude how they can align new unstructured disparate data sources with their existing data to derive actionable decisions [1–3]. With the recent advances in mobile computing and wireless networking, mobile advertising is now becoming popular because of the effective platform that the mobile devices can offer. Thus, mobile-based Big Data Analytics provides new opportunities of inputs to advertising.

Presently, the approaches involved in advertising processes use behavior targeting (BT) technology to provide static services. This is obsolete and very poor scenario as compared to the fast and real-time expectation and requirements in the upcoming scenario of Big Data Analytics. It is the need of the hour to develop a new service in advertising based on Big Data Analytics techniques. The objective of such initiatives will be to provide, real time and static on-demand services for advertisers and publishers on 'when,' 'what,' 'how,' to advertise, identify customer behavior patterns that are collected by data. This becomes essential to develop models for advertising recommendations and trend-setting statements.

C. S. R. Prabhu et al., *Big Data Analytics: Systems, Algorithms, Applications*,
https://doi.org/10.1007/978-981-15-0094-7_12

12.2 What Role Can Big Data Analytics Play in Advertising?

Targeted personalized campaigns can be created to save money and increase efficiency by targeting the right people with the right product by gathering data and learning user behavior [3].

The digital footprint of a customer is highly valuable in today's era of personalized marketing and advertising. There is so much valuable information available with every Google search, every Facebook or Twitter posting, all online actions, a consumer's social media and digital world will be flooded with advertisements of various products that the customer may be willing or be interested to buy.

The information about live online communities can be targeted after due understanding of the patterns in user behavior. The customer motivations can be better understood with the details of such behavior. Advertising agencies can obtain information and thereby measure accurately the customer interests, and with subgroups level analysis, they will be able to track and measure the customers' impressions about specific latest trends or products.

12.3 BOTs

We already know that about a quarter of all advertisements are being shown only to Bots, not humans. Therefore, advertisement frauds can be seen with humans being not exposed to advertisements.

12.4 Predictive Analytics in Advertising

Huge advertisement fraud can be overcome by applying predictive analytics techniques for Big Data problems. Big Data Analytics techniques make it possible to define accurately the types of customers being targeted, thus enabling effective, efficient and least cost mechanism to have reach and impact on specific targets. Optimove is a platform for marketing automation that uses predictive analytics to prioritize within the existing customers instead of acquiring new customers by additional investment. For achieving customer retention, targeted deals and services are offered to specific customer groups based on their requirements as understood by using this predictive analytics platform.

12.5 Big Data for Big Ideas

New and big ideas can be conceived and delivered by advertisers by working with Big Data companies, at a pace fast than conventional methods of working with ideas and pitches through various departments. Today, advertisers can provide new ideas and campaign very fast, deploying Big Data technologies.

12.6 Innovation in Big Data—Netflix

Netflix advertises TV shows and movies based on what the customers have previously watched by collecting data on TV shows watched, time spent on each show, preferences of actors, etc. With all this, Netflix is able to calculate the worthiness of a customer for advertisers.

12.7 Future Outlook

Since the fact that a lot of business are sitting on a lot of data of their customers but lack infrastructure or capability to understand analyze it, the need for better and new technology infrastructure and analytics will continue to grow.

Thus, by using the power of data analytics, advertisers can identify emerging trends and provide real-time live options of advertisements: Predictive analytics based on Big Data technology can help reach the right audience at the right time, the goal of all advertising.

This provides an opportunity for companies to mine their data to improve both bottom level and customer service, instead of blindly sitting on a gold mine of data of customers.

12.8 Conclusion

In this chapter, we have analyzed the possible role of the Big Data Analytics in advertising. We have presented the role of predictive analytics in advertising with the example of Netflix. We finally concluded by providing the future outlook for a role of Big Data Analytics in advertising.

12.9 Review Questions

1. Explain Internet advertising in contrast to traditional advertising.
2. How existing advertising approaches use Behavior Targeting. What is their drawback?
3. How Big Data Analytics will beneficial?
4. What is the role Big Data Analytics can play in Advertising?
5. How predictive analytics can help in advertise?
6. Explain how Netflix functions?
7. What is the future outlook for Big Data Analytics in Advertising?

References

1. Y. Freund, R. Iyer, R.E. Schapire, Y. Singer, An efficient boosting algorithm for combining preferences. J. Mach. Learn. Res. **4**, 933–969 (2003)
2. C. Schmitt, D. Dengler, M. Bauer, Multivariate preference models and decision making with the MAUT machine, in *Proceedings of the 9th International Conference on User Modeling (UM 2003)*, pp. 297–302 (2003)
3. J.L. Herlocker, J.A. Konstan, L.G. Terveen, J.T. Riedl, Evaluating collaborative filtering recommender systems. ACM Trans. Inf. Syst. **22**(1), 5–53 (2004)

Chapter 13
Big Data Analytics in Bio-informatics

13.1 Introduction

Bio-informatics is an interdisciplinary science, which provides life solutions in the discipline of biology and health care by combining the tools available in various disciplines such as computer science, statistics, storage, retrieval and processing of biological data. This interdisciplinary science can provide inputs to diverse sectors such as medical, health, food and agriculture.

The practice of bio-informatics usually involves:

1. Collection of biological data.
2. Building a computational model.
3. Solving a computational modeling pattern.
4. Testing and evaluation of computational algorithms.

The above steps can be performed with the help of several bio-informatics tools and databases.

Bio-informatics refers to the application of information technology the study of biological systems.

The data sizes in bio-informatics are rising dramatically in the last five years. For example, the European Bio-informatics Institute (EBI) has more than 40 PB of data about genes, proteins and small molecules and their total size doubles every year [1]. Therefore EBI has established the Hinxton center with 17,000 servers and 74 TB of RAM for processing data, with computing power being upgraded every month. Similarly, other organizations such as National Center for Biotechnology Information (NCBI) of USA and National Institute of Genetics (NIG) of Japan also have comparable repositories of Bio-informatics data.

C. S. R. Prabhu et al., *Big Data Analytics: Systems, Algorithms, Applications*,
https://doi.org/10.1007/978-981-15-0094-7_13

13.2 Characteristics of Problems in Bio-informatics

While the availability of such data is helpful for more accurate analysis, the big data characteristics and challenges in Bio-informatics research are quite different from those in other application (such as CERN data on physics or satellite data at NASA/NRSA/ISRO (Bhuvan.nrsc.gov.in) open data archive). Being not only largely spread geographically the data is largely nonuniform and non-homogenous. In order to perform analysis and inferences in bio-informatics area, we have to deal with several heterogeneously independent databases. Also, as many different and uncontrolled organizations generate the data, the same data types are represented in different forms and formats at their respective sources. Secondly, the bio-informatics data is massive and expanding naturally in global geographical distribution as well as the number of instances. Also, only some parts of this data are transferable over the Internet, while other cannot be transferred. Such data cannot be transferred due to their huge size, cost, privacy and also regulatory and ethical issues [2]. Therefore, it becomes necessary to process remotely before results are shared. Thus, the big data problems in bio-informatics are characterized by the standard Vs of volume, velocity and variety but also by physical and geographic distribution across the globe.

13.3 Cloud Computing in Bio-informatics

Cloud computing technology comes to be great rescue with success in this context. The cloud can be deployed for both purposes of data storage and for computations [2]. Beijing Genomics Institute (BGI), a prominent genome sequencing centers of the world, has installed a cloud-based Gaea (workflow system), for analytics on Hadoop which can perform large-scale genomic analysis. Similarly at Stanford University and University of California, Berkeley a hardware component called Bina box (www.bina.com) helps the users perform genome analytics on preprocessed data. Bina box also reduces the size of genome data for their efficient transfer to the cloud component, thereby improving the efficiency and throughout of genome analytics by several orders of magnitude [3].

13.4 Types of Data in Bio-informatics

There are many freely available web-based and computer-based tools, through which a bio-informatician can extract valuable information from biological data. There are some of these tools to analyze data coming from diverse sources. Scientists from various disciplines can focus on their respective research areas using them.

Characterizing genes, study of the properties of proteins and perform simulation to identify how the behavior of biomolecule in a living cell can be performed with the help of computational tools. Although these tools can generate information in accordance with experiments, these are inexpensive and less time-consuming [4–8].

Sequence Analysis

Sequence analysis refers to analysis of features and functions of biomolecules (protein or nucleic acids). In the other words, it explains the functional properties of different biomolecules. For this, we need to retrieve the sequence of a biomolecule from the corresponding database. There are many databases which are freely available on the Internet to retrieve the sequence data. Once refined, predictive tools can be used for predicting the features related to structure, functions, history of this evolution or homolog identification.

We have integrated database of Entrez, which is a data retrieval tool. The Entrez system can provide views of gene and protein sequences and chromosome map. Tools like this help in comparing the genome sequences between the organisms or between the species. Tools like BLAST (Basic hold Alignment Search Tool) and Clustal W can be used to compare sequences of gene or proteins to investigate about their origin or history of their evolution. Gene View, Tree Viewer Gene Graphs allow the analysis by graphic description of data. Techniques such as clustering regression analysis sequence mining, hidden Markov models and artificial neural networks are used.

Phylogenetic Analysis and Databases

Phylogenetic analysis is aimed at reconstructing the relationships of evolutionary nature (between molecules or organisms), which may be related for predicting certain features of a molecule where functions may be unknown for tracking flow of genes and for concluding relationships of genetic nature.

A genealogy tree represents the genetic relationships between organisms. The similarity of organisms decides their position in the tree. Methods such as unweighted pair method, neighbor joining or unweighted pair group method are available for constructing a phylogenetic tree.

Algorithms are also used in the construction of the phylogenetic trees. Creating, developing and evaluating the algorithm result in phylogenetic tree with similarities. These are also used in the comparative analysis.

Sequence Databases

In a sequence database, the data of large molecules like proteins, polymers and nucleic acids are stored and each molecule is identified with a unique key. This information is essential for sequence analysis. Due to advances in technology in this domain, we have reached a scale of a whole genome, thus generating massive amounts of data daily. Large databases have been created due to this large data turn over. Each such database is an autonomous representation of molecular unit of life.

We have databases classifiable into primary, secondary or composite categories. Data in primary databases is directly observed in experimentations such as XRD or NMR approaches. Data in secondary databases comprises of data derived from primary databases.

Primary databases: Uni Prot, Swiss Prot, PIR, GenBank, EMBL, DDBI and Protein data bank.

Secondary databases: SCOP, CATH, PROSITE, cMOTIF.

While primary databases are archives, secondary databases are curated.

Composite databases also exist, like PDB, SWISS PROT, PIR, PRF, Uni Prot PIR-SPD, TrEMBL, WWPDB, RCSB, PDB, MSD, PDB (Composite of 3D Structures).

Databases of Sequence of Genomes

NCBI [National Center for Biotechnological Information] built GenBank. It has data for 250,000 species. NCBI contains almost all the sequences of species of organisms. It is the store house of the sequences with Accession number, Accession Id, length of amino acid and types of Nucleotides.

Searches of similarity of sequences (based on previous annotations) will help provide annotations. Sequences so annotated (based on similarity to others) are deposited in a database. Such database will become the basis for future annotations.

Micro-array databases are from sources such as Array Express (www.ebi.ac.uk) from European Bioinformatics Institute (EBI) Gene Expression Omnibus from National Center for Bioinformatics (NCBI) (www.ncbi.nlm.gov/geo) and also the Stanford Microarray Database (smd.princeton.edu).

The applications of DNA sequencing include the studying of the association of diseases and phenotypes with genomes and proteins. This will lead to drug discovery, understanding evolutionary biology and identification of microspecies in samples.

In order to perform sequence analysis including the sequences of DNA, RNA and peptides, we can identify their features, functions, structures and evolution.

Sequencing of RNA can be an alternative for micro-arrays. In addition, it can be deployed for other purposes such as identification of mutations, also for identifying post-transcriptional mechanisms and also for detecting or identifying viruses, exogenous RNAs, and also Polyadenylation.

Sequence analysis is superior and more effective than microarray analysis as sequence data embeds better and richer information.

Sequence analysis requires more sophisticated analytic tools and computing infrastructures capable of dealing with the massive sizes of sequencing data. Examples of sequence databases are DNA Databank of Japan, (www.ddbj.hig.ac.jip) RDP (rdp.cme.msu.edu) and mirbase (www.mirbase.org) [9–13].

Protein Sequence Databases

Protein sequence databases are: SWISS PROT, TrEMBL, Uni Prot, etc. These databases provide complete information regarding the sequence of the protein. These protein sequences may help in building the three-dimensional structure of the protein complex. In addition to this, it plays a crucial role in drug designing, docking of the molecules. Proteins being classified into different categories; similarly, databases are also classified accordingly.

Predicting Protein Structure and Function

In a living cell, the interactions and interactomics are having PPI (protein–protein interactions) as intrinsic to them. Therefore, diseases such as Alzheimer and even cancer are traceable to anomalous PPIs which are of interest of study in bio-informatics biochemistry, quantum chemistry, molecular dynamics, resulting in high volume of heterogeneous data pertaining to interactions. Some of the PPI repositories are DIP (dip.doe-mbi.ucla.edu), STRING (string.embl.de), BioGRID (thebiogrid.org), etc.

Pathway analysis is essential to understand molecular basis of a disease.

Gene ontologies are available in gene ontology or G.O. data base (www.geneontology.org).

13.5 Big Data Analytics and Bio-informatics

Bio-informatics research became a big data problem. Big data is characterized by the 3Vs; Volume, Velocity and Variety. In addition, incremental data and distributed data are the characteristics of bio-informatics big data. These characteristics or properties call for deploying machine learning techniques. The same machine learning techniques as used in Big Data Analytics have been deployed in bio-informatics too. However, the challenges are still to be solved or overcome: Traditional machine learning techniques were not developed for big data scenario. The volume, velocity and variety of data to be analyzed in big data context were not taken into account in traditional machine learning. Velocity of data in bio-informatics makes it different. Traditional database is arrayed in set schema. Data warehouses store data following ETL operations. In Big Data systems, the data could be unstructured or semi-structured, data in high speed and large variety. Also, traditional machine learning techniques are inefficient with big data. Appropriate improvements are to be made on machine learning algorithms for Big Data. In addition, privacy, very important in bio-informatics, has to be protected in the Big Data scenario.

13.6 Open Problems in Big Data Analytics in Bio-informatics [14]

We can identify seven categories of problems to be solved in big data analytics in bio-informatics:

(1) **Micro-array Analysis**

In the context of ever-increasing sizes and datasets of microarray data (due to widespread use of microarray experiments), there exists a great opportunity for deploying big data analytics techniques.

Microarray experiments capture changes in values of gene expressions proportionate to various stages of disease. It will enable identification of genes that are affected.

The identification of cancer from gene expression data and then grouping cancer patients into high- or low-risk groups has given direction to many researchers to study the application of machine learning (KL) methods. ML techniques have been utilized with an aim to model the progression and the treatment of cancerous conditions. The ML tools have the ability to detect key features from complex datasets. A variety of ML methods have achieved success, including support vector machines (SVMs), genetic algorithm (GA), K-nearest neighbor (KNN), random forest, artificial neural networks (ANNs) and Bayesian networks (BNs). These methods have been widely applied in cancer research for the development of predictive models, resulting in effective and accurate decision making. However, there are limitations in ML methods to find hidden expression from high dimensional data.

In recent years, deep learning techniques have been applied to variety of problems such as computer vision, speech recognition and natural language processing. In many of these applications, algorithms based on deep learning have outperformed over ML methods. Y. Bengio et al., originally proposed deep learning, a method to learn a multiple hierarchical neural networks by training. Deep learning can be expressed as a deep nonlinear network which can result in complex function approximation. According to the need of different learning tasks like feature learning, classification and recognition, numerous deep learning methods are proposed. Deep Learning methods are widely used in the field of bio-informatics, especially in genomics.

Each gene expression dataset consists of huge number (more than thousands) of genes. It is extremely difficult to analyze such a large set of gene expression data. Moreover, for regulation of gene expression level, only small set of genes takes part. These small set of genes are called as characteristic genes. These characteristic genes are related to specific biological process of various cancer types. Identifying such genes from large arrays of gene expression dataset is an important research topic.

In the traditional analysis of gene complexes, the time dimension was not there. When time dimension is added, the computational complexity increases and big data analytics techniques will come handy.

Molecular Interactions

Bio-informatics also clearly explains about the chemical bonding interactions between the molecules like hydrogen bonding, number of hydrogen atoms present in the molecule, structure and shape of the molecules, number of amino acid molecules present in the molecule and about the active sites of the bonding molecules. In addition to check the bond energies, bond affinities are also essential for molecular interactions. Another special field of science known as the structural biology helps in predicting the structural and functional properties of a molecule. This information effectively helps in molecular interactions. There are many tools and databases which help in predicting the above properties.

In this way, bio-informatics performs the required action in a very less span of time with accurate results.

Bio-informatics being an interdisciplinary science; it also includes other sciences like computational biology and structural biology. These fields also make use of computer technology for performing the biological activities.

Computational biology can be broadly described as the science which includes the study of biology, applied mathematics, statistics, biochemistry, biophysics, molecular biology, genetics and computer sciences. It is an interdisciplinary science which involves computer science and bioengineering, which is very much similar to bio-informatics.

In this science of bio-informatics, we develop computer models for evaluating the biological results developed from the biological data. In addition to this, there is another field known as structural biology. This field also comes under bio-informatics. Structural biology studies about the structure and functional features of different biomolecules. We use bio-informatics databases and tools for predicting the structure and functional properties of a biological molecule. Mostly, computational tools are used for performing the above activities.

Systems Biology

Systems biology is an interdisciplinary field for mathematical and computational modeling of biological systems with a focus on complex interactions, by adopting a holistic approach (instead of the conventional reduction approach). Systems biology became successful and popular since the year 2000. Human genome project is an example of applied systems thinking in biology. The properties and functions of cells, tissues, organs and organisms are modeled is a single system viewed with a systems biology approach. It can also be viewed as a study of interaction between components, how such interactions result in functions and behavior of that system, as a whole. For conducting research, a cycle comprising a theory, hypothesis (about a biological system), experimentation and the consequent validation of that theory/hypothesis so as to utilize the newly acquired data to validate or improve the existing hypothesis or theory or model. Thus, quantitative data can be collected and used in the construction and validation of the models in fields such as transcriptomics, metabolomics and proteomics.

Other aspects are:

- Models for biological processes.
- Information extraction and text mining.
- Database development of online databases and repositories.
- Syntactically and semantically strong models for representing biological models.
- High dimensional data of genomes can be analyzed using network-based approaches. Network analysis with correlation (with weights) is used for locating intramodular hubs as also in cluster location.
- For omics data analysis, pathway-based methods are deployed for identifying pathways which exhibit differential activity of their members whether they are genes, proteins or metabolic members See [15–20].

13.7 Big Data Tools for Bio-informatics

All the tools developed in bio-informatics area before the advent of big data analytics techniques are stand alone and were not designed for very large-scale data. In recent years, even cloud-based tools and platform such as Galaxy [21] and cloud blast [22] have been developed for a variety of problems in bio-informatics area as:

(i) ***Tools for large-scale micro-array data analysis***
(ii) ***Tools for gene-gene network analysis***
(iii) ***Tools for PPI data analysis*** [11–13]
(iv) ***Tools for sequence analysis***
(v) ***Tools for pathway analysis***.

For sequence analysis problems, several tools have been developed on top of the Hadoop MapReduce platform to perform analytics on large scale sequence data. BioPig [23] is a notable Hadoop-based tool for sequence analysis that scales automatically with the data size and can be ported directly to many Hadoop infrastructures. SeqPig [24] is another such tool. The Crossbow [25] tool combines Bowtie [26] an ultrafast and memory-efficient short-read aligner, and SoapSNP [27] an accurate genotyper, to perform large-scale whole genome sequence analytics on cloud platforms or on a local Hadoop cluster. Other cloud-based tools that have been developed for large-scale sequence analysis are Stormbow [28], CloVR [29] and Rainbow [30]. There exist other programs for large-scale sequence analysis that does not use big data technologies, such as Vmatch [31] and SeqMonk [32].

13.8 Analysis on the Readiness of Machine Learning Techniques for Bio-informatics Application [14]

In this section of this chapter, we address the issues involved in applying the techniques of Big Data Analytics to bio-informatics domain. Bio-informatics data is growing in size, very large and very fast. It has large dimensionality in addition to large size and large variety. It is spread geographically all over the globe. As an example, the well-known MapReduce Hadoop fault tolerant mechanism. Stream-based application platforms such as Apache Spark were required to be deployed. Therefore, what is the appropriate Big Data architecture for bio-informatics application domain? There is a dire need to come out with a fully satisfactory big data architecture for bio-informatics.

Conventional ML techniques like clustering and classification were originally developed for small data not only for big data. Some attempts were made subsequently in extending clustering techniques. Incremental, parallel and multiview clustering methods were also developed for bio-informatics applications.

The following characteristics are essential for applying a machine learning technique to Big Data: (a) Scalability. (b) Velocity robustness with low time complexity for processing the incoming high velocity stream data. (c) Variety: One indication of Big Data characteristics is variety of data. This means the data to be processed and analyzed could be structured data or unstructured data or semi-structured data or poly-structured data. On the other hand, the conventional machine learning algorithms could be usually capable of handling only single structure of data with one schema of data. But the new scenario demands variety as a necessity. The machine learning algorithm shall handle successfully multiple data types coming with different schemas from different sources, respectively. Further, the data so arriving could be incremental instead of conventional static data expected. It could be partial data and merging of all partial datasets may be executed. Distributed Processing:

BioPig [23] and Crossbow [25] are the few tools for performing sequence analysis on Hadoop and MapReduce platform. Bio-informatics tasks such as Protein–Protein interactions (PPI) network analysis or disease network analysis (DNA) are still to come out with readiness for Big Data scenario (See [33–40] for guidance for future developments).

13.9 Conclusion

To conclude, we can state that considering the big data boom in bio-informatics and the research opportunities that emerge out of that boom, the open and pending problems in bio-informatics research need to be addressed from the perspective of big data technologies, platforms and machine learning techniques for coming up with appropriate solutions, a task that is still open.

In this chapter, we have presented the different use cases of Big Data Analytics in the bio-informatics domain, types of data in bio-informatics, the open and pending issues for research in the application of big data analytics and machine learning techniques to bio-informatics domain, tools for analytic in bio-informatics and finally concluded highlighting the desirable characteristics of machine learning algorithm for being capable to apply to Big Data scenarios in the context of bio-informatics.

13.10 Questions and Answers

1. What are the characteristic challenges of problems of Big Data Analytics in bio-informatics area?
2. How cloud computing can play a role in bio-informatics?
3. What are the various types of data in bio-informatics area? Explain them.
4. What are the open problems for research in the content of Big Data Analytics in bio-informatics?
5. Explain open problems of Big Data Analytics micro-array analysis.
6. Explain open problems of Big Data Analytics in gene–gene network analysis.

7. Explain open problems of Big Data Analytics in (1) PPI data analytics and (2) sequence analytics.
8. Explain open problems of Big Data Analytics in evolutionary research.
9. Explain open problems of Big Data Analytics in pathway analytic and disease network analytics.
10. What are tools available for micro-array analytics?
11. What are tools available for gene–gene network analytics?
12. What are tools available for PPI data analytics?
13. What are tools available for sequence analytics?
14. Is machine learning fully ready for bio-informatics domain?
15. What are the issues and open problems for research?
16. System improvements in machine learning algorithm for handling bio-informatics.

References

1. EMBL-European Bioinformatics Institute, EMBL-EBI annual scientific report 2013 (2014)
2. V. Marx, Biology: the big challenges of big data. Nature **498**(7453), 255–260 (2013)
3. S.Y. Rojahn, Breaking the genome bottleneck. MIT Technology Review (May 2012)
4. A. Nekrutenko, J. Taylor, Next-generation sequencing data interpretation: enhancing reproducibility and accessibility. Nat. Rev. Genet. **13**(9), 667–672 (2012)
5. M. Kanehisa, S. Goto, KEGG: kyoto encyclopedia of genes and genomes. Nucleic Acids Res. **28**(1), 27–30 (2000)
6. D. Croft, G. OKelly, G. Wu, R. Haw, M. Gillespie, L. Matthews, M. Caudy, P. Garapati, G. Gopinath, B. Jassal et al., Reactome a database of reactions, pathways and biological processes. Nucleic Acids Res. **39**, D691–D697 (2010)
7. E.G. Cerami, B.E. Gross, E. Demir, I. Rodchenkov, O. Babur, N. Anwar, N. Schultz, G.D. Bader, C. Sander, Pathway commons, a web resource for biological pathway data. Nucleic Acids Res. **39**(suppl 1), D685–D690 (2011)
8. J. Mosquera, A. Sanchez-Pla, Serbgo: searching for the best go tool. Nucleic Acids Res. **36**(suppl 2), W368–W371 (2008)
9. T.H. Stokes, R.A. Moffitt, J.H. Phan, M.D. Wang, Chip artifact CORRECTion (caCORRECT): a bioinformatics system for quality assurance of genomics and proteomics array data. Ann. Biomed. Eng. **35**(6), 1068–1080 (2007)
10. J.H. Phan, A.N. Young, M.D. Wang, ominBiomarker a web-based application for knowledge-driven biomarker identification. IEEE Trans. Biomed. Eng. **60**(12), 3364–3367 (2013)
11. M. Liang, F. Zhang, G. Jin, J. Zhu, FastGCN: a GPU accelerated tool for fast gene co-expression networks. PLoS one **10**(1), e0116776 (2014)
12. D.G. McArt, P. Bankhead, P.D. Dunne, M. Salto-Tellez, P. Hamilton, S.D. Zhang, cudaMap: a GPU accelerated program for gene expression connectively mapping. BMC Bioinform. **14**(1), 305 (2013)
13. A. Day, J. Dong, V.A. Funari, B. Harry, S.P. Strom, D.H. Cohn, S.F. Nelson, Disease gene characterization through large scale co-expression analysis. PLoS One **4**(12), e8491 (2009)
14. H. Kashyap, H.A. Ahmed, N. Hoque, S. Roy, D.K. Bhattacharyya, Big data analytics in bioinformatics: a machine learning perspective
15. A. Day, M.R. Carlson, J. Dong, B.D. O'Connor, S.F. Nelson, Celsius: a community resource for Affymetrix microarray data. Genome Biol. **8**(6), R112 (2007)

16. P. Langfelder, S. Horvath, WGCNA: an R package for weighted correlation network analysis. BMC Bioinform. **9**(1), 559 (2008)
17. C.G. Rivera, R. Vakil, J.S. Bader, NeMo: network module identification in cytoscape. BMC Bioinform. **11**(Suppl 1), S61 (2010)
18. G.D. Bader, C.W. Hogue, An automated method for finding molecular complexes in large protein interaction networks. BMC Bioinform. **4**(1), 2 (2003)
19. T. Nepusz, H. Yu, A. Paccanaro, Detcting overlapping protein complexes in protein protein interaction networks. Nat. Methods **9**(5), 471–472 (2012)
20. B.P. Kelley, B. Yuan, F. Lewritter, R. Sharan, B.R. Stockwell, T. Ideker, PathBALST: a tool for alignment of protein interaction networks. Nucleic Acids Res. **32**(suppl 2), W83–W88 (2004)
21. J. Goecks, A. Nekrutenko, J. Taylor et al., Galaxy: a comprehensive approach for supporting accessible, reproducible, and transparent computational research in the life science. Genomic Biol. **11**(8), R86 (2010)
22. A. Matsunaga, M. Tsugawa, J. Fortes, Cloudblast: combining MapReduce and virtualization on distributed resources for bioinformatics applications, in *eScience'08 IEEE Fourth International Conference on IEEE*, 2008, pp. 222–229
23. H. Nordberg, K. Bhatia, K. Wang, Z. Wang, BioPig: a hadoop based analytic toolkit for large-scale sequence data. Bioinformatics **29**(23), 3014–3019 (2013)
24. A. Schumacher, L. Pireddu, M. Niemenmaa, A. Kallio, E. Kotpelainen, G. Zanetti, K. Heljanko, SeqPig: simple and scalable scripting for large sequencing data sets in hadoop. Bioinformatics **30**(1), 119–120 (2014)
25. B. Langmead, M.C. Schatz, J. Lin, M. Pop, S.L. Salzherg, Searching for SNPs with cloud computing. Genome Biol. **10**(11), R134 (2009)
26. B. Langmead, C. Trapnell, M. Pop, S.L. Salzberg et al., Ultrafast and memory-efficient alignment of short DNA sequences to the human genome. Genome Biol. **10**(3), R25 (2009)
27. R. Li, Y. Li, X. Fang, H. Yang, J. Wang, K. Kristiansen, J. Wang, SNP detection for massively parallel whole-genome resequencing. Genome Res. **19**(6), 1125–1132 (2009)
28. S. Zhao, K. Prenger, L. Smith, Strombow: a cloud-based tool for reads mapping and expression quantification in large scale RNA-Seq studies. Int. Sch. Res. Not. **2013** (2013)
29. S.V. Angiuoli, M. Matalka, A. Gussman, K. Galens, M. Vangala, D.R. Riley, C. Arze, J.R. White, O. White, W.F. Fricke, CloVR: a virtual machine for automated and portable sequence analysis from the desktop using cloud computing. BMC Bioinform. **12**(1), 356 (2011)
30. S. Zhao, K. Prenger, L. Smith, T. Messina, H. Fan, E. Jaeger, S. Stephens, Rainbow: a tool for large-scale whole-genome sequencing data analysis using cloud computing. BMC Genom. **14**(1), 425 (2013)
31. S. Kurtz, The vmatch large scale sequence analysis software. Ref Type: Computer Program, pp. 4–12 (2003)
32. www.bioinformatics .bbsrc.ac.uk
33. A.C. Zambon, S. Gaj, I. Ho, K. Hanspers, K. Vranizan, C.T. Evelo, B.R. Conklin, A.R. Pico, N. Salomonis, GO-Elite a flexible solution for pathway and ontology over representation. Bioinformatics **28**(16), 2209–2210 (2012)
34. M.P. van lersel T. Kelder, A.R. Pico, K. Hanspers, S. Coort, B.R. Conklin, C. Evelo, Presenting and exploring biological pathways with PathVisio. BMC Bioinform. **9**(1), 399 (2008)
35. P. Yang, E. Patrick, S.X. Tan, D.J. Fazakerley, J. Burchfield, C. Gribben, M.J. Prior, D.E. James, Y.H. Yang, Direction pathway analysis of large-scale proteomics data reveals novel features of the insulin action pathway. Bioinformatics **30**(6), 808–814 (2014)
36. P. Grosu, J.P. Townsend, D.L. Hartl, D. Cavalieri, Pathway processor: a tool for integrating whole-genome expression results into metabolic networks. Genome Res. **12**(7), 1121–1126 (2002)
37. Y.S. Park, M. Schmidt, E.R. Martin, M.A. Pericak-Vance, R.H. Chung, Pathway PDT: a flexible pathway analysis tool for nuclear families. BMC Bioinform. **14**(1), 267 (2013)

38. W. Luo, C. Brouwer, Pathview: an R/Bioconductor package for pathway based data integration and visualization. Bioinformatics **29**(14), 1830–1831 (2013)
39. S. Kumar, M. Nei, J. Dedley, K. Tamura, MEGA: a biologist-centric software for evolutionary analysis of DNA and protein sequences. Brief. Bioinform. **9**(4), 299–306 (2008)
40. M.S. Barker, K.M. Dlugosch, L. Dinh, R.S. Challa, N.C. Kane, M.G. King, L.H. Rieseberg, EvoPipes net: bioinformatic tools for ecological and evolutionary genomics. Evol. Bioinform. Online **6**, 143 (2010)

Chapter 14
Big Data Analytics and Recommender Systems

14.1 Introduction

Recommender systems are designed to augment human decision making. The objective of a recommender system is to suggest relevant items for a user to choose from a plethora of options. In essence, recommender systems are concerned about predicting personalized item choices for a user. Recommender systems produce a ranked list of items ordered in their order of likeability for the user. To capture the likeability, recommender system feeds on multiple patterns of data starting from user's past interaction history on the platform, his profiles of likes and dislikes, demographics and behavioral pattern of other users on the same platform. The interesting bit about recommender systems is the way all these data or parts of it are combined to predict likeability of an item for a user. The underlying dynamics of recommender systems also extend to multiple other similar applications like search engines where based on search query the engine ranks results in the order of their relevance to the query. So, understanding recommender systems provides a strong base for understanding other similar applications.

14.2 Background

In the era of Internet, for any user on popular platforms like Amazon or Netflix or Flipkart, surfing through the numerous listings on the web site can be a daunting task. Apart from the challenge of information overdose and subsequent challenges in processing so much information, the user is also constrained by the amount of time he would want to spend on finding and buying the correct item. Hence, from a user's point of view, recommender system helps them with time optimization by suggesting personalized options.

For an organization like Amazon or Netflix, a recommender system helps in increasing sales of their products by enhancing the customer's interaction on the

C. S. R. Prabhu et al., *Big Data Analytics: Systems, Algorithms, Applications*,
https://doi.org/10.1007/978-981-15-0094-7_14

portal as recommenders make the shopping experience smooth for the customer. Also, a well-designed recommender system helps in increasing discovery of niche products. The recommenders also help the organization to gain more understanding about user's behavior on the portal along with the performance of different products in different user segments.

Before we proceed further with the ideas of how a recommender system is built, let us define some of the terms that we would be using in this chapter subsequently.

Portal
Portal is the platform on which users come and interact with variety of items. Most of the times, portal tends to be an e-commerce nature. But it could be completely different nature as well like YouTube/Facebook. The core idea of the portal being that it lists a lot of items (in case of YouTube/Facebook, the items are content posted by other users) all of which cannot be experienced by a single user in finite time. So, there is a need to recommend items, personalized to the user's tastes in order to make his decision making easier on the portal.

Users
Users include all the registered/non-registered users of a portal. They could be logged in the portal (in which case the recommender system has access to his past interactions with the portal) or could be just browsing the portal. The assumption for any recommender system being the user is not biased by any external agencies, free willing and rational. Users are characterized by their demographics as well as interaction on the portal and sometimes interaction on other web portals as well.

Active User
Active user is the user for whom the recommendation is to be made. He is the user-in-question whose profile demographics, interaction history and other allied information are processed by the recommender to suggest him the list of likeable items.

Items
Item or product is the medium through which the user interacts with the portal. The items could be movies in case of Netflix, videos in case of YouTube or products for an e-commerce Web site. Items, much like users, can be characterized by details regarding the product itself or interaction of users with similar items. Also for most of the recommender systems, it could be noticed that there is an inherent grouping of items. For example, on Netflix, movies can be grouped by genres like thriller, comedy, etc. In case of Amazon, products could be grouped into utility categories like laptops, apparel, etc.

Seen and Unseen Items
All the items that a user has already interacted and by our assumption rated are called seen items for that user. The set of items which are yet to be discovered by the user is called unseen items. It is to be noted here that the job of a recommender is to suggest items from unseen items that the user is likely to rate highly.

Ratings

Ratings are the means of describing the user's likeability toward an item on the portal. The ratings could be on a discrete scale of 1–5 or 'good' or 'bad' or on a continuous scale. It is imperative to note here that ratings are just one measure of recording user's interaction with an item. There could be multiple such ways to record the interaction. For example, on Netflix apart from the user's rating for a movie—other measures which could serve as proxy for user's likeability toward a movie could be length of movie watched if it was watched in multiple sessions or single session and so on.

14.3 Overview

Recommender systems started off back in early 2000s, on a simple idea that users often mimic choices of other users, similar to them while making choices on a daily basis. For example, individuals tend to rely on film reviews while deciding on movies to watch. Similarly, users tend to depend on what other similar users on the portal have rated highly. This is the underlying essence across all recommender systems.

To make this more tangible, let us assume we have n users on the platform u_1, u_2, ..., u_n and m items i_1, i_2, ..., i_m which the portal hosts. There is a possibility interaction between users and items, and it can be assumed that each time a user interacts with item, he rates the item on a score 1–5. Then for each user, we have multiple items which the user has not interacted with also called the unseen items. In the figure, it can be understood that u_1 has not interacted with i_2, i_5, and similarly u_2 has not interacted with i_1 and so on and these cells are marked as '?'.

The job of a recommender is for each active user generating a list of say five items, ordered in the descending order of their likeability for that particular user. The underlying algorithm of all recommenders is same—for each active user the system would try to predict the rating of the unseen items. Once the rating of unseen items is predicted, they are sorted in the descending order of rating, and top n items from the list are given out as output of the recommender system.

Again, to illustrate this, let us consider the fictitious case of user u_1 who has not interacted with items i_1, i_5, i_8, i_{10}. All recommender systems would try and predict $u_1{}'$ s rating for all the four items. Let us say the four ratings be (u_1, i_1) be 2, (u_1, i_5) be 3, (u_1, i_8) be 1 and (u_1, $i1_0$) be 4, then the items sorted in the order of predicted ratings would be [i_{10}, i_5, i_1, i_8]. Then, if the recommender system needs to predict top three recommendations—they would be [i_{10}, i_5, i_1].

The recommender systems differ from each other in the way they predict ratings of an item for active user. That is exactly what we cover in the next part of the chapter.

14.3.1 Basic Approaches

The most naïve form a recommender engine is often referred to as popularity-based recommender. The popularity-based recommender, as the name suggests, recommends the most popular items on the platform. The metric to be used for judging popularity can be different across recommenders. These kinds of recommenders may seem naïve at the outset but in case of portals where no or very less user history is available for active user, popularity-based recommendations are the most commonly used. Similar kind of recommenders can be seen being deployed on any portal when the user is not logged in [1–6].

Highest Frequency of Rating
One of the most approaches to decide on popularity is the frequency of an item rated. The frequency of rating can be counted as such or calculated with exponential weighting to give more importance to the latest trend. Exponential weighting would mandate the items that are most recently rated multiple times get more weights. The items with the highest frequency of rating are the ones which are controversial or polarizing items.

Highest Rating
Highest rating of items or highest rating weighted exponentially can be used as another measure for popularity. The highest rated one historically may not be always significant but highest rated with exponential weights can indicate the most loved items in recent times.

Other Rules
Other rule-based approaches include measuring popularity by other metrics like profitability to user in monetary terms, most viewed, etc. Also, there could be customized business rules like recommending multiple items from different categories or different time horizons that could be used for constructing popularity recommenders.

Hybrid
Hybrid approaches in case of basic recommender systems include algorithms which combine some or all of the above approaches to build the final recommender systems. The hybrid approach is akin to the way ensembles of machine learning models are constructed for final prediction. Outcomes from multiple rule-based recommenders are taken and combined to dish out final recommendations.

Drawbacks with all these approaches is the lack of personalization in the recommendations. As noted earlier, these are useful systems in cases where enough data about a user's history is not available for the recommender to make personalized choices. In order to make slightly more personalized recommendations, we head on to the next section of the chapter.

14.3.2 Content-Based Recommender Systems

Content-based recommender systems use the item-related information to make inferences about user's likely rating of an item based on his past history of interaction. The content-based recommender systems can broadly be divided into two categories—unsupervised or IR-based methods and supervised or model-based methods. The underlying problem statement is same for both the approaches—given a user's past history and his ratings, predict the rating for unseen items based on how similar are unseen items as compared to seen items. All these methods leverage on item-based information to measure similarity across multiple items. The item-related information could be things like release date, cast, language, pre-bookings, etc., if the items concerned are movies. If the items are like products on an e-commerce portal, then item-related information could include features like seller, manufacturer brand, price, warranty, etc., which are specific to each product in the portal.

14.3.3 Unsupervised Approaches

Unsupervised approaches have almost similar approach as information retrieval (IR) to solving the similar items puzzle. The algorithms in this class of recommenders use multiple ways to represent item-related information of the seen item(s) as vector (v_u). The vector could simply represent the item-related information of the most highly rated item. It could be weighted average of item-related information of the top three most highly rated items with the weights being the ratings of those items. It could also be weighted combination of all item-related information of all seen movies combined using weights as ratings for each item. The final outcome here is that v_u should represent item-related information as best as possible for the most highly rated items of the active user.

Once v_u is computed, the next part of all these algorithms constitutes finding distance of v_u from each unseen item and ranking them in the order of least distance. The distance metric could be anything like cosine distance or Euclidean or Manhattan distance. The idea being the items with least distance to v_u bear most resemblance to active user's highly rated items, and hence, should be a reasonable approximation to his likeability. As discussed, there are multiple variants of this idea with different choices for representing v_u as well as different choices for measuring similarity.

14.3.4 Supervised Approaches

Supervised approaches, on the other hand, rely on machine learning models to predict the likely rating of an active user for an unseen item given the model is trained on active user's history. For any user, the history consists of item information as independent features and rating as dependent feature. The model, when trained on this data for an active user would be able to predict the rating of an unseen item which gives its information as independent features.

The supervised model, trained on a user's history would learn the prediction pattern of the user which may be a nonlinear complex model. Such nonlinear complex dependencies would not be captured in the unsupervised approaches described above. Unsupervised approach tends to calculate similarity just based on distance between items. In cases, where user's behavior tends to be a complex function of item-related information, supervised approaches could be helpful in building a better recommender system.

The major drawback with supervised approach, however, is twofold and most of the times could be expensive for most organizations. Firstly, the number of such models would increase linearly with the number of users which could be computationally inefficient. In such cases, an easy solution is bucketing of users into user groups and having models trained on collective interaction data for each bucket. Secondly, a major drawback of such supervised models when trained on individual level user history is the lack of enough data to train robust models. Again, the solution to overcome this problem is the bucketing of users into different user groups so that the collective user interaction history data is enough to train robust supervised machine learning models.

All of these approaches described above leverage on item-related information to draw inferences about similarities across items. There is an even better way to measure similarities across items by leveraging how the items were ranked by different users. This is the concept we deal at length in the next section of the chapter.

14.3.5 Collaborative Filtering

A central idea that kick started research on recommendation systems was the realization that users prefer to rely on item recommendations by other users. This central idea is referred to as collaborative filtering. As the name suggests, the recommendation is basically a filtering of all items on the portal based on collaborative suggestions from other users. Ratings from other users for unseen items can be used for a proxy for recommendation by other users.

Collaborative filtering can be implemented using two major approaches [7–12]. The first is neighborhood-based approach, and the second one is using dimensionality reduction techniques. Neighborhood-based approaches can further be divided into groups. One is user–user collaborative filtering which is finding items recommended by similar users. The other group of ideas is item–item collaborative filtering where items similar (as deemed from other user's ratings) to user's seen and highly rated items are suggested.

14.3.5.1 User–User Collaborative Filtering [10, 12]

As mentioned above, user–user collaborative filtering builds on the idea of finding similar users to active user. Active user in case of user–user collaborative filtering is defined as vector representing his ratings of all the items on the portal (u_i). The distance of u_i from all other users $u_1, u_2, \ldots, u_n$ is measured, and the users are sorted in the order of increasing distance. The distance metric could be one of distance metrics that we have mentioned earlier. The user with least distance to u_i occurs first, and user with maximum distance from u_i occurs last. Now from this sorted list of most similar users, top n most similar users are chosen and their ratings are used to approximate user's ratings for the unseen items.

So, for better understanding let us assume user u_1 is similar to u_2 and u_3 based on item ratings. Ratings of unseen items by u_i can be considered as average of the ratings for those items as given by u_2 and u_3. Instead of simple average, it could also be a weighted average with the weights being inverse of distance of u_2 and u_3 from u_1. Instead of just using two most similar users—there could be any number of users chosen based on robustness and efficiency required from the recommender system.

14.3.5.2 Item–Item Collaborative Filtering

Item–item collaborative filtering is almost exactly same as content-based recommender systems we discussed above. The only difference between the concepts is the way item–item similarity is computed. In case of content-based recommendation systems, item–item similarity is computed based on item features like price, release date, cast, etc. In case of collaborative filtering, item–item similarity is computed based on ratings of items by users. Two items would be similar if for majority of users either have rated both the items high or rated both of them low. Contrarily, two items would be dissimilar if majority of users have rated one item high and another item low. Item–item similarity as computed based on rating by different users is the only underlying concept behind item–item collaborative filtering.

Rest of the steps remains same once item–item similarity is computed. To illustrate once again, for every active user, based on his interaction history and items he has highly rated, similar but unseen items are looked up from item–item similarity matrix. The looked-up items are then suggested in the form of a recommender system.

14.3.5.3 Dimensionality Reduction Techniques

A major setback for collaborative filtering is that due to the presence of numerous items and even higher number of users—the user-item rating matrix tends to be extremely sparse. So, two items can end up being extremely similar if they were viewed by just one user who happened to rate both the items highly. Sparsity affects similarity calculations drastically.

This is where dimensionality reduction techniques come into play. The underlying concept behind these approaches encapsulate the idea that a user had liking toward an inherent grouping of items rather than linking toward specific items. This can be clearly understood in case of movies, where user would tend to like a genre of movies rather than particular movies. The grouping of items could be dictated by business rules or can be inferred from user-item rating matrix. Dimensionality reduction techniques like matrix factorizations are useful for capturing inherent grouping among items as well as users and capturing user-genre rating matrix instead of user-item rating matrix.

User-item rating matrix can be broken down using matrix factorization techniques into the following three segments as explained in the figure. The first matrix is user-genre rating matrix, the next is genre–genre distribution, and the final is item-genre constitution matrix. The item-genre constitution matrix describes the amount of contribution of each genre in a particular item.

In a simplistic setting for understanding the concept clearly, given active user's liking toward a particular genre, unseen items that constitute mostly of that particular genre can be easily looked up and recommended to him.

The most of industrial recommender systems tend to be hybrid recommenders which combine multiple recommendations from both collaborative filtering methods as well as content-based recommendation methods. In fact, there tends to be multiple different types of collaborative filtering systems as well incorporated in such hybrid recommender systems. This concludes our basic understanding of recommender systems and we proceed to explore the evaluation techniques of recommender systems.

14.4 Evaluation of Recommenders

Recommender systems are, at the end of the day, devised to increase business-related metrics like sales, viewership as well make the user experience smoother. So, the conventional techniques like A/B testing are the ultimate test of whether a recommender system is working fine or not. In conventional A/B tests, two different versions of recommender systems are used for two different sets of consumers, and whichever set of customers shows better engagement or higher conversion would indicate likely better performance of the recommender system deployed for that group.

Apart from this, there are some other techniques to evaluate recommenders in a stand-alone way which we discuss specifically in this segment.

Error Measures

Recommender systems, as we understand, concern with predicting likely rating of active user for unseen items. This group of metrics tries and measures the deviation of predicted ratings from actual ratings for hold -out validation dataset. The entire dataset of users is segregated into training and validation set. The recommender is built and fine-tuned based on data from training set. For every pair of user and item in the validation set where actual rating is available, the recommender assumes the item is unseen by the user and predicts the likely rating. The deviation of predicted rating from actual rating can be captured using different metrics like

MSE, MAE, RMSE, etc.
$\text{MAE} = \text{Sum}\ (|ri - <ri>|)/n$
$\text{MSE} = \text{Sum}\ ((ri - <ri>)^2)/n$
$\text{RMSE} = \text{Sqrt}(\text{MSE})$
Lower the error, better the recommendation system is.

Relevancy Levels
This group of performance metrics concern with the idea if the one of the items seen and rated highly by the user is forcibly changed to an unseen item, would it be picked up recommender system as one of the recommendations. The relevancy can be measured in multiple metrics like precision, recall and MAP.

To illustrate the metrics, let us consider the case where recommender system returns N items for recommendations for active user where some of the items he had seen and rated highly are turned to unseen items forcibly. Let us call the N recommendations as recommendation set and the seen and highly rated items forcibly converted to unseen as favorite set. Precision and recall are calculated as

Precision = (# items common to both recommendation and favorite set)/ (# items in recommendation set)
Recall = (# items common to both recommendation and favorite set)/ (# items in favorite set).

Higher precision implies majority of items in the recommended list are relevant for the active user. Higher recall implies majority of relevant items for the active user have been recommended to the user. The higher the precision and recall, better the recommender system overall is.

Diversity Scores
This group of metrics checks for the diversity of the items recommended to the active user. The idea here is that if recommender systems are recommending items which are similar to each other then there is not much of a value to that recommender system.

Diversity score is measured as average of dissimilarity between all pairs of items recommended to the active user. This can be easily calculated from item–item similarity matrix that was used in item–item collaborative filtering and content-based recommender systems. Higher the diversity score, better the recommender system is.

14.5 Issues

In the final section of the chapter, we discuss some of the common practical issues that are faced in the industry while building recommender systems. We have already discussed solutions to some of the issues while discussing the concept. For better understanding, let us gloss over the issues once again along with their engineered solutions.

Sparsity
In case of industrial grade collaborative filtering-based recommender system with numerous users along with numerous items as we mentioned above, sparsity is a major drawback in those cases. Sparsity can lead to incorrect conclusion about similarity of items since the number of interaction on the item could be very low. This could be overcome in two ways. First, we use dimensionality reduction techniques like matrix factorization to reduce user-item matrix to user-genre matrix. Secondly, we can use hybrid recommenders which capture item–item similarity not only based on collaborative filtering but also item features as is done for content-based recommendation systems.

Cold Start
Cold start problem refers to the issues that arise when a new user joins the portal or a new item is added to the portal. In case a new user is added, the lack of interaction history makes it incredibly tough to personalize recommendations to the user. The issue here can be mitigated using two different approaches. The first approach is to use a popularity-based recommender system to suggest him popular items on the platform. The second approach concerns around his profile and demographic user information to find similar users and then recommending items highly rated by similar users. In case of Twitter, for example, they have engineered the solution where they make every new user compulsorily follow different topics (similar to the concept of genre we have discussed).

The other cold start problem consists of issues when new item is added to the portal. In collaborative filtering, the lack of interaction on the new item would make it tough to find similar items to the new item. This kind of problems can be solved using hybrid recommenders where item–item similarity can also be computed using item-related features in case user ratings on item is not abundant to calculate similarity.

Fraud
Most of the portals which deploy recommender systems usually see a huge influx of fake users which, programmatically or otherwise, tend to rate items randomly. In case of user–user collaborative filtering, the results could be detrimental since active user could be matched to fake users and recommendation based on fake user rating would not be suitable for active user. In such cases, content-based recommender systems would not be affected at all since they are focused on finding similar items based on item features and not user ratings. So, hybrid recommenders combining

both collaborative filtering as well content-based recommenders can be an effective solution to tackle the fraud problem.

14.6 Conclusion

So, it is evident that there are multiple ways to build a functional recommender system. However, the 'best' way to build a recommender system is dependent on a lot of criteria as we have discussed throughout the entire chapter. For example, if there are lot more items than users, then it is better to choose a collaborative filtering based on users as that would be computationally efficient. In some other cases, if the platform is new, then it would imply user-item matrix would be sparsely populated in which case dimensionality reduction techniques would be useful for constructing a recommender system. Alternately, if there is a lot of reliable item-related information then item–item similarity should be computed from item-related information [11, 13–23]. Also, a lot of advanced techniques [24–32] can be used for more reliable item-user similarity matrix based on state-of-the-art deep learning methodologies [33–39].

14.7 Review Questions

(1) What is the kind of recommenders that you would use if there is a lack of data (in terms of user-item interaction) on the platform?
(2) How is YouTube recommender different from Amazon e-commerce recommender?
(3) How to tackle for cold start problems?
(4) How can you build a search engine based on your understanding of recommender systems?
(5) Can user–user similarity matrix be used for recommendation and if yes, what is the technique known as from previous description?
(6) If user identities get swapped—that is user A is now user B and user B is user A- would that affect the predictions of recommender system?

References

1. Y. Freund, R. Iyer, R.E. Schapire, Y. Singer, An efficient boosting algorithm for combining preferences. The J. Mach. Learn. Res. **4**, 933–969 (2003)
2. C. Schmitt, D. Dengler, M. Bauer, Multivariate preference models and decision making with the MAUT machine, in *Proceedings of the 9th International Conference on User Modeling (UM 2003)*, pp. 297–302

3. J.L. Herlocker, J.A. Konstan, L.G. Terveen, J.T. Riedl, Evaluating collaborative filtering recommender systems. ACM Trans. Inf. Syst. **22**(1), 5–53 (2004)
4. S. Schiaffino, A. Amandi, Intelligent user profiling, in *Artificial Intelligence An International Perspective*, ed by M. Bramer. Lecture Notes in Computer Science, vol 5640 (Springer, Berlin, Heidelberg, 2009)
5. R. Baeza-Yates, C. Hurtado, M. Mendoza, Query recommendation using query logs in search engines, in ed by W. Lindner, M. Mesiti, C. Türker, Y. Tzitzikas, A.I. Vakali, *Current Trends in Database Technology—EDBT 2004 Workshops. EDBT 2004*. Lecture Notes in Computer Science, vol 3268 (Springer, Berlin, Heidelberg, 2004)
6. P. Bellekens, G.-J. Houben, L. Aroyo, K. Schaap, A. Kaptein, User model elicitation and enrichment for context-sensitive personalization in a multiplatform TV environment, in *Proceedings of the 7th European Conference on Interactive TV and Video (EuroITV'09)*. (ACM, New York, NY, USA), pp. 119–128 (2009). https://doi.org/10.1145/1542084.1542106
7. K. Lang, Newsweeder: learning to filter netnews, in *Proceedings of the 12th International Conference on Machine Learning* (Morgan Kaufmann publishers Inc.: San Mateo, CA, USA, 1995), pp. 331–339
8. S. Aciar, D. Zhang, S. Simoff, J. Debenham, Informed recommender: basing recommendations on consumer product reviews. IEEE Intell. Syst. **22**(3), 39–47 (2007)
9. S.-H. Yang, B. Long, A.J. Smola, H. Zha, Z. Zheng, Collaborative competitive filtering: learning recommender using context of user choice (2011)
10. U. Shardanand, P. Maes, Social information filtering: algorithms for automating 'word of mouth', in *Proceedings Conference Human Factors in Computing Systems, 1995*
11. P Melville, R.J. Mooney, R. Nagarajan, Contentboosted collaborative filtering for improved recommendations. In *Proceedings of the Eighteenth National Conference on Artificial Intelligence (AAAI02)* (Edmonton, Alberta, 2002), pp. 187–192
12. R. Burke, B. Mobasher, R. Bhaumik, C. Williams. Segment-based injection attacks against collaborative filtering recommender systems, in *ICDM'05: Proceedings of the Fifth IEEE International Conference on Data Mining* (Washington, DC, USA, IEEE Computer Society, 2005), pp. 577–580
13. A.M. Rashid, I. Albert, D. Cosley, S.K. Lam, S.M. McNee, J.A. Konstan, J. Riedl, Getting to know you: learning new user preferences in recommender systems, in *Proceedings International Conference Intelligent User Interfaces, 2002*
14. M.J. Pazzani, A framework for collaborative, content-based and demographic filtering. Artif. Intell. Rev. **13**(5–6), 393–408 (1999)
15. M. Balabanovic, Y. Shoham, Fab: Content-based, collaborative recommendation. Commun. Assoc. Comput. Mach. **40**(3), 66–72 (1997)
16. J.J. Rocchio, Relevance feedback in information retrieval, in *SMART Retrieval System—Experiments in Automatic Document Processing*, ed by G. Salton, Chapter 14 (Prentice Hall, 1971)
17. G. Salton, *Automatic Text Processing* (Addison-Wesley, 1989)
18. R.J. Mooney, L. Roy, Content-based book recommending using learning for text categorization, in *Proceedings ACM SIGIR'99 Workshop Recommender Systems: Algorithms and Evaluation, 1999*
19. J.S. Breese, D. Heckerman, C. Kadie, Empirical analysis of predictive algorithms for collaborative filtering, in *Proceedings of the Fourteenth Conference on Uncertainty in Artificial Intelligence*, Madison, WI (July 1998)
20. X. Su, T.M. Khoshgoftaar, A survey of collaborative filtering techniques. Adv. Artif. Intell. **2009**, 1–20 (2009)
21. B. Sarwar, G. Karypis, J. Konstan, J. Reidl, Item-based collaborative filtering recommendation algorithms, in *WWW'01: Proceedings of the 10th International Conference on World Wide Web* (ACM, New York, NY, USA, 2001), pp. 285–295
22. G. Linden, B. Smith, J. York, Amazon.com recommendations: Item-to-item collaborative filtering. IEEE Internet Comput. **7**(1):76–80 (2003)
23. N. Srebro, T. Jaakkola, Weighted low-rank approximations, in *International Conference on Machine Learning (ICDM), 2003*

24. J. Rennie, N. Srebro, Fast maximum margin matrix factorization for collaborative prediction, in *International Conference on Machine Learning, 2005*
25. D.D. Lee, H.S. Seung, Learning the parts of objects by non-negative matrix factorization. Nature **401**(788) (1999)
26. P. Cotter, B. Smyth. PTV: intelligent personalized TV guides, in *Twelfth Conference on Innovative Applications of Artificial Intelligence, 2000*, pp. 957–964
27. M. Claypool, A. Gokhale, T. Miranda, Combining content-based and collaborative filters in an online newspaper, in *Proceedings of the SIGIR-99 Workshop on Recommender Systems: Algorithms and Evaluation, 1999*
28. X. Su, R. Greiner, T.M. Khoshgoftaar, X. Zhu, Hybrid collaborative filtering algorithms using a mixture of experts, in *Web Intelligence, 2007*, pp. 645–649
29. G. Shani, A. Gunawardana, Evaluating recommendation systems, in *Recommender Systems Handbook*, ed. by F. Ricci, L. Rokach, B. Shapira, P. Kantor (Springer, Boston, MA, 2011)
30. S. Vargas, P. Castells, Rank and relevance in novelty and diversity metrics for recommender systems, in *Proceedings of the fifth ACM conference on Recommender systems (RecSys'11)*. ACM, New York, NY, USA, 109–116 (2011). http://dx.doi.org/10.1145/2043932.2043955
31. G. Adomavicius, A. Tuzhilin, Toward the next generation of recommender systems: a survey of the state-of-the-art and possible extensions. IEEE Trans. Knowl. Data Eng. **17**(6), 734–749 (2005). https://doi.org/10.1109/TKDE.2005.99
32. S.K. Lam, J. Riedl, Shilling recommender systems for fun and profit, in *WWW'04: Proceedings of the 13th international conference on World Wide Web* (ACM, New York, NY, USA, 2004), pages 393–402
33. In *Proceedings of the 34th International ACM SIGIR Conference on Research and Development in Information Retrieval (SIGIR'11)*. (ACM, New York, NY, USA), pp. 295–304. http://dx.doi.org/10.1145/2009916.2009959
34. R. Bell, Y. Koren, C. Volinsky, Matrix factorization techniques for recommender systems. IEEE Comput. **42**(8), 30–37 (2009)
35. J. Li, O.R. Zaïane, Combining usage, content, and structure data to improve web site recommendation, in *Proceedings Fifth International Conference Electronic Commerce and Web Technologies (EC-Web'04)*, pp. 305–315 (2004)
36. O.R. Zaïane, J. Srivastava, M. Spiliopoulou, B. M. Masand (eds.), *J. Proceedings WEBKDD 2002—Mining Web Data for Discovering Usage Patterns and Profiles*, 2003
37. S.E. Middleton, N.R. Shadbolt, D.C. de Roure, Ontological user profiling in recommender systems. ACM Trans. Inf. Syst. **22**(1), 54–88 (2004)
38. W. Hill, L. Stead, M. Rosenstein, G. Furnas, Recommending and evaluating choices in a virtual community of use, in *Proceedings Conference Human Factors in Computing Systems, 1995*
39. P. Resnick, N. Iakovou, M. Sushak, P. Bergstrom, J. Riedl, GroupLens: an open architecture for collaborative filtering of Netnews, in *Proceedings 1994 Computer Supported Cooperative Work Conference, 1994*

References

24. J. Rennie, N. Srebro, Fast maximum margin matrix factorization for collaborative prediction, in *International Conference on Machine Learning*, 2005
25. D.D. Lee, H.S. Seung, Learning the parts of objects by non-negative matrix factorization. Nature 401(755) (1999)
26. D. Leung, B. Smyth, PTV: intelligent personalized TV guides, in *Twelfth Conference on Innovative Applications of Artificial Intelligence*, 2000, pp. 957–964
27. M. Claypool, A. Gokhale, T. Miranda, Combining content-based and collaborative filters in an online newspaper, in *Proceedings of the SIGIR-99 Workshop on Recommender Systems: Algorithms and Evaluation*, 1999
28. X. Liu, K. Aberer, LM Khoshgoftaar, X. Zhu, Hybrid collaborative filtering and ultimate using a mixture of experts, in Web Intelligence, 2012, pp. 645–649
29. G. Shani, A. Gunawardana, Evaluating recommendation systems, in *Recommender Systems Handbook*, ed. by F. Ricci, L. Rokach, B. Shapira, P. Kantor (Springer, Boston, MA, 2011)
30. ... of the ... 2011

Chapter 15
Security in Big Data

15.1 Introduction

In the previous chapters, we have seen how techniques of the Big Data Analytics can be applied to various application domains such as Social Semantic Web, IOT, Financial Services and Banking, Capital Market and Insurance. In all these cases, the success of such application of the techniques of Big Data Analytics will be critically dependent on security. In this chapter, we shall examine how and to what extent it is possible to insure security in Big Data.

The World Economic Forum recently dubbed data as 'the new oil.' There is a new age gold rush in which companies such as IBM, Oracle, SAS, Microsoft, SAP, EMC, HP and Dell are aggressively organizing to maximize profits from the Big Data phenomenon [1]. Since the new oil, the most valuable resource is data now and those who are in possession of the greatest amounts of data will have enormous power and influence. Thus, companies such as Face book, Google and Acxion are creating the largest datasets about human behavior, ever created in history and they can leverage this information for their own purposes, whatsoever they may be, whether for profit, surveillance or medical research.

Similar to other valuable resources, this new most valuable resource, 'data' should be adequately protected and safeguarded with due security provisions, as called for similar resources. Unfortunately, this is missing; we do not have adequate security mechanisms presently to adequately safeguard this most valuable resource. The millions or trillions of records stored say in a supermarket database or data warehouse of a supermarket are not having any serious security for protection. The database storing such data is vulnerable and can be accessed and hacked by illegal or criminal elements. This is also true for the Big Data stored in companies like Face book and Google, with possibly a better situation, but still vulnerable for access, hacking and misuse or abuse. When the 'big data' is being stored in a vulnerable manner, our ability to capture and store information is greatly outfacing our ability to understand it or its implications. Even though the cost of storing the Big Data is coming down

C. S. R. Prabhu et al., *Big Data Analytics: Systems, Algorithms, Applications*,
https://doi.org/10.1007/978-981-15-0094-7_15

drastically, the social costs are much higher, posing huge future liabilities for Society and our World.

The more data we produce and store, the more organized crime is happy to consume.

This situation is very similar to a bank which stores too much money in one place and this will be of very much greater interest to robbers and thieves who will have much easier and better opportunity to rob. Eventually, our personal details will fall (if not already fallen) into the hands of criminal cartels, the competition or even the foreign governments. Examples of this scale of leakages are: 2010 wiki-leaks debacle, (which leaked millions of classified diplomatic information), Snowden leaks of National Security Agency (NSA) (of classified security files).

15.2 Ills of Social Networking—Identity Theft

Social media provide ready-made provisions for identity theft since all the information that the criminals are looking for is readily available online: date of birth, mother's maiden name, etc. in every Face book account. Contrary to the trust of the subscribers, the criminals will have free access to all this information in the Face book account. While resetting the TOS, Face book can override privacy options given by the subscriber and make available all the information to anyone, such as advertisers and therefore to the data brokers, including criminals. With more than 600,000 Face book accounts compromised daily, anyone's account is as vulnerable as anyone else's. In fact, criminals have created specialized tools such as targeted viruses and Trojans to take over the personal data in Face book and other social media accounts, without any personal permission. In reality, about 40% of all the social networking sites have been compromised and at least 20% of the email accounts have been compromised and taken out by criminals, without anybody's permissions. A technique called 'Social Engineering' is used by criminals posing as friends or colleagues and thereby illegally exploits the trust we repose on our trusted friends and colleagues. In a single click on a masquerading, friend 'request' or 'message' will lead virus spreading across. Sensational news stories with their links being clicked can also be misleading into viruses. For example, Koob face is a targeted Face book worldwide virus [2–8].

15.3 Organizational Big Data Security

Individual organizations also may maintain their own Big Data repositories and yet they may not have made adequate arrangements from the security perspective. If the security of Big Data is breached, it would be resulting in substantial loss of credibility and its consequent effects from the provisions of law; more than whatever is the immediate damage.

In today's new era of Big Data, various companies are using the latest technology to store and analyze petabytes of data about their own company business and their own customers. As a result, the classification of information becomes even more critical. For insuring that Big Data becomes secure, techniques such as encryption, logging, honey pot detection must be necessarily implemented. Big Data can play a crucial role in detecting fraud, in banking and financial services sectors. We can also deploy techniques for analyzing patterns of data originating from multiple independent sources to identify or anomalies and possible fraud [9–12].

15.4 Security in Hadoop

In the highly popular Hadoop framework (a Java-based distributed parallel processing system), significant security vulnerabilities can be identified. Specific techniques for handling such security issues in Hadoop environment are suggested below:

1. The W3C had identified SPARQL for protecting data originating from divergent sources. A 'secured query' concept was proposed for privacy protection.
2. Jolene proposed that processing of queries may be performed in accordance with the service provider's security policy. This will insure that only those queries which are acceptable according to the security policy will be processed, while the others will be not processed, for security reasons.
3. Access control for XML documents [13] was proposed by Bertino by adopting techniques from cryptography and digital signatures [14]. Another approach proposed by IBM researchers is that query processing of queries may be performed in a secured environment, using the mechanism 'Kerberos' (of MIT). 'Kerberos' uses an encryption technology along with a trusted third party, an arbitrator, to be able to perform a secure authentication on an open network. 'Kerberos' uses cryptographic tickets to prevent transmission of plain text passwords over the network (and 'Kerberos' is based on Needhan Shouder Protocol).
4. Airavat [15] is an access control mechanism (by Roy et al.) along with privacy, which aims at preventing leakage of information beyond the security policy of the data provider [16–21].

15.5 Issues and Challenges in Big Data Security

Data security involves not only encryption of data as a primary requirement but it shall also depend upon the enforcement of security policies for access and sharing. Also, it is required to provide security for the algorithms deployed in memory management and allocation of resources.

In industry sectors as telecom, marketing, advertising, retail and financial services, Big Data security becomes crucial.

In e-governance sector also the issues of security in Big Data scenarios assume great importance. Data explosion in the Big Data scenario will make life difficult for many industries if they do not take adequate measures of security.

15.6 Encryption for Security

Since the data is present in the clusters of Hadoop environment, it is possible for the critical information stored in it to be stolen by a data thief or a hacker. Encryption of all the data stored will be insuring security. Keys used for encryption should be different for different servers and the key information may be stored centrally, under the protection of firewall.

15.7 Secure MapReduce and Log Management

Both mappers and data are required to be accessed in the presence of an entrusted mapper.

For all MapReduce jobs which may manipulate the data, we may maintain logs along with individual user ID's of those users who executed those jobs. Auditing the logs regularly helps protecting the data.

15.8 Access Control, Differential Privacy and Third-Party Authentication

It is effective to integrate differential privacy along with access control to achieve better security. The owners or providers of Big Data sources will define the security policy and control privacy violations if they take place. Thus, the users able to perform the execution of their jobs without any data leakage and S.E Linux (Security-Enhanced Linux) [22] can be deployed for prevention of data leakages.

Security policy can be specified and supported using the Linux Security Module (LSM). By modifying the Java Virtual Machine (JVM) and MapReduce framework, it is possible to enforce differential privacy. In a cloud service, the user identity pool can be stored, so that individual identities for each application will not be required to be stored.

In addition to the above, third-party authentication is also supported by cloud service provider. The third party will be trusted by both cloud service provider and the user who is accessing the data offered in the cloud service. This third-party authentication will add an additional layer of security to the cloud service. Third-party publication of data required for outsourcing of data also is for external publication purposes. The machine itself serves and plays the role of a third-party publisher when the data is stored in the cloud.

15.9 Real-Time Access Control

Operational control within a database in the cloud can be used to prevent configuration drift and/or unauthorized changes to the application. For this purpose, the parameters such as IP address, time of the day, authentication methods—all can utilize. It will also be better to keep the security administrator different from a database administrator. For protecting sensitive data, label security method can be implemented by affixing data labels or by classifying data as public, confidential or sensitive. The user will also have labels affixed to them similarly. When the user attempts to access, the user's label can be matched with data classification label and only then the access can be permitted to the user. The prediction, detection and prevention of possible attacks can be achieved by log tracking and auditing. Fine-grain auditing (such as column auditing) also is possible by deploying appropriate tools (such as those offered in DBMSs such as Oracle).

15.10 Security Best Practices for Non-relational or NoSQL Databases

Non-relational databases or NoSQL databases are not yet evolved fully with adequate security mechanisms. Robust solutions to NoSQL injunction are still not matured, as each NoSQL database is aimed at a different modeling, objective, where security was not exactly a consideration. Developers using NoSQL databases are usually dependent on security embedded in the middleware only, as NoSQL databases do not explicitly provide for support for enforcing security.

15.11 Challenges, Issues and New Approaches Endpoint Input, Validation and Filtering

Many Big Data systems acquire data from endpoint devices such as sensors and other IOT devices. How to validate the input data to create trust that the data received is not malicious and how to filter the incoming data?

Real-Time Security Compliance Monitoring
Given the large number of alerts that may be generated by security devices, real-time security monitoring is a challenge. Such alerts correlated or not may lead to

many false positives which may be ignored or 'clicked away' by humans who cannot cope up with the large numbers. This problem is going to be serious in Big Data scenario where the input data streams are large and are incoming with high velocity. Appropriate security mechanisms for data stream processing are to be evolved.

Privacy-Preserving Analytics
Big Data can be viewed as big brother, invading privacy with invasive marketing, decreased civil freedom and increased state control. Appropriate solutions are required to be developed.

15.12 Research Overview and New Approaches for Security Issues in Big Data

The security research in the context of the Big Data environment can be classified into four categories according to NIST group on Big Data security: Infrastructure security, data privacy, data management and integrity/reactive security. In the context of infrastructure security for Big Data, the Hadoop environment becomes the focus. There is a proposal for G-Hadoop, an extension of the MapReduce framework to run multiple clusters that simplifies user authentication and offer mechanisms to protect the system from traditional attacks [23]. There are also new proposals for a new scheme of [24], a secure access system [25] and encryption scheme [26]. High availability is proposed for Hadoop environment [27] wherein multiple active node names are provided at the same time. New infrastructures of storage system for improving high availability and fault tolerance are also provided [27, 28]. Alternative architectures for Hadoop file system which when combined with network coding and multimode reading enable better security [29]. By changing the infrastructure of the nodes and by the deploying certain specific new protocols, better secure group communication in large scale networks is achieved by Big Data systems.

Authentication
An identity-based sign cryption scheme for Big Data is proposed in [30].

In the context of the Big Data, the access control problem is addressed and techniques are proposed for enforcing security policies at key, value level [31] and also a mechanism of integrating all access control problem features is proposed [32].

In the context of data management, security provision can be made at collection or storage. One solution proposed [33] suggests that we can divide the data stored in Big Data system into sequenced parts and storing them in different cloud storage providers.

In the context of integrity or reactive security, the Big Data environment is characterized by its capacity to receive streams of data from different origins and with distinctive formats whether structural or unstructured. The integrity of data needs to be checked that it can be used properly. On the other hand, Big Data itself can be applied for monitoring security so as to detect whether a system is newly attacked or not.

Traditionally, integrity is defined as the maintenance of consistency, accuracy and trustworthiness of data. It protects the data from unauthorized alteration during its life cycle.

Security comprises of integrity, confidentiality and availability. While insuring integrity is critical, the management of integrity in Big Data scenario is very difficult. Proposals have been made for external integrity verification of the data [34] or a framework to insure it during a MapReduce process [35].

In the context of the possible attacks on Big Data systems by malicious users, where detection [36] can be made by provenance data related to the MapReduce process [37].

Recovery from disaster in a Big Data system also is an important problem to solve by providing adequate mechanisms for recovery.

15.13 Conclusion

In this chapter, we have identified the security vulnerabilities and threats in Big Data and also summarized the possible techniques as remedial measures.

15.14 Review Questions

1. How the Big Data scenario in the context of social networking is vulnerable and what are the security risks?
2. Is there adequate protection insured for data in Big Data?
3. Explain the problems of Identity theft in social networks.
4. Explain organizational Big Data security threads and protection mechanisms.
5. Explain social engineering thread.
6. Explain security provisions in Hadoop.
7. Explain 'Kerboros.'
8. Explain the role of encryption in Big Data security.
9. How can we deploy secure MapReduce and log management?
10. Explain access control, deferential privacy and third-party indication.

References

1. M. Goodman, *Future Crimes* (Bantam Press, 2015)
2. H.S. Rekha, C. Prakash, G. Kavitha, Understanding trust and privacy of Big Data in social networks—a brief review. In *Proceedings of the 2014 3rd International Conference on Eco-Friendly Computing and Communication Systems (ICECCS 2014)*, Bangalore, India, 18–21 December 2014, pp. 138–143
3. A. Mantelero, G. Vaciago, Social media and Big Data, in *Cyber Crime and Cyber Terrorism Investigator's Handbook* (Syngress: Boston, MA, USA, 2014), pp. 175–195
4. V. Estivill-Castro, P. Hough, M.Z. Islam, Empowering users of social networks to assess their privacy risks, in *Proceedings of the 2014 IEEE International Conference on Big Data*, Washington, DC, USA, 27–30 Oct 2014, pp. 644–649
5. H. Ren, S. Wang, H. Li, Differential privacy data aggregation optimizing method and application to data visualization, in *Proceedings of the 2014 IEEE Workshop on Electronics, Computer and Applications (IWECA 2014)*, Ottawa, ON, Canada, 8–9 May 2014, pp. 54–58
6. L. Xu, C. Jiang, Y. Chen, Y. Ren, K.J.R. Liu, Privacy or utility in data collection? A contract theoretic approach. IEEE J. Sel. Top. Signal Proc. **9**, 1256–1269 (2015)
7. A.S. Weber, Suggested legal framework for student data privacy in the age of big data and smart devices, in *Smart Digital Futures*, vol. 262 (IOS Press: Washington, DC, USA, 2014)
8. D. Thilakanathan, R. Calvo, S. Chen, S. Nepal, Secure and controlled sharing of data in distributed computing, in *Proceedings of the 16th IEEE International Conference on Computational Science and Engineering (CSE 2013)*, Sydney, Australia, 3–5 Dec 2013, pp. 825–832
9. J.B. Frank, A. Feltus, The widening Gulf between genomics data generation and consumption: a practical guide to Big Data transfer technology. Bioinf. Biol. Insights **9**(Suppl. 1), 9–19 (2015)
10. J.J. Stephen, S. Savvides, R. Seidel, P. Eugster, Program analysis for secure big data processing, in *Proceedings of the 29th ACM/IEEE International Conference on Automated Software Engineering*, Vasteras, Sweden, 15–19 Sept 2014, pp. 277–288
11. J. Chen, Q. Liang, J. Wang, Secure transmission for big data based on nested sampling and coprime sampling with spectrum efficiency. Secur. Commun. Netw. **8**, 2447–2456 (2015). [CrossRef]
12. V. Chang, Towards a Big Data system disaster recovery in a private cloud. Ad Hoc Netw. **35**, 65–82 (2015). [CrossRef]
13. E. Bertino, S. Castano, E. Ferari, M. Mesiti, Specifying and enforcying access control policies for XML documents sources 139–151 (2004)
14. E. Bertino et al., Specifying and enforcing security polices in XML document sources, 139–151. Open Circus Summit (OCS), 2012 seventh, Beijing, 19–29 June 2012). For imposing one additional trusted security layer, authentic third party distribution of XML documents was also propos [3]
15. A. Kilzer, E. Witchel, I. Roy, V. Shmatikov S.T.V. Setty, Airavat security and privacy for map reduce
16. C.-T. Yang, W.-C. Shih, L.-T. Chen, C.-T. Kuo, F.-C. Jiang, F.-Y. Leu, Accessing medical image file with co-allocation HDFS in cloud. Future Gener. Comput. Syst.**43–33**, 61–73 (2015). [CrossRef]
17. Z. Wang, D. Wang, NCluster: using multiple active name nodes to achieve high availability for HDFS, in *Proceedings of the 2013 IEEE 10th International Conference on High Performance Computing and Communications & 2013 IEEE International Conference on Embedded and Ubiquitous Computing (HPCC_EUC)*, Zhangjiajie, China, 13–15 Nov 2013, pp. 2291–2297
18. J. Kepner, V. Gadepally, P. Michaleas, N. Schear, M. Varia, A. Yerukhimovich, R.K. Cunningham, Computing on masked data: a high performance method for improving big data veracity, in *Proceedings of the 2014 IEEE High Performance Extreme Computing Conference (HPEC)*, Waltham, MA, USA, 9–11 Sept 2014, pp. 1–6
19. Z. Quan, D. Xiao, D. Wu, C. Tang, C. Rong, TSHC: trusted scheme for Hadoop cluster, in *Proceedings of the 2013 Fourth International Conference on Emerging Intelligent Data and Web Technologies (EIDWT)*, Xi'an, China, 9–11 Sept 2013, pp. 344–349

20. M. Kuzu, M.S. Islam, M. Kantarcioglu, Distributed search over encrypted Big Data, in *Proceedings of the 5th ACM Conference on Data and Application Security and Privacy*, San Antonia, TX, USA, 2–4 March 2015 pp. 271–278
21. A. Irudayasamy, L. Arockiam, Scalable multidimensional anonymization algorithm over big data using map reduce on public cloud. J. Theor. Appl. Inf. Technol. **74**, 221–231 (2015)
22. *Security Enhenced Linux*. Security Enhanced Linux N.P. Web 13 Dec 2013
23. J. Zhao, L. Wang, J. Tao, J. Chen, W. Sun, R. Ranjan, J. Kolodziej, A. Streit, D. Georgakopoulos, A security framework in G-Hadoop for big data computing across distributed Cloud data centres. J. Comput. Syst. Sci. **80**, 994–1007 (2014). [CrossRef]
24. Y.-S. Jeong, Y.-T. Kim, A token-based authentication security scheme for Hadoop distributed file system using elliptic curve cryptography. J. Comput. Virol. Hacking Tech. **11**, 137–142 (2015). (Cross Ref)
25. B.A. Kitchenham, D. Budgen, O. Pearl Brereton, Using mapping studies as the basis for further research—A participant-observer case study. Inf. Softw. Technol. **53**, 638–651 (2011). [CrossRef]
26. J.C. Cohen, S. Acharya, Towards a trusted HDFS storage platform: mitigating threats to Hadoop infrastructures using hardware-accelerated encryption with TPM-rooted key protection. J. Inf. Secur. Appl. 19, 224–244 (2014). [CrossRef]
27. M.A. Azeem, M. Sharfuddin, T. Ragunathan, Support-based replication algorithm for cloud storage systems, in *Proceedings of the 7th ACM India Computing Conference*, Nagpur, India, 9–11 Oct 2014, pp. 1–9
28. P. Meye, P. Raipin, F. Tronel, E. Anceaume, Mistore: a distributed storage system leveraging the DSL infrastructure of an ISP, in *Proceedings of the 2014 International Conference on High Performance Computing & Simulation (HPCS)*, Bologna, Italy, 21–25 July 2014, pp. 260–267
29. Y. Ma, Y. Zhou, Y. Yu, C. Peng, Z. Wang, S. Du, A novel approach for improving security and storage efficiency on HDFS. Procedia Comput. Sci. **52**, 631–635 (2015). [CrossRef]
30. G. Wei, J. Shao, Y. Xiang, P. Zhu, R. Lu, Obtain confidentiality or/and authenticity in Big Data by ID-based generalized signcryption. Inf. Sci. **318**, 111–122 (2015). [CrossRef]
31. H. Ulusoy, P. Colombo, E. Ferrari, M. Kantarcioglu, E. Pattuk, GuardMR: fine-grained security policy enforcement for MapReduce systems, in *Proceedings of the 10th ACM Symposium on Information, Computer and Communications Security*, Singapore, 14–17 April 2015, pp. 285–296
32. P. Colombo, E. Ferrari, Privacy aware access control for Big Data: a research roadmap. Big Data Res. **2**, 145–154 (2015). [CrossRef]
33. H. Cheng, C. Rong, K. Hwang, W. Wang, Y. Li, Secure big data storage and sharing scheme for cloud tenants. China Commun. **12**, 106–115 (2015). [CrossRef]
34. C. Liu, C. Yang, X. Zhang, J. Chen, External integrity verification for outsourced big data in cloud and IoT. Future Gener. Comput. Syst. **49**, 58–67 (2015). [CrossRef]
35. Y. Wang, J. Wei, M. Srivatsa, Y. Duan, W. Du, Integrity MR: integrity assurance framework for big data analytics and management applications, in *Proceedings of the 2013 IEEE International Conference on Big Data*, Silicon Valley, CA, USA, 6–9 Oct 2013, pp. 33–40
36. Z. Tan, U.T. Nagar, X. He, P. Nanda, R.P. Liu, S. Wang, J. Hu, Enhancing big data security with collaborative intrusion detection. IEEE Cloud Comput. **1**, 27–33 (2014). [CrossRef]
37. C. Liao, A. Squicciarini, Towards provenance-based anomaly detection in MapReduce, in *Proceedings of the 2015 15th IEEE/ACM International Symposium on Cluster, Cloud and Grid Computing (CCGrid)*, Shenzhen, China, 4–7 May 2015, pp. 647–656

Chapter 16
Privacy and Big Data Analytics

16.1 Introduction

While the personal information in the realm of Big Data [1] is considered the most valuable resource of the twenty-first century, the 'new oil', [2] its analytics, i.e., Big Data Analytics can be described as the 'new engine' for economic and social value creation. At the same time, the threat of loss of privacy of the personal data looms large. Enterprises or other organizations keen to harness the power of Big Data with its vast potential are also cautious and hence recognizing their responsibility to protect the privacy of personal data collected and analyzed in the Big Data framework [3].

Therefore, in any Big Data initiative, the mechanisms to govern and protect the privacy [4] in the context of risk, and thereby maintaining the adequate mechanisms to mitigate the same assume importance. The delicate balance between realizing the benefits on one side and optimizing the risk levels and resource levels on the other side is to be carefully carved out by using business frameworks such as COBIT [1].

16.2 Privacy Protection [4]

The velocity, volume and variety features of Big Data demand that the enterprises seek new ways to adequately effectively address legal, regulatory, business and operational requirements and needs. To obtain adequate measurable ROI (Return on Investment) on Big Data initiatives on the data stored in online or offline repositories, enterprises are performing analytics tasks for correlating, aggregating and statistically analyzing huge chunks of terabytes and petabytes of data in real time. Also, when the enterprises are deciding to move data into the cloud and to use cloud-based analytics services using Massively Parallel Processing (MPP) or Symmetric Multiprocessing (SMP) analytical databases, the issues of cloud security become reinforced with the issues and laws of data privacy protection. Currently, each region

C. S. R. Prabhu et al., *Big Data Analytics: Systems, Algorithms, Applications*,
https://doi.org/10.1007/978-981-15-0094-7_16

(EU or USA) is handling privacy in a different way, enterprises are forced to reconsider the methods they adopt to handle and protect the privacy of individuals and the information collected about them and how they implement cloud-based Big Data solutions. This, in turn, has impact on the software project execution and delivery. The growth of Big Data has led to distributed and disparate storages of personally identifiable health records and also online transaction such as credit card transactions. The processing of such data is liable to regulatory privacy mechanisms as Payment Card Industry Data Security Standard (PCI DSS), 1998 UK Data Protection act, US Health Insurance Portability and Accountability Act (HIPPA) COBIT offers a systematic comprehensive framework for compliance.

16.3 Enterprise Big Data Privacy Policy and COBIT 5 [1]

In any enterprise, Big Data is a given asset, and the enterprise has a responsibility to maintain privacy of data and individual as much as its commitment to derive and deliver value from its own Big Data assets. Big Data is an asset of the enterprise that will be required to fit within the domain of COBIT 5 principle, i.e., meeting the stakeholder needs COBIT 5 distinguishes clearly between corporate governance and corporate management. The corporate governance is the responsibility of the board. The needs of stakeholders of Big Data are ensured and maintained at a high level in the enterprise. The 'RACI' chart of COBIT 5 framework represents the respective responsibilities of the Board and Management with the respective roles as R (Responsible), A (Accountable), C (Consulted) and I (Informed). 'R' indicates 'taking responsibility' and takes the main operational stakes in fulfilling the activity list and creating the intended outcomes. 'A' indicates the role that is 'accountable' for the success of the task given; 'C' indicates 'Consulted' or provided impact; 'I' or 'Informed' receiving information about achievements and/or deliverable of the task. Each role (one of the R, A, C, I,) is assigned to each stakeholder. For example, Board has all others accountable ('A') to it while CEO has the responsibility ('R') to execute under consultation ('C') with CFO while giving information ('I') to the Business Process Owner and Chief Information Security Officer.

The following questions are important to be answered in the context of privacy of Big Data in the enterprises.

1. Are the sources of Big Data trust worthy?
2. What structure and skills are existing in the enterprises to govern Big Data privacy?
3. Are there the right tools to maintain Big Data privacy requirements?
4. How to ensure and maintain authenticity of data?
5. How and in what way the information will be used by whom?
6. How to improve the processes of acquiring data?
7. How to protect the sources of the data?
8. What are the insights we need to derive from Big Data in the enterprise?

9. Which are the legal and regulatory requirements containing or affecting the enterprise for which the particular data is collected?
10. Which trends are we creating which can be exploited by the rivals of the enterprise?
11. How to assure the secrecy and privacy of data and protect it from employees?

16.4 Assurance and Governance

The main steps required to drive assurance in an enterprise are:

1. Ensuring that interested parties are provided with positive substantiated opinions in governance and management of the enterprise-IT in accordance with the objectives of Assurance.
2. Defining clearly the objectives of Assurance in consistence with the Enterprise objectives so as to maximize the value of Assurance initiatives.
3. Ensuing regulatory and contractual requirement for Enterprise to provide Assurance over their own IT management.

To achieve the above objectives, the Assurance Professionals should be made the part of Big Data initiatives in the enterprise right from their inception. Such Assurance Professionals should have adequate knowledge about the business and also have expertise in using analytics tools such as R, Hadoop, Greenplum, Teradata, etc., along with the skills to interpret the data correctly to the stakeholders concerned. They shall keep abreast with the new development in tools and techniques of Big Data and also keep the management and audit teams updated on the tools to be used. The Assurance Professionals shall also ensure that Big Data privacy, and security solutions are implemented and also that adequate Big Data privacy governance exists by taking action in the following directions:

1. That data annoymization/sanitization or de-identification is ensured.
2. Adequate and relevant privacy policies and processes/procedures and supporting structures are implemented for Big Data scenarios.
3. Involve senior management and ensure their commitment to implementing the above.
4. Define and implement clear cut data destruction policies including comprehensive data management policy, data disposal, ownership and accountability.
5. Ensuring legal and regulatory compliance on Big Data and privacy assurance requirement.
6. Continuous education and training on Big Data policies, process and procedures.

Data governance is the exercise and enforcement of authority over management of data assets and performance of data functions. It comprises of pragmatic and practical steps to formalize accountability for the management of data assets: deciding who should do what and ensure that they do so; identifying assigning data stewards; a 3-d approach: De Facto, Discipline and Database.

Data stewardship is formalizing accountability for management of data resources. Data stewards need to be provided with knowledge, tools, forums and processes to become more effective and efficient in data management.

Metadata plays an important role–metadata is for management of data recorded in IT tools that improve the business and technical understanding of data and data-related processes.

Data governance requires commitment to enforcing behavioral improvements focused on managing the effectiveness and efficiency with which data will be managed. Data governance formalizes the processes for providing proactive and reactive data issue escalation and resolution. All activities of data governance should become part of the daily work process. Best practices of data governance are to be defined, and the standard processes are to be adopted for implementation. The assessment of gap between the best practices on one hand and the existing practices on the other has to be identified along with risks associated with the identified gap. Based on the gap assessment, the nature of the implementation plan for the best practices can be identified. Accordingly, an action plan can be drawn for delivering data governance program.

The best practices for implementing data governance and data stewardship include:

1. Commitment and support of the management of the organization after duel assessment and understanding by sponsoring the activities of data governance program and endorsing the efforts of the data governance team.
2. Management should ensure that the data governance team shall focus on the issues that have been identified and planned for resolution.
3. The responsibilities of data stewards should be identified and recognized.
4. The goals, scope, expectations and measurement of success of the data governance program should be well defined and communicated to all business units, functional teams, project teams and IT areas/departments of the organization for compliance.
5. It should be realized that data governance is not a single process or discipline or change of behavior.
6. Accountability towards Management about the data definition, production and usage will be the responsibility every individual identified with one or more responsibilities.
7. The data governance team will be given the responsibility to implement the planned program; it will be applicable to business data, technical data and metadata, consistently across the organization.

16.5 Conclusion

Big Data's depth represents its value and its challenge. By collecting data from different sources into a single source, Big Data promises new answers to new questions [5] hitherto never asked. But it is also fundamentally challenging regulations based on collection, identification and aggregation. At the same time, we also need to focus on transparency, expert review, efficacy, security, audit and accountability. In this chapter, we surveyed the scenario of enterprises Big Data privacy policy definition and management implementation. We have also surveyed the processes involved in enterprise data governance.

16.6 Review Questions

1. What is the threat of privacy in Big Data? How it can be instigated.
2. How privacy protection is possible in Big Data?
3. Explain enterprises Big Data privacy and COBIT.
4. Explain RACI profile.
5. Explain Assurance and governance in enterprises.
6. Explain what are the challenges and issues in privacy protects.
7. What is Trusted Scheme for Hadoop Cluster (TSHC)?
8. What is anonymization?
9. Is privacy possible in social networking scenario?

References

1. *Privacy and Big Data*, An ISACA White Paper, Aug 2013
2. D. Hirsch, The glass house effect: why Big Data is the new oil and what to do about it? *Big Data and Privacy: Making Ends Meet Digest*, pp. 44–46. ISACA White Paper, Aug 2013
3. *Big Data Impacts and Benefits*, ISACA log, March 2013
4. M. Birnhack, S-M-L-XLDATA: Big Data as a new international privacy paradigm. *Big Data and Privacy: Making Ends Meet Digest*, ISACA White Paper, Aug 2013
5. S. Freiwald, Managing the muddled mass of Big Data. *Big Data and Privacy Making Ends Meet Digest*, pp. 31–33. ISACA White Paper Aug 2013

Chapter 17
Emerging Research Trends and New Horizons

17.1 Introduction

The upcoming new horizons and recent research trends in Big Data Analytics frameworks, techniques and algorithms are as reflected in research papers recently published in conferences such as ACM International Conference on Knowledge Discovery and Data Mining (ACM SIG KDD), SIAM International Conference on Data Mining (SDM), IEEE International Conference on Data Engineering (ICDE) and ACM International Conference on Information and Knowledge Management (CIKM). In this chapter, we shall survey the research trends and the possible new horizons coming up in Big Data Analytics.

Current research areas include data mining, pattern recognition, natural language processing, business intelligence, collective intelligence, machine intelligence and web intelligence.

17.2 Data Mining

Data mining problems of interest include outlier detection, community detection, sequential pattern mining, network clustering, feature extraction, causal inference, parallel and distributed mining and predictive analytics. The problem of representing complex training data in a simple form suitable for statistical analysis permeates a host of applications, including network traffic and medical and biological data analysis, social web mining, e-Commerce, recommender systems and computational advertising.

C. S. R. Prabhu et al., *Big Data Analytics: Systems, Algorithms, Applications*,
https://doi.org/10.1007/978-981-15-0094-7_17

17.3 Data Streams, Dynamic Network Analysis and Adversarial Learning

The literature taking into account the management of uncertain data at scale focuses on the management of streaming and dynamic data, likely to be changing in time. In this chapter, we shall survey the state-of-the-art in adversarial machine learning and dynamic network analysis for Big Data samples [1–3].

Adversarial learning [4–7] accounts for changes to data by an intelligent adversary. Dynamic network analysis accounts for changes to data modeling real-world phenomena as complex networks in time. The dynamic data usually considered is either physical measurements or the World Wide Web data at different levels of granularity and complicity. With the emergence of Big Data systems in computer science, research into adversarial learning and dynamic networks borrow ideas from a wide spectrum of subjects such as mathematics, statistics, physics, computer science and social science. Thus, the research method evaluating the computational machine learning, data mining and data analytics models varies by research area. However, the common elements of Knowledge Discovery in Databases (KDD) process and algorithm design are seen in all the models. Here, the algorithmic design is concerned with reducing the computational complexity of implicit patterns found in the data. Such patterns are compressed in terms of computer programs with corresponding inputs and expected outputs. The program design is then specified in terms of approaches and improvements to existing algorithms and systems used in data analytics.

17.4 Algorithms for Big Data

The existing algorithms designed to handle very large volumes of data by indexing and search are not optimum for tasks requiring more precise and accurate analysis. In a static context, within time windows and tasks concern the study and analysis of incremental, heterogeneous data over search spaces issued from complex random processes.

17.5 Dynamic Data Streams

Dynamic data streams impose new challenges on the analytic methodologies since it is usually impossible to store an entire high-dimensional data stream or to scan it multiple times due to its tremendous volume and changing dynamics of the underlying data destruction over time.

17.6 Dynamic Network Analysis

Dynamic networks are an emerging topic of research in the area of graph mining. The edges and vertices of a graph can change over time. This is called a dynamic network which needs to be analyzed. The network can also be modeled as properties computed over one or more of weighted or un-weighted, directed or undirected; homogenous or heterogeneous networks found in the real world. Common sources and examples for dynamic networks are biological networks, Internet networks, bibliographic networks, email networks, health networks and road networks.

A variety of classical data mining techniques can be adapted for dynamic network analysis. Such techniques include classification, clustering, associations and inference under supervised and unsupervised learning categories.

Applications of dynamic network analysis can be found extensively in system diagnosis, financial markets, industrial surveillance, computer networks, space telemetry and information networks.

Research challenges in applying dynamic networks techniques include the algorithmic ability to deal with multidimensional data analysis, temporal data analysis, complex data type recognition and supervised classification learning.

17.7 Outlier Detection in Time-Evolving Networks

Outlier detection in the time-evolving networks, a special case of dynamic networks is a focus area in research. A data structure to summarize the changes in a time-evolving network is provided in [8]. In [9], an algorithm that evaluates the changes in a time-evolving network is proposed. In [8], stochastic guarantees on the performance of the data structure are presented by designing mathematical theorems on set properties and hash properties. In [9], a search algorithm is proposed which uses local optimization algorithm for searching dense blocks of tensors. The search criteria in the algorithm are defined in terms of certain statistics that are of interest in a real-world tensor. The input data may be represented as both a complex network and a complex tensor. Therefore, for dynamic network analysis, the graph search techniques based on optimization criteria become very much relevant. Both [8] and [9] deal in networks with a notion of change over time periods or ontologies or both. The accuracy of the data structure proposed in [8] depends on its ability to summarize the underlying data stream. For the definition of outlier given in terms of edge cuts in the graph, [8] provides approximate guarantees on the performance of the proposed data structure. The performance guarantees vary and depend upon sampling bias in reservoir sampling procedure and hashing procedure. The accuracy of algorithm prepared in [8] depends on the suitability of proposed statistical evaluation metrics for various complex networks. There, metrics may be readily applicable to data mining methods from tensor decompositions but not for data mining methods derived from graph search. Thus, the reliability of statistics in [9] may be guaranteed only after significant amount of data preparation.

Adopting static network analysis techniques, static sequence analysis techniques and static set analysis techniques to a dynamic network is an emerging area of research. With the increasing emergence and availability of dynamic networks in application areas such as social networks or bio informatics networks, the relevance of techniques of mining and analysis of dynamic networks is only increasing gradually.

17.8 Research Challenges

The emerging research challenging includes the generalizing of the data mining methods for various data types combining structured and unstructured data, developing online versions of data mining methods and also formulating network analysis problems in the context of various application domains. The deployment of parallel and distributed data mining algorithms suitable for large changing datasets synthesized from multiple structured data sources may be useful in addressing these above-identified research problems. Examples of such data sources include Microsoft Academic Graph Network, Internet Movie Database, Wikimedia Commons, Web Data Commons, Apache Foundation Mail Archives and Git Hub Code Repositories.

17.9 Literature Review of Research in Dynamic Networks

We will now review the existing research literature on dynamic network analysis and adversarial learning. The papers reviewed in this section have been ordered by the ideas relevant for Knowledge Discovery in Databases (KDD) with an emphasis on dynamic network analysis and adversarial learning. KDD is a research methodology and thought process recommended for developing a data mining algorithm and solution. KDD has been formalized as the specification of Cross-Industry Standard Process for Data Mining (CRISP-DM). In its simplest form, KDD has three steps—preprocessing, modeling, post-processing—that must be addressed by any data mining method.

17.10 Dynamic Network Analysis

For dynamic network analysis, the properties or dynamics of the network to be modeled need to be defined. This is possible by sampling the dynamic network to be able to represent changes in the network. The changes are statistically quantified by certain validation metrics defined on the sample. The ideas of sampling, change detection, validation metrics affect the preprocessing and post-processing phases of the KDD process. Once a suitable sample is available, we need to define the data mining model to be built on the current sample that is then validated on future samples. The elements of the data mining model include the type of the input complex network (such as a labeled network) and the data mining problem (such as evolutionary clustering and block modeling) that is being solved. The proposed solution of the data mining problem is then packaged and presented for real-world usage (through event mining, for instance). Thus, this literature review progresses along the lines of preprocessing with sampling and validation metrics, post-processing for change detection and event mining over labeled graphs that is then modeled as the data mining problems of evolutionary clustering and block modeling.

We conclude the literature review with a discussion on the existing data mining algorithms suitable for dynamic network analysis by summarizing the research gaps discussed in survey papers and books.

17.11 Sampling [8]

In [8], the authors discuss a sampling procedure to compress structural summaries underlying data stream of networks. The data used in the paper is from co-authorship networks and social media networks. The sampling is a modification of the well-known reservoir sampling procedure. In compressing the data, outlier detection is defined as identifying graph objects which contain unusual bridging edges.

This definition of outlier detection generalizes to unusual connectivity structure among different nodes found with respect to the historical connectivity structure. The algorithm maintains the information about the connected components dynamically during stream processing. The connected components are dynamically tracked by using the spanning forests of each of the edge samples.

Each outlier is assumed to satisfy a general condition on the structural criteria referred to as set monotonicity. Set monotonicity is determined as stochastic stopping criteria by examining the behavior of the edges in the individual graph objects. Set monotonicity is suitable for processing stream objects that can be examined at most once during the computation. The algorithm not only dynamically maintains node partitioning for the graph stream but also maintains a statistical model for outlier determination. Reservoir sampling is used to create data partitions that are then sorted by hashing with fixed random hash function. A current hash threshold is maintained to make decisions on whether or not incoming elements are to be included in the reservoir. The algorithm processes the edges in the reservoir in decreasing order of the hash function value until we are satisfied that the resulting reservoir is the smallest possible set which satisfies the stopping constraint. In the algorithm, current sample is the only set we need for any future decisions about reservoir sample maintenance.

The output partitions represent densely connected nodes with a small number of bridge edges. A cross-validation approach is used to model the likelihood statistics in a more robust way that avoids over fitting of the likelihood statistics to the particular structure induced by a reservoir sample. The paper presents first known real-time and dynamic method for outlier detection in graph streams with structural statistics. The problems addressed in the paper are comparable to distance-based or density-based methods for outlier detection in multidimensional data. Adaptations of such research work are suitable for Big Data structures constructed around subgraphs. A subgraph is stated to be more suitable for Big Data mining. While it may not be possible to construct a statistically robust model on the basis of edge presence or absence between individual nodes, it may be possible to create a statistically more robust model for groups of nodes which correspond to the partitions. The paper's focus is on graph partitioning and reservoir sampling by estimating moments of data stream. Both partition sizes and interaction sizes are considered for estimating moments. Thus, possible extensions to the paper are in the direction of sampling with ensemble methods and estimation of higher-order moments. Furthermore, the paper uses only co-authorship networks. The methods in paper can be extended toward various kinds of heterogeneous networks, by and dynamic networks considering semantic features also in addition to other features.

17.12 Validation Metrics [9]

The authors in [9] discuss a tensor search algorithm applied to pattern discovery applications. The tensor is constructed with an intention of modeling suspicious behavior in twitter networks. The search algorithm finds dense blocks across modes and dimensions of the tensor. The tensor is used to represent high-dimensional features in the original dataset. The outlier detection algorithm then computes dense subgraphs or labeled subsets of tensor. The output of algorithm is used for anomaly or fraud detection. The search algorithm does a greedy search of the tensor by parameterizing the tensor dimensions in terms of validation metrics defined on subgraph densities. The notion of density is defined in terms of mass and sparsity of the tensor dimensions. The subgraph density is assumed to be generated from an underlying probability model generating a random graph. Specifically, the Erdos-Renyi model is assumed to be the ideal probability density function for tensor density. Then discrete data distribution actually found in the tensor is measured against the ideal probability density function using KL divergence. The local greedy search method estimating density is an alternating maximization procedure. The procedure searches each mode of the tensor for changes while assuming the remaining modes in tensor are fixed. The dynamics of the network are measured by the output of the search method on both un-weighted and weighted graphs.

The statistics proposed in the paper are useful for feature extraction, evolutionary clustering, event mining on labeled graphs and heterogeneous networks. The statistics are also useful for parallel data mining on the cloud provided issues in parallelizing the sparse matrix representations in search algorithm are well understood. The tensor search method in the paper is comparable to tensor decomposition methods where high-order singular values represent the importance of the dense block found in the tensor. The statistical metrics of mass and size proposed in the paper are related to notion of singular values in tensor decomposition.

However, all the statistics in the paper put together generalize low-rank matrices found in tensor decompositions. The method proposed in paper is suitable for Big Data algorithms. The algorithm is suitable for dynamic, big, distributed processing of streaming data. The algorithm can be extended by constrained optimization and multiobjective optimization methods. To apply the method to a particular domain of application, we need to define the features, patterns and labels in the dataset that can act as constraints on static and streaming data. The alternating maximization procedure is comparable to coordinate descent in optimization methods. That is why, the proposed local search can be considered to be one step in the more general expectation–maximization procedure for searching tensors. To evaluate various block modeling methods with the proposed statistics, alternatives to KL divergence can be studied to measure distance between probability distributions.

17.13 Change Detection [10]

The authors of [10] propose a graph compression algorithm on a series of graphs where each graph is represented as an adjacency matrix. The algorithm takes into account both the community structures, as well as their change points in time, in order to achieve a concise description of the data. Standard benchmark graph data is used to evaluate the algorithm. The compression algorithm is derived from the Maximum Description Length (MDL) principle. Intuitively, the encoding objective tries to decompose graph into subgraphs or cliques that are either fully connected or fully disconnected. Specifically, MDL is used to define the objective function quantifying the cost of encoding an adjacency matrix as a binary matrix compression problem. The changes across compressed graphs are summarized by a cross-association method. The cross-association method incrementally summarizes tensor streams (or high-order graph streams) as smaller core tensor streams and projection matrices. The compression algorithm is a lossy compression that results in loss of information while capturing changes across graphs. The fundamental trade-off is between the number of bits needed to describe the communities and the number of bits needed to describe the individual edges in the stream. This algorithm belongs to the class of graph compression algorithms that use community discovery and optimization algorithms to compress the changes in graphs. Thus, the main challenge in the algorithm is to define objective function over a series of graphs and select corresponding objective function, search method in a parametric model. The objective function estimates the number of bits needed to encode the full graph stream (partition, community and edge) so far in space and time.

An alternating minimization method coupled with an incremental segmentation process performs the overall search over space and time. Such algorithms can be adapted into parallel, distributed algorithms over complex, dynamic networks. Such an adaptation would have to be benchmarked for the often competing objectives—accuracy, speed and scale. If the compression algorithm defines the compression objective in terms of both rows and columns of the adjacency matrix, the algorithm can be adapted for information-theoretic correlation and biclustering-driven compression. Thus, various methods for grouping communities and time segments constitute future work in the paper. A data stream management system can also aid with data retrieval in the algorithm.

17.14 Labeled Graphs [11]

In [11], the authors discuss structural anomaly detection using numeric features. A TCP/IP access control database and a ATM bank transaction database are used as the input data. The anomalous security patterns investigated by the algorithm include temporal patterns, repetitive access patterns, displacement patterns and out-of-sequence patterns. The distribution of edge weights in a graph is used to compute anomaly scores. The edge weights, in turn, are computed from a structural function defined on the numeric labels on vertex or edge in a labeled graph. An anomaly is defined with reference to infrequent substructures in the graph. The infrequent substructures are computed by matching against frequent subgraphs. The anomaly scores are computed as parametric statistics from a Gaussian mixture model. In general, the anomaly scores can be computed by various data mining models defined in the context of various graph structures. The models can be determined by tradeoffs in accuracy and efficiency of the algorithm. The proposed data mining model is useful for forensic analysis of graph transaction databases and online detection of anomalies using dynamic graphs. The proposed algorithm is extensible to dynamic graphs and Big Data. For extending the algorithm, a variety of feature extraction techniques in general and unsupervised discretization methods, in particular, can be defined for vertex and edge attributes on the graph. Both parametric and non-parametric anomaly scoring models can also be investigated as an extension to the algorithm. As an extension to the model, a distance and density-based clustering can be used to snapshot clustering quality, social stability and sociability of objects computed from the anomaly scores.

17.15 Event Mining [12]

In [12], the authors define community dynamics and event mining in a time-evolving network. Events are summarized in terms of the heaviest dynamic subgraphs computed over a time sequence. The heaviest dynamic subgraph is supposed to indicate the behavior of an edge among neighbors in a graph. A search heuristic is used to compute local and global anomaly scores for an edge. In the search heuristic, edge behavior conditioned by neighborhood is generalized to a heavy subgraph. Then heavy subgraph behavior is studied over a time sequence. The computed network can be correlated with external events for semantic event mining in time-evolving networks. By mining the full history of a large road network, a typical application of such event mining would compute a comprehensive summary of all unusual traffic congestions and their time of occurrence, and report it to the police or to urban planners.

Community dynamics and evolution in event mining is an active area of research. Naive adaptations of existing methods for dynamic network analysis are inefficient on large problem instances. So, future work would involve designing efficient ways to summarize the heaviest dynamic subgraphs. In such a subgraph, significant anomalous regions would involve very large-scale neighborhood search for both the heaviest subgraph and the maximum score subsequence. Thus, data mining issues include the need to reduce network scans and generate effective initial solutions (seeds) for the search heuristic to converge onto a globally optimum solution. Rather than a minimum spanning tree, an arbitrary shaped subgraph may also be defined to summarize the current graph in search algorithm.

17.16 Evolutionary Clustering

The authors of [13] propose an evolutionary bibliographic network summarization algorithm. The algorithm does iterative ranking and agglomerative evolutionary clustering according to an underlying probabilistic generative model. In the probabilistic generative model, an expectation–maximization estimates prior probabilities while maximum likelihood algorithm estimates posterior probabilities. A heterogeneous bibliographic information network is input to the algorithm where vertices correspond to different entities in the information network. For example, in DBLP Graphs the vertices are taken to be author, conference, paper and term. In general, the approach is suitable for star schema-based diagnosis of heterogeneous information networks. To introduce a temporal smoothness approach into the graph analysis, the core idea in proposed approach toward agglomerative evolutionary clustering is to cluster the entire entity or group of related objects as a whole, rather than clustering of individual types separately.

Such evolutionary clustering is combined with careful evolutionary diagnosis and metrics to determine merges and splits of different topical areas, authorship evolution and topical evolution. The new vector space model proposed in the algorithm can be leveraged for similarity computation and object assignment. Algorithmic tradeoffs exist between clustering quality and clustering consistency across time. The paper summarizes a sequence of graphs as a sequence of trees such that trees represent high-quality clusters. The maximum likelihood-based tree estimation model is only suitable for log-linear probabilities found in the dataset. This approach can be extended to nonlinear estimation of probabilities using probabilistic graphical models. Specifically, conditional random fields and dynamic Bayesian networks are well suited for mixing multiple probability generation mechanisms. The cluster evolution metrics may also be adapted for quantifying the appearance and disappearance rate of objects across different time granularities. Also, to be adapted for complex network analysis, the approach needs to incorporate variable number of clusters across different time periods. This would be possible by diagnosing of features in the clusters across time. Looking into the clustering methods for data streams is a direction for improving the proposed algorithm. Commonly used clustering methods include partitioning clustering, hierarchical clustering, model-driven clustering, biclustering and multilevel clustering. Another possible extension is to integrate the evolutionary clustering with research into mixed membership stochastic block models used for iterative search and optimization.

17.17 Block Modeling [14]

Block modeling methods are a generalization of community detection and graph clustering methods. Reference [14] gives a block modeling method suitable for vertex, edge, subgraph data distributions found in a temporal graph. The block modeling is done with respect to validation metrics on weighted graphs as well as un-weighted graphs. The metrics are defined as objective functions measuring data distribution. These objectives are defined over subspaces and clusters to consider both local and global optimization criteria. The solution to the optimization problem is found by tensor decomposition methods. The solution of optimization can be used to measure changes in data distribution as well as compare changes across multiple levels of data. As future work, the proposed validation metrics on dynamic networks can be integrated with static graph mining algorithms. For example, graph distance metrics can be used to sequentially compare graphs by creating a time series of network changes from adjacent periods. The time series then act as the data distribution for the proposed validation metrics. Sensitivity functions and reconstruction errors in the data distribution can also be used as optimization objectives. These various objectives can be combined using multiobjective optimization measures. The solution to multiobjective optimization can then be found in terms of metaheuristic search that converges onto a globally optimum solution. Another topic for investigation is the encoding, indexing and retrieval of spatiotemporal networks as high-dimensional time series data distributions.

17.18 Surveys on Dynamic Networks

References [1–3] provide a comprehensive survey of the most recent developments in dynamic network analysis, evolutionary network analysis and temporal data analysis. Presently, the focus is on discussing the problems of outlier detection methods over complex networks with reference to these survey papers. Reference [3] classifies anomaly detection algorithms in dynamic networks by the types of anomalies they detect—nodes, edges, subgraphs and events. For each of these methods, the available graph mining models are community-based models, compression-based models, decomposition-based models, clustering-based models, probability-based models, sequence-based models and time-based models. The above papers summarized fall into the categories of decomposition-based models, compression-based models, clustering-based models and probability-based models. Such models are an interesting extension to the emerging methods on graph behavior, change detection, event mining, outlier detection and evolution over dynamic networks. The features detecting spatiotemporal properties for machine learning over dynamic networks are commonly obtained from samples, trees, clusters, wavelets, kernels, splines, nets, filters, wrappers and factors in data series, sequences, graphs and networks. Reference [1] overviews the vast literature on graph evolution analysis by assuming the dynamic networks are composed of data streams. In data stream mining, all data mining models must be adapted to the one-pass constraint and high dimensionality of data streams. Thus, evolution analysis on data streams focuses on either modeling the change in data stream or correcting for results in data mining, or both. Direct evolution analysis is closely related to the problem of outlier detection in temporal networks because temporal outliers are often defined as abrupt change points. By this definition of outlier detection, most data mining problems such as clustering, classification, association rule mining can be generalized to network data. In the context of evolutionary graphs, additional graph-specific problems suitable for outlier detection include link prediction, preferential attachment, counting triangles, spectral methods, shortest path methods, graph matching, graph clustering, dense pattern mining and graph classification. Aggarwal and Subbian [1] also give an interesting summary of application domains for complex network analysis. These application domains include World Wide Web, Telecommunication Networks, Communication Networks, Road Networks, Recommendations, Social Network Events, Blog Evolution, Computer Systems, News Networks, Bibliographic Networks and Biological Networks. Future work in these domains includes content-centric analysis, co-evolution of the content with network structure, collective classification and domain problems to streaming networks. In this context, system structure and network topology can be studied by outlier detection methods. Gupta et al. [2] are focused on outlier detection for temporal data. Thus, the data under discussion may or may not be graph data. The temporal data collection mechanisms are classified into data streams, spatiotemporal data, distributed streams, temporal networks and time series data, generated by a multitude of applications. In this context, outlier detection has been studied on high-dimensional data, uncertain data, streaming data, network data and time series data. A common characteristic in all these problem formulations is that temporal continuity has key role in formulations of anomalous changes, sequences and patterns in the data. The paper then goes on to summarize the various outlier detection methods by four facets important to data mining and computational algorithms: Time series versus Multidimensional Data Facet, the Point versus Window Facet, the Data Type Facet and the Supervision Facet. The paper also provides an extensive section on applications and usage scenarios of outlier detection in dynamic networks. These scenarios span Environmental Sensor Data, Industrial Sensor Data, Surveillance and

Trajectory Data, Computer Networks Data, Biological Data, Astronomy Data, Web Data, Information Network Data and Economics Time Series Data. Thus, the existing survey papers on dynamic network analysis are focused on particular application domain or on a single research area. The assumptions made by various data mining models in each data mining model for dynamic network analysis are critical for design and analysis of computational algorithms.

17.19 Adversarial Learning—Secure Machine Learning [4–7, 15, 16]

Many algorithms in use today are not provably secure. Adversarial learning is concerned with making machine learning algorithms secure. Here, security is measured in terms of an objective function which also considers the reliability, diversity and significance of the algorithm. Thus, the objective of adversarial learning is to show that a specific machine learning algorithm or a data mining model or an analytics solution can be made insecure only with an impractical amount of computational work. Adversarial learning assumes that the underlying pattern in the data is non-stationary when training or testing this algorithm. In the context of such non-stationary data, the standard machine learning assumption of identically distributed data is violated.

Furthermore, it is assumed that the non-stationary data is caused by an intelligent adversary who is attacking the algorithms. In other words, the feature test data is the data generated at application time to minimize a cost function.

Thus, adversarial learning incorporates defense mechanisms into the algorithmic design and action. Thus, in literature, the analysis frameworks for adversarial learning specify the learning and attack processes in terms of the corresponding objective functions.

The attack processes also specify the attacker's constraints and optimal attack policy. The learning processes also specify the algorithm's gain and attacker's gain under the optimal attack policy.

Adversarial learning has application in areas such as spam filtering, virus detection intrusion detection, fraud detection, biometric authentication, network protocol verification, computed advertising, recommender systems, social media web mining and performance monitoring.

Improvements to the existing status of adversarial learning are concerned with dynamic features and regularized parameters computed while processing high-dimensional data for correlations. Such correlations may exist between one and more of samples, features, constraints, indicators and classes in computational algorithms. With changing spatiotemporal data, these correlations also change within the purview of conflicting goals of data mining algorithms such as accuracy, robustness, efficiency and scalability. Such conflicting properties of the algorithms are cast in the garb of a search and optimization framework finding the best (linear and nonlinear) decision boundaries between classes of interest. The objective function of the optimization algorithm may have either open or closed-form solutions. The objective of optimization can be either one or multiple objectives functions. Deep learning over time has been found to be susceptible to adversarial examples. Thus, applying the ideas of adversarial learning to make robust deep learning algorithms is an interesting area of research.

In terms of machine learning, the supervision facet of deep learning algorithms connects the data analytics output of dynamic networks and adversarial learning since single or multiple adversaries must learn about the classifier using some condition of prior knowledge, observation and experimentation formulated as cost functions or loss functions or utility functions or payoff functions.

The supervision over time facet of deep learning is also observed in related areas of research like dictionary learning, representation learning, semi-supervised learning, similarity learning, ensemble learning, online learning and transfer learning.

To discuss computational complexity of such kinds of machine learning, adversarial learning and dynamic network analysis, we can use ideas from game theory, linear algebra, causal inference, information theory and graph theory. The features (of networks, sets, sequences and series) for adversarial learning can also be obtained from complex networks, dynamic networks and knowledge graphs.

Ideas for feature extraction can be obtained from research into data mining problems like biclustering, structure similarity search, quasi cliques mining, dense subgraph discovery, multilevel graph partitioning, hypo-graph partitioning, agglomerative and divisive graph clustering.

17.20 Conclusion and Future Emerging Direction

In this chapter, we have surveyed in detail the current research trends and also identified the future emerging trends of research in Big Data Analytics. Evidently, some of the future emerging, critical research directions are in Adversarial Machine Learning in the context of dynamic networks.

17.21 Review Questions

1. What are the various data mining techniques and what are their application domains?
2. What is data streaming?
3. What is Adversarial Machine Learning?
4. What is dynamic network analysis? Explain.
5. How Adversarial Machine Learning and dynamic network analysis are related?
6. What algorithms of data mining can be adopted for dynamic network analysis?

7. What are the application domains for (a) dynamic network analysis? (b) What are the research challenges in applying them?
8. What are the various outline detection and analysis techniques in dynamic networks?
9. How states network analysis techniques, state sequence analysis techniques and state set analysis techniques are applicable to dynamic network analysis problems?
10. What is practical real-life illustration of dynamic networks?
11. What are the existing research challenges in Big Data Analytics?
12. Explain salient features of approaches in literature for research in dynamic network analysis.
13. Explain how dynamic network analysis can be made? What are the salient features?
14. Explain sampling.
15. Explain validation metrics.
16. Explain block modeling.
17. How secure machine learning can be adopted in adversarial conditions?
18. What is Adversarial Security Mechanism?

References

1. C. Aggarwal, K. Subbian, Evolutionary network analysis: a survey. ACM Comput. Surv. **47**(1):10:1–10:36 (2014)
2. M. Gupta, J. Gao, C.C. Aggarwal, J. Han, Outlier detection for temporal data: a survey. IEEE Trans. Knowl. Data Eng. **26**(9), 2250–2267 (2014)
3. S. Ranshous, S. Shen, D. Koutra, S. Harenberg, C. Faloutsos, F.N. Samatova, Anomaly detection in dynamic networks: a survey. **7**, 223–247 (2015)
4. I.J. Goodfellow, J. Shlens, C. Szegedy, Explaining and Harnessing Adversarial Examples (2014). ArXiv e-prints
5. W. Liu, S. Chawla, J. Bailey, C. Leckie, K. Ramamohanarao, AI 2012: advances in Artificial Intelligence: 25th Australasian Joint Conference, Sydney, Australia, 4–7 Dec, 2012, in *Proceedings, Chapter An Efficient Adversarial Learning Strategy for Constructing Robust Classification Boundaries* (Springer, Berlin, Heidelberg, 2012), pp. 649–660
6. N. Papernot, P. McDaniel, I. Goodfellow, S. Jha, Z. Berkay Celik, A. Swami, Practical Black-Box Attacks against Deep Learning Systems using Adversarial Examples (2016). ArXiv e-prints
7. M. Vidyadhari, K. Kiranmai, K.R. Krishniah, D.S. Babu, Security evaluation of pattern classifiers under attack. Int. J. Res. **3**(01), 1043–1048 (2016)
8. C.C. Aggarwal, Y. Zhao, P.S. Yu, Outlier detection in graph streams, in *Proceedings of the 2011 IEEE 27th International Conference on Data Engineering, ICDE'11*. IEEE Computer Society Washington, DC, USA, 2011, pp. 399–409
9. M. Jiang, A. Beutel, P. Cui, B. Hooi, S. Yang, C. Faloutsos, A general suspiciousness metric for dense blocks in multimodal data, in *2015 IEEE International Conference on Data Mining, ICDM 2015*, Atlantic City, NJ, USA, 14–17 Nov 2015, pp. 781–786
10. J. Sun, C. Faloutsos, S. Papadimitriou, P.S. Yu, Graphscope: Parameter-free mining of large time-evolving graphs, in *Proceedings of the 13th ACM SIGKDD International Conference*

on Knowledge Discovery and Data Mining, KDD'07 (New York, NY, USA. ACM, 2007), pp. 687–696
11. M. Davis, W. Liu, P. Miller, G. Redpath, Detecting anomalies in graphs with numeric labels, in *Proceedings of the 20th ACM International Conference on Information and Knowledge Management, CIKM'11* (ACM, New York, NY, USA, 2011) , pp. 1197–1202
12. M. Mongiov, P. Bogdanov, R. Ranca, E.E. Papalexakis, C. Faloutsos, A.K. Singh, *NetSpot: Spotting Significant Anomalous Regions on Dynamic Networks*, Chapter 3, pp. 28–36
13. M. Gupta, C.C. Aggarwal, J. Han, Y. Sun, Evolutionary clustering and analysis of bibliographic networks, in *2011 International Conference on Advances in Social Net-Works Analysis and Mining (ASONAM)*, pp. 63–70
14. J. Chan, N.X. Vinh, W. Liu, J. Bailey, C.A. Leckie, K. Ramamohanarao, J. Pei, Advances in Knowledge Discovery and Data Mining: 18th Pacific-Asia Conference, PAKDD 2014, Tainan, Taiwan, 13–16 May 2014, in *Proceedings, Part I, chapter Structure-Aware Distance Measures for Comparing Clusterings in Graphs* (Springer International Publishing, Cham, 2014) pp. 362–373
15. F. Wang, W. Liu, S. Chawla, On sparse feature attacks in adversarial learning, in *2014 IEEE International Conference on Data Mining*, 2014, pp. 1013–1018
16. H. Xiao, B. Biggio, B. Nelson, H. Xiao, C. Eckert, F. Roli, Support vector machines under adversarial label contamination. J. Neuro Comput., Spec. Issue Adv. Learn. Label Noise (2014 in press)

Case Studies

Case Study 1: Google

Big Data and Big Business go together. Every second endless amount of information that is being produced in the world is attempted to be utilized by the world's largest businesses.

Google is one of the most innovative organizations in the world, doing things ahead of time, which other companies will do in the future. Google produced all the core components of Big Data Analytics: MapReduce, BigQuery, Google File System (GFS). Four billion queries per day are processed by Google on a database of 20 billion web pages. Google search is the core business. In addition, Google performs many innovations, far beyond the search process itself. The 20 billion web pages are refreshed and upgraded daily. The bots of Google crawl the web, copy down what is found as new in the web and then add on to the Google Index database. Google Search spans greater and wider data than all other competitor search engines.

PageRank

PageRank included information about Websites that linked to a particular Website in the index, to help evaluate the importance of that Website globally. Previously, all other search engines worked on the principle of matching keywords in the query input with the keywords in the Website. PageRank revolutionized the search process by incorporating other elements, in addition to the keywords in the search process. Moving forward from keyword search, now semantic search is taking importance, where the 'objects' behind keywords or words or phrases in the query will be identified. This means the query will be better matched with the semantic content of the web pages.

Starting from 2007, Google launched Universal Search, which pulls in data from hundreds of data sources such as language databases, weather forecasts, historical

C. S. R. Prabhu et al., *Big Data Analytics: Systems, Algorithms, Applications*,
https://doi.org/10.1007/978-981-15-0094-7

data, geographical data, financial data, travel information, currency exchange information, sports data and also a database of mathematical functions.

In 2012, Google produced the Knowledge Graph which displays information on the subject of search from a large variety of sources directly into search results. After that, the information about the search history all the individuals who is searching/browsing, his/her Geographical information, his/her profile (from Google+) and his/her Gmail messages. It also planned to interact with user in the native local language, as a friend, and provide everything, every information that the individual may need from time to time. What will Google get in return? Google makes money from the information it gathers about the individual users, anyone who performed a Google search. Google builds up a vast amount of information base about the individuals who are using Google's services in any manner, such as Google search, Gmail or Google+ social networking. After that, Google matches up companies with potential customers through its ÄdSense algorithm. Companies pay handsomely to Google for the information about the individual. Companies will then approach the individuals through their advertisements during the browser sessions and interactive sessions of Google search or Gmail. This is how Google makes the money to become very rich, by selling data about the individuals.

In 2010, Google launched the 'BigQuery' for allowing companies to store and analyze big datasets on its cloud platform. Customer companies which use this service will pay for the storage and for computer time consumed for executing the queries on the data. Google Maps and Street View provide accurate satellite images for navigation purposes.

Google Car is a self-driven car with sensor data inputs, camera inputs, tracking devices all integrated with the real-time analysis from Google Maps and Street View, onboard, so as to enable the self-driving car drive successfully without any accidents.

Google also aims at crowd prediction technique of prediction based on multiple sources of data or Big Data.

Case Study 2: General Electric (GE)

GE is a specialist in industrial Internet applications: the M2M or IoT-based Smart Industry Internet Ecosystem. Sensors connected to all the machines in a plant or a factory enable every aspect of the industrial operations to be monitored and tracked for optimal performance and for reduced downtime.

Data pertaining to every aspect of machine operation is being monitored through IoT devices. In 2012, GE established a one billion dollar Analytics Center at California. In aviation industry, to improve fuel economy, reduce maintenance cost, optimize flight scheduling, minimize cancellation of flights, etc., GE offered intelligent operation technology, in partnership with Accenture. Etihad Airways has adopted this technology for their operations. Large amount of data was being captured in real time from every aircraft and every aspect of ground operation was being captured, targeting recovery from disruptions and returning to regular schedule.

In 2015, GE launched a Hadoop cluster-based database system on the cloud, to enable industrial customers to move to the cloud. It offers 'Predictivity Services' for real-time automated analysis. This means, order placement for new parts will be made and also minimizing of downtime will be achieved to prevent huge wastages due to unpredicted down time.

For ecological improvements, 22,000 wind tunnels across the globe are attached with the sensors which stream constantly the captured input data to the Cloud and that can be used by the operators, to be able to remotely fine-tune the pitch, the speed and direction of the blades, so as to capture as much energy from the wind as possible.

Each turbine will speak to the others around it for allowing automated responses such as adopting to mimic more efficient neighbors, pooling up resources (such as wind speed monitors) if the device fails.

Similarly, electricity meters at homes can be monitored for power consumption, so as to be able predict power cuts, using social media data and also weather data.

As above, GE has been very successful will deploying Big Data Analytics through the deployment of IoT-based smart devices.

Case Study 3: Microsoft

Founded in 1975 by Bill Gates, Microsoft has always been in the forefront and highly successful in bringing the new and disruptive technological developments into the hands and home of the masses. Microsoft remains a super brand, market leader in business and home computer operating systems, office productivity software, web browsers, games consoles and search engine (Bing is ranked no. 2 in search engine market). Microsoft Cloud offerings with Azure are highly successful. Microsoft aims to be a big player in Big Data Analytics tools by offering a suite of services and tools, including hosting and offering analytics services based on Hadoop, to business users.

Microsoft Kinect device for Xbox aims to capture more data than ever from our own living rooms. An array of sensors captured minute hand movements and is capable of monitoring heart rate of users as well as their activity levels. Future applications possible are about monitoring of television viewers and thereby provide interactive environment for content delivering through TV.

In business applications, Microsoft is aiming at offering Big Data Analytics services to the users of business enterprises.

Similar to Google AdWords, Microsoft Bing Ads provides pay per click advertising services targeting precise audience segment, identified through data collected about individual browsing habits.

Similar to Amazon and Google, Microsoft aims to provide 'Big Data in business' solution services, combining open source with proprietary software, to offer large-scale data analytics operations to all different sizes of business.

Microsoft Analytics Platform System marries Hadoop with industry-standard SQL Server DBMS, while Office 365 will make data analytics available to lay audiences, by joining with Power BI, adding basic analytics to office productivity software.

Microsoft offers IoT gateway in Azure cloud to gather data from millions of sensors in online-enabled industrial and domestic devices from manufacturing scale to personal bedroom scales.

Microsoft is bound to bring Big Data to its heart of business operations and yet provide simple solutions to the end users.

Case Study 4: Nokia

Nokia has been around for more than 150 years in various sectors of business. Today, Nokia is a leading mobile phone manufacturer, connecting 150 billion people globally. Nokia was successful in turning resources into useful products—now data is the big resource. Nokia is now embarking on the third phase of mobility which is on leveraging digital data to make it easier to navigate the physical world. Toward this goal, Nokia wanted to find the technological solution for the collection, storage and analysis of data which is virtually unlimited in data types and in volume.

For improving the user experience with their mobile phones and their location data, effective collection and use of data became a central concern of Nokia. The company leveraged on the data by data processing and by performing complete analysis in order to build maps with predictive traffic and layered elevation models for sourcing points of interest around the world, to improve the quality of service and much more.

For this purpose, Nokia deployed Teradata's enterprise data warehouse (EDW), several Oracle and MySQL data marts, visualization technologies and in its core, the Hadoop cluster. Nokia has 100 terabytes (TB) of structured data on Teradata's EDW platform and 0.5 petabytes (PB) of multistructured data on Hadoop Distributed File System (HDFS) of the Hadoop cluster. Data warehouse and the data marts continuously stream multistructured data into the multitenant Hadoop environment, allowing its 60,000 employees to access the data. Hundreds of thousands of 'Scribe' processes run each day to move data from various nodes all over the globe to the Hadoop cluster in their UK Data Center. 'Sqoop' is used to move data from HDFS to Oracle and /Terra Data. HBase is used for sewing the data from Hadoop cluster.

Benefits include highly improved and integrated user experience with integrated query retrieval and processing system without having to be bogged down with many limited individual query and information systems.

Hadoop support was enabling unprecedented scale to build 3D digital images of the globe, using which geo-spatial applications were built by Nokia.

Case Study 5: Facebook

Facebook, the largest social network in the world, contains substantial amounts of data about the personal lives of more than a billion of its subscribers, which they intend sharing with their friends and family members only. However, Facebook became one of the richest businesses in the world by selling the personal data of its subscribers to anyone who pays for it.

Of late, the concerns of privacy started to be raised and hence the need for data and privacy protection. There have also been complaints of another type: conducting unethical psychological research-effectively experimenting on its subscribers, without seeking their explicit permission to do so. Facebook has been accused of breaking many ethical guidelines when it recorded and measured the observations of its users or subscribers by showing specific parts with either positive or negative vibes. Such allegations cannot be dismissed; such risks do exist with all social media. While the benefits of social networking and social media have been immense, the risks of misuse and breach of privacy are there surely. The original and main business model of social networks like Facebook was to sell advertisements as per the individual or group interests. Advertisers benefit immensely from the detailed profiles of individuals and groups which they build up over time and they will use the features that attract them as per their own specific requirements of the individual profiles.

Facebook became the most successful among all the social networks due to its simple and effective interface. Facebook profiles of individuals are exact, identified and more accurate than other profiles such as plain profiles of Google. Therefore, Google initiated a new social networking model Google+ as an innovation. Facebook made acquisitions such as Netflix and WhatsApp which came in with their own subscriber base. Facebook countered its critics about privacy by claiming anonymization. Even this anonymization is being challenged. Facebook claims that information is recorded from subscribers along with permissions from the owner of the information. But the mechanism to seek permission is too complex. Too much battered by the targeted advertisement, subscribers wanted to delete permanently their personal information from Facebook, a feature not existing previously, but now Facebook has made provision for the same.

Facebook has a data science team which handles everything from strategy to operations. They have their own Website and often post-updates on the habits of millions who browse Facebook.

Case Study 6: Opower

Opower is an electric power and electrical energy management company that offers a cloud-based platform on Big Data and behavioral sciences to help 93 utility companies offering electric power to their customers globally. The objective is to reduce the electrical energy consumption and operation cost and also reduce the carbon dioxide emissions. These 93 utility companies offer energy services to 33 million customer households, globally. Thus, Big Data Analytics techniques deployed by Opower influence and benefit 32 million customer households, globally.

Opower gathers data of electrical energy consumption from 7 million data points using IoT-based smart motors and thermostats every day and provides analytical reports to the respective utility companies. These analytics reports are not only analyzed by the utility companies, but they are also presenting them to the customers also, so as to motivate them to conserve and save energy, in comparison with their neighbors.

Opower was previously using MySQL for database infrastructure but soon realized that MySQL queries are not capable of analyzing the data as fast as was required and also much of the data collected was not even being fully utilized. Even after deploying 60 instances of MySQL, it was not becoming possible for them to get the required results either in coverage or in speed or in analytics.

Therefore, Opower switched over to more advance and the latest technologies of Big Data. They created Energy Data Hub, by migrating their data comprising of 200 tables from MySQL infrastructure to Hadoop infrastructure and used their proprietary platform 'Datameer' for Analytics. Today, Opower is able to answer customer queries directly—without any assistance from the IT department. For example, to understand the patterns in energy consumption they can run analytics algorithms. What is the end result? Opower could dramatically reduce the time required to access the data for analytics and also empower the energy managers which could result in reducing the energy consumption by 500 million dollars and also reduce the carbon dioxide outputs by 7 billion pounds.

Case Study 7: Kaggle

Kaggle is one company (recently acquired by Google) which incorporates all the principles and aspects of Big Data Analytics—crowdsourcing, predictive modeling and gamification. This organization, based in San Francisco, offers awards and cash prizes to 'citizen scientists' who compete to untangle Big Data challenges of all shapes and sizes. Kaggle projects stretch wide, from applying crowdsourcing techniques to Data Analytics, from applications ranging from basic sciences to applied sciences and technologies, from molecular biology research on HIV to cosmology and tracing of dark matter.

Hal Varian, Chief Scientist of Google, described the operations Kaggle as: 'A way to organize the brainpower of the world's most talented data scientists and make it accessible to every organization of every size.' When every organization in the world wants to avail the Big Data analytics solutions but does not get appropriate manpower, Kaggle offers them the requisite solutions as services by crowdsourcing.

With an acute shortage of trained and skilled manpower, Kaggle offers 150,000 of them ready to farm out to the highest bidder.

By charging up to 300 dollars per hour for consultancy work for top companies including Amazon, Microsoft, Facebook and Wikipedia, Kaggle also offers competitions with gamification which is a motivation for all youngsters who take it as a challenge to participate in the game and there are no prerequisites.

After acquiring the requirement of the company through a middle man, Kaggle will package the same as a challenge and emulate it with simulated datasets. In such competitions, all the entrants will be challenged to provide an optimal solution, including the algorithm, better than the existing ones. Kaggle will offer a good reward to the winner.

In addition to such crowdsourcing challenges, Kaggle offers 'Kaggle in class' solution that offers tools and simulated challenges to academic institutions as colleges which are trying to produce future generation of data scientists.

Case Study 8: Deutsche Bank

Being well aware of the transforming potential of Big Data Analytics in banking sector, the Deutsche Bank, Germany, made significant investments in all its sectors of banking activity toward the deployment of this new technology.

Deutsche Bank has currently set up many Hadoop cluster platforms within the Bank. Originally, it was the Global Technology and Operation Division and Finance Division of the Bank which adopted Big Data technologies. After that, it became well recognized all over the bank that Big Data is able to provide benefits to all departments and businesses of the bank. Thereafter, a data laboratory (connected with innovation laboratory) was set up, cutting horizontally across the different businesses of the bank, providing for each of them the necessary internal data and the associated consultancy. For example, if any specific business division or function has a new idea which has clear business value for specific way of exploring or analyzing its data, the Data Lab will come forward to set up the infrastructure required and implement that particular idea and its requirements. It will serve as an Incubator to materialize that idea into practice. They also provide technology services and staff to work along, side by side with the businesses, in proving the data solution.

Many such ideas became materialized successfully and became full pledged projects in the Bank.

Risk Management Systems have been built on Big Data platforms. The data extracted from the Bank's systems can be spanning over P&L Risk, Market Risk and Volcker Key Performance Indicators (KPI).

The Volcker Rule requires the bank to keep the archive of a data for a long period and also be able to pull out the requisite information quickly, when required. Luckily, the Big Data technologies have enabled the Bank to store the proprietary trading data for 10 years and also ensure easy accessibility.

Another important Big Data Algorithm used by the Deutsche Bank is the matching algorithm which enables the businesses to gain data visibility on its performance. Other Big Data Analytics solutions include profiling of data to identify abnormal information through rule-based algorithms by training a machine learning algorithm. A machine learning algorithm will be trained on what is normal and what is abnormal, so that it can quickly flag errors and minimize false positives. This is applied in activity monitoring and anti-money laundering processes. In the context of reporting capabilities, the Big Data technology makes a possible to generate reports directly from the data without the need to develop any new software for this purpose.

Inside the Global Transaction Banking (GTB), Product Management and CIO Divisions, the operating model (OM) is structured around three pillars: (1) productization; (2) product management framework (PMF); (3) lean thinking. All of which drive the bank toward its primary goal of delivering client value by being product centric and market driven.

Many highlighted projects of Big Data Analytics have been completed successfully for individual GTB businesses across regions such as Cash Management, Germany, and Cash Management, USA, with a focus on product lead financial and volume of data.

All the projects in Big Data Analytics have to be executed with careful planning with transparency and systematic implementation to prevent a risk failure.

Case Study 9: Health Sector Analytics

Health sector organizations have massive amounts of patient data, insurance data and billing data, etc. Regulatory requirements make it challenging for health sector organizations to achieve compliance on integrated reporting by leveraging on their existing divergent and disparate data sources. The data landscape has been evolving from simple to complex: Volumes grew exponentially and data types have grown substantially from traditional structured data types to unstructured data types such as electronic medical record or electronic health record (EMR/EHR) data, sensor data, physician prescriptions, patient Websites. How to handle all the volume and all the varieties for the data types for achieving better health care, resource efficiency, financial reporting and regulatory compliance?

Some health organizations have been studying patient health history which is crucial for evolving more effective treatments and prescriptions. Some other health organizations have performed data mining of electronic medical record (EMR) and blended with other sources of data to improve quality of health care. While doing so, the challenges faced are poor quality of data, data stored in multiple sources which cannot be integrated and also lack of effective analytics tools.

Oklahoma University at Tulsa faced the same problems. The data was in silos and could not be analyzed to monitor the effects of treatment or to track vital statistics such as blood pressure, blood sugar and cholesterol. By deploying Big Data Analytics tools (Pentaho, in this case), they were able to gather vital statistics data and provide reporting of the same, post-treatment. They were thus able to ingest, manage and increase the vitality of these underline data sources by using an analytics tool that was easy to be used by the physicians, medical students and other staff. They could then see how the patients are responding to each treatment and accordingly come up with the subsequent treatments or other modified interventions such as the frequency of clinical visits by the patients.

In the case of Loma Linda University, more than 500,000 patients and 600 doctors faced similar challenges in data availability and reporting across all departments for deriving insights into data for achieving improved patient management.

For achieving a single version of truth, the University set up a central data warehouse and performed business analytics over it. They could integrate SAP data with their patient data in their EPIC system. By blending data from disparate systems (using tools as Pentaho), a more holistic view of the quality care metrics could be achieved. They could add further analysis steps on parameters such as visit times and patient satisfaction scores. When these metrics are blended with physician productivity scores, they could make better evaluation of performance.

St. Antanio Hospital, Netherlands wanted to integrate the data of over 500,000 patients and then create a self-service analytics culture across the hospital so that all staff in the hospital could access analytics so as to make recommendations to stream line the hospital processes and improve the patient care. They could integrate data from different silos into one and also integrate Analytics into their existing hospital systems. This could enable the hospital staff improve operational efficiency and reduce the turnaround time in the emergency room.

In addition to improving the operational efficiency, the healthcare units required to maintain compliance and regulatory requirements and third-party requirements. They also needed to have timely availability of data for insurance compliance. Data needs to be integrated so that specific data points can be mined and reported for compliance or for obtaining performance-based funding from the Government for research purposes at the universities. This requires accurate tracking of clinical metrics and aggregating data from disparate systems.

A single view of genomic sequencing data, demographic data and EMR data was also attempted to be created to provide it as a resource for scientific research. The challenges were with data accuracy, data security and the difficulties in Hadoop skills for ingesting and retrieving data. Using other tools (such as Pentaho), a leading hospital could ingest both structured and unstructured data such as EMR data, oncology data and genotype data into a Hadoop data lake. This data could be extracted by Pentaho and was made available to researchers and analysts who could then have access to secure, blended data and thereby identify the correlations.

Case Study 10: Online Insurance

A Fortune 500 insurance company wanted to out perform the competition and achieve sales in a new world of online insurance. It wanted to find out exactly what strategic steps will help it sustain and grow by out-beating the competition in an online insurance sales path. It wanted to switch over to a direct distribution operating model, to support the future growth of online sales of insurance.

The company wanted to get strategic advice from a consulting company (PWC) to help estimate the potential for selling individual life insurance through its direct channel and forecast sales for the next three to five years along with the strategic advice on the steps to be implemented to help achieve the growth forecast.

PWC, the Advisor to the Insurance Company, performed the analysis of data by sourcing third-party data from three large datasets to answer the following questions: (1) How could the new regulations of health care and the proliferation of electronic medical records (EMR) impact the online sales? (2) How much marketing effort will be required to make consumers comfortable shopping life insurance in this new online way? and (3) What kind of upcoming new technology changes will make online sales more viable?

For the purpose of analysis, the third-party large datasets utilized were macroeconomic data, consumer data, technology advancement data, all modeled for 5–10 years. The following three potential barriers to market growth were identified by analysis: (1) Insurance applications essentially require some or other kinds of medical underwriting (as medical records and samples); (2) the fact that consumers will be reluctant to share their most personal medical information online; and (3) the complexity of insurance products.

If only these barriers were largely overcome, if not completely demolished, with more patient information available electronically and therefore more easily accessible to the insurance company, with the consumers more willingly coming forward to share it online, only then the switch over to online sales could occur quicker and sooner.

Who could the most prospective sales target be? By obtaining data on who already have life insurance and of what type, who do not have, but qualify what is their net worth, what demographic categories they fall into, how much digital savvy they are, how much time they spend online, etc., it could be predicted fairly accurately and confidently who could be the target for online sale of insurance.

The technological readiness and underpinnings required as prerequisites were also identified for the company to take up online sales by developing and deploying a direct distribution system which could analyze a customer application, write a policy, confirm it, all with online speeds, as is normal in any online sales experiences.

Finally, the sales forecast for 3–5 years could be made.

Motivated by the forecast, the company actually implemented the strategy for online direct distribution of insurance policies through online sales successfully, aiming to capture its market share in insurance in the next few years.

Case Study 11: Delta Airlines

Delta Airlines, headquartered at Atlanta, Georgia, USA, is a well-known major international airline, transporting 160 million passengers annually across 64 countries. Delta has its domestic hubs located within USA and the international hub located at Paris Airport.

With continuously increasing customer base and income from its customers, Delta has deployed customer relationship management (CRM) based and dependant heavily on its Big Data Analytics infrastructure.

Delta's business model is heavily depended on its relationships with its customers. Therefore, Delta's CRM based on Big Data is very critical for its business because:

- focus on retaining high-value customers,
- maximize customer knowledge in terms of customer value and customer needs,
- develop personalized services and improve customer service efficiency and
- increase marketing efficiency.

CRM is significant because it is essential to retain customer loyalty to withstand competition.

Customer loyalty is based on customer satisfaction. Many customers stated that for them getting personalized service is essential and very important. If they have satisfaction, then the chances of customer churn are reduced.

The CRM of Delta is controlled through multiple avenues: mobile application, Website and call center, from pre-flight to post-flight, as follows:

1.	Pre-flight:	– Website
		– Call center
		– Ticket office
		– Customer mail box
		– Sales staff
2.	Departure at Airport:	– Check in/priority check in
		– Multipurpose automation
		– Lounge
		– Gate
3.	In flight:	– Outbound crew
		– In-flight entertainment system
		– Transfer desk
4.	Arrival at Airport:	– Baggage claim
		– Baggage service
		– Arrival lounge
		– Transfer desk
6.	Post-flight:	– Website
		– Call center
		– Customer mail box
		– Feedback or surveys

By meticulously following the above process efficiently, Delta can successfully create a financial, social, structural and personalized service for its passenger base.

Big Data in CRM of Delta Airlines

How does Big data help Delta Airlines in its business operations, CRM, business development and profit maximization?

Big Data as understood today comprises high volume, high variety, high velocity, high veracity information assets which have and can create high economic value, help improve the operations, improve decision-making quality and speed, improve risk management and improve customer services. As stated by IBM, Big Data is essential for the evolution of data management and information management.

Big Data Analytics has been used for increasing sales in e-Commerce companies such as Amazon or eBay by understanding customer purchase patterns and accordingly promoting products for profit maximization.

In the case of Airlines industry, Big data Analytics can be deployed for:

1. Greater value creation for customers and
2. Increased personalization.

Both the above objectives will evidently improve customer satisfaction levels and hence improved revenues through customer retention and attracting new customers.

Delta Airlines had deployed Big Data Analytics for achieving the following opportunities:

1. utilizing real-time input of information about customer preferences and needs creating value by offering targeted individualized services,
2. for more effective baggage handling purposes, more advanced tools of baggage data collection and its analysis have been deployed.

Delta staff at airports was better enabled to track baggage losses, misplacements and mishandlings so that more improved solutions can be implemented.

By integrating real-time data into its baggage system, Delta is now able to alert baggage handlers while connecting bags that need to be transferred directly to another aircraft instead of putting through the luggage sorting system of the airport.

3. Delta has made use of advanced data to better engage with customers and generate customer loyalty. The airline combines customer data generated from flight ticket purchases, routes flown, credit profiles and demographic profiles of the customers, travel habits and spending abilities and habits of the customers and even their employer company profiles. Smartly making use all this data, Delta airlines carefully prepares tailor-made promotion schemes and target customers who are identified to have more opportunity for action. The only threat possible in using Big Data is the objections by customers about infringing their data privacy.

Case Study 12: LinkedIn

LinkedIn is the world largest HR-related social network. It deploys Hadoop-based big data analytics stack which allows data scalability. Machine learning specialists draw insights and build product features from the massive amounts of data available in the LinkedIn.

The above is inclusive of interfaces, both input and output with the existing online systems and production systems and processes. This enables an easy interface for data analysis, without getting affected by the internals of distributed system functions and concerns.

This entire ecosystem is used to solve all kinds of problems, ranging from recommendations to news feed updates to e-mail digesting or distributed dashboards for internal users.

In the related work of machine learning workflows, Twitter had to develop an extension to Pig for the purpose of workflows. Facebook uses Hive for its analytics and warehousing platform but not much is known on how Twitter has put into production its machine learning applications. For input interface, ETL is deployed. Twitter's transport mechanisms use Scribe (developed originally at Facebook). Yahoo has similar log ingestion system Chukwa.

Coming to OLAP, it is well known in data warehousing and is deployed for a long time. It is still not clear how to use MapReduce for cubing in a multidimensional cube and how to perform the subsequent query processing.

The data pipeline and end-to-end pipeline for data mining applications also are not well known.

In LinkedIn, the member data and activity data generated by the online portion of the Website flow into the offline system of LinkedIn for building various derived datasets, which is then pushed back into the online serving side.

HDFS is used for sinking all the data coming in which is either (1) activity data or (2) core database snapshots.

The database change logs are periodically compacted into time-stamped snapshots for easy access. The activity data consists of streaming events generated by the services handling requests on LinkedIn.

Once the data is available into an ETL HDFS instance, it is replicated as two Hadoop instances, one for production and another for development.

Azkaban, the open-source scheduler for LinkedIn, manages these workflows. It is a general-purpose execution framework and supports diverse job types such as native MapReduce, Pig, Hive, shell scripts and others. The production workflows of LinkedIn are in Pig, even though native MapReduce is also used sometimes for performance reasons.

Ingress:

1. Data coming in as (a) database and (b) event data is loaded into Hadoop.
2. Without manual intervention, it is attempted to (i) loading data should be scalable to cover large volumes, (ii) diverse data and (iii) data schemas are evolving as functionality of this site changes rapidly.

Kafka:

For activity data, Kafka system is the infrastructure in LinkedIn. The large volume of activity data of several orders of magnitude larger than database data is to be handled.

The persistence layer is optimized, to allow for fast linear reads and written, and data is batched end-to-end to avoid small I/O operations.

Workflows:

The data stored in HDFS is processed by many of the chained MapReduce jobs to form a workflow, a directed acyclic graph of dependencies.

Workflows can be built using a variety of tools such as Hive, Pig and native MapReduce; there are three primary processing interfaces.

For supporting data interoperability, Avro is used as serialization format.

In case of Hive, partitions are created for every event data topic during the data pull, so as to enable users to run queries within partitions.

A wrapper helps restrict data in Pig and native MapReduce being read to a certain time range based upon the parameters specified.

For managing all complex and simple workflows, LinkedIn uses Azkaban, an open-source workflow scheduler. Three Azkaban instances are maintained, one corresponding to each of three Hadoop environments.

Egress:

The output results of these workflows are pushed to other systems, either back for online serving or as derived dataset for further consumption.

This is achieved by workflows adding an extra job (of one-line command for data deployment) at the end of their pipeline for data delivery out of Hadoop.

Three mechanisms are available depending upon the needs of the application. Key value: The derived data can be accessed as an associative array or their collection. Streams: Data is published as a changelog of data tuples. OLAP: Data is transferred offline to multidimensional cubes for online analytical queries to be processed.

Applications:

Most application features at LinkedIn depend on the data pipeline either explicitly (data being the product) or implicitly (derived data infusing into the application).

Many applications leverage Kafka ingress inflow and use Azkabanas their workflow and dependency manager to schedule computation at regular intervals (say every 4 h) based on a trigger.

Key value:

Key value access using Voldemort from Hadoop is the most common Egress mechanism.

Over 40 products use this mechanism and account for 70% of all Hadoop deployment activity in LinkedIn.

Case Study 13: Traffic Management

In Zhejiang, China, traffic developed too fast and too big to handle, due to its rapid economic growth and affluence of its residents going up.

The city government felt that it was necessary to monitor and manage local traffic in order to provide better transportation services to the public.

By taking a data-driven approach, the local traffic department and city police installed more than 1000 digital monitoring devices such as cameras in the checkpoints all over the city. These monitoring devices which are cameras capture images and video continuously. One terabyte of data is compiled each month. What are the objectives and requirements? The following were the objectives and requirements:

1. Centralized management of traffic data: Access was required to be provided to image and video traffic data in a centralized location—otherwise, the data was distributed in different data centers in different divisions of the city. Centralized access to traffic management facilities, equipment and application systems was required to be made available.
2. Optimized utilization of massive data: The vehicle monitoring data was to be stored for as long as possible, to provide information support for departments such as public security, criminal investigation, in addition to front-line Police department.
3. Improve traffic flow across the city: Dispatch capability was required to be enhanced for better dealing with various kinds of emergencies and also accurately forecast the traffic patterns.

Solutions Provided

1. A Unified Centralized Data Center: All individual distributed data centers located across the city were merged into one single centralized data center by deploying high-end (Intel Xeon E-5) servers and a 198-terabyte storage space for centralized storage of digital traffic information.

2. Hadoop Deployment: HDFS and HBase were deployed to provide permanent storage and seamless expansion of vehicle and traffic violation image data accumulated in the last 24 months (2 years). Hadoop was used for retaining the data in real time.
3. Data Mining and Analysis: Intel's Trust Way Open Platform was deployed for data mining and analysis.

Technical Features

- Apache Hadoop was provided as a mass data storage solution with high fault tolerance and high throughput.
- Powerful I/O processing function was provided: Intel Xeon E-5 series servers provided enhance I/O processing. A single server can enable synchronous transmission of 500 KB picture with an average speed of 250 times per second or asynchronous continuous storage of 2000 times.
- High-performance HBase database provided answers to complex queries on data in vehicle monitoring system. In less than one second, a query search can be performed for a vehicle plate number or driving record of a vehicle from a database of 2.4 billion records.

Business Value Outcomes

The following outcomes could be observed to enhance business value:

1. Improved traffic case detection ability: With 24 months' video data stored in Hadoop, the traffic police department can easily retrieve the vehicle information such as color, model and license plate in real time along with other relevant information such as historical behavior, driving routes, operating company of the vehicle and the identity of the driver.
2. Improved traffic supervision of motor vehicles by the police department: Traffic police can easily retrieve the vehicle plate numbers and the driving traits of a moving vehicle from 2.4 billion records.
3. Easy access to relevant vehicle analysis data: Complex enquiries involved in investigation of traffic cases such as data from multiple checkpoints or multiple vehicles can be answered in 10 s.

Case Study 14: Cisco

Cisco is a world leader in networking, enabling effective collaboration, by connecting people and organizations together. Cisco manages 38 global data centers, with about 90% virtualization all over.

Business intelligence can be derived by Cisco from its vast resource of structured data about its customers, products and their networking activity, all comprising of structured data. Similarly, the unstructured data in Cisco comprising of web logs, video, e-mail, documents and images is vast in quantity. Therefore, both the structured data and unstructured data together comprise the big data ecosystem in Cisco.

For processing and harnessing the Big Data, Cisco adopted the open-source Hadoop framework, which serves the purpose of an affordable super-computing platform. Hadoop enables computation at the same location where data is stored, thereby preventing disk I/O, and also provides almost linear scalability. Hadoop enabled the consolidation data spread as islands throughout the enterprise.

Before offering big data analytics solution services to the end-user business teams within the company, Cisco needed to develop an enterprise-level platform that could quote and support service-level agreements (SLAs) the availability and performance of the platform.

The requirements were scalability and enterprise-class availability and multitenant support so that multiple teams in the company could use the same platform at the same time and also integrate with the IT support processes.

Scalability was achieved by deploying linearly scalable hardware with very large online storage capacity as follows:

Cisco's Hadoop platform is comprised of Cisco UCS C240 M3 High-Performance Rack Servers 2RU server power by two Intel Xeon E%-2600 series Processors 256 GB RAM 24 TB of local storage of which 22 TB is for Hadoop system and 2 TB is for operating system.

The system architecture comprised of four racks, each containing of 16 server nodes supporting 384 TB of raw storage per rack.

This configuration could be scaled up to 160 servers in a single management domain supporting 3.8 PB of raw storage capacity.

Connectivity: Cisco UCS 6200 Series Fabric Interconnects offers high-speed and low latency connectivity for servers and central management of all connected devices with UCS Manager.

Deployment of redundant pairs offers the full redundancy performance (active–active) and scalability to very large number of nodes required in Big Data clusters. Each rack connects to fabric interconnects through a redundant pair of Cisco Nexus @2232PP Fabric Extensions.

Administration: Cisco IT Server Administrators manage all elements of Cisco UCS: servers, storage access, networking and virtualization from a single Cisco UCS Manager Interface. Even upon scaling with large clusters, the administration is easy and effective.

Cisco uses Map R distribution for Apache Hadoop which speeds up MapReduce jobs with an optimized shuffle algorithm, direct access to the disk, built-in compression and code written in advanced C++ instead of Java.

The traditional database management tools such as Oracle and Teradata continue alongside with Hadoop. Hadoop acts as a complement and helps in processing unstructured data and also very large datasets very fast at low cost.

HDFS aggregates the storage on all UCS C240 MB servers in the cluster to create one large logical l unit. After that, it splits data into chunks to accelerate processing by eliminating time-consuming ETL (extract, transform, load) operations. The failure tolerant features in-built in HDFS ensures reliability through redundancy, because Hadoop makes multiple copies of every data element, distributing them across several servers in the cluster. If any node fails, there will be no data loss. On detecting node failure, Hadoop automatically creates another copy of the data, distributing it across the remaining nodes.

Job scheduling: Cisco uses its own job scheduler TES as more user-friendly alternative to the native Apache job scheduler Oozie. The built-in TES connectors to Hadoop components eliminate manual steps such as writing Sqoop code to download data into HDFS and executing a command to load data to Hive. Thus, using TES saves a lot of time on each job, compared to Oozie because reducing the number of programming steps means reducing time for debugging. Further, the added advantage of TES is that it can operate on mobile devices, thus enabling Cisco and users to execute and manage big data jobs from anywhere, using their own mobile phones.

Big Data Analytics: The aim of Cisco in performing analytics on its Big Data is to identify the hidden opportunities for partners to sell Cisco's servers. For this purpose, previously the traditional data warehousing techniques were being deployed, to analyze the install base that identified the opportunities in the next four quarters. Such an analysis took 50 h and hence could generate reports once in a week only. Further, the data was originally spread across many data marts. The new Big Data Analytics Solution harnessed the power of Hadoop on Cisco UCS CPA within one-tenth of the time.

The data formulation included the following items of data: Those technical service contracts Cisco which are either ready for renewal or expiring in the next five quarters:

- opportunities to either activate or up-sell software subscriptions in the next five quarters,
- business rules and management interface to identify new types of opportunity data and
- partner performance measure (ratio of opportunity to bookings conversion).

Other modules of Hadoop deployed: In addition to Cisco's own TES, the following were the other modules of Hadoop ecosystem which were deployed:

Hive: SQL-like interface for analysis of large datasets.

Pig: Data flow language that enables the processing of the big data without writing of MapReduce programs.

HBase: Column database built on top of HDFS for low latency read and write operations.

Sqoop: Tool for importing and exporting data between traditional databases and HDFS.

Flume: Tool to capture and load log data to HDFS in real time.

ZooKeeper: Coordination of distributed processes so as to provide high availability.

Results: The following results were achieved by Cisco operating the Big Data Analytics program form their own UCS.

Common Platform Architecture (CPA):

1. Partner sales and service opportunities could be completed in one-tenth of the time at one-tenth of the cost.
2. Processing 1.5 billion records daily, new service opportunities could be identified in order to help achieve targets.
3. Consolidating all data of customers, partners and employees into a single version of truth.
4. Providing a customer base data opportunity for various groups in Cisco to analyze.
5. To make procuring IPR easier, Cisco is replacing static, manual tagging of Websites with dynamic tags based on user feedback, deploying the techniques of machine learning Cisco is also creating Smart Business Architecture (SBA) for offering to other companies by implementing content auto-tagging.
6. Hadoop platform analyses the logs from collaboration tools such as Cisco's Unified Communications, e-mail, Cisco TelePresence, Cisco Webex, Cisco WebEx Social, Cisco Jabber to reveal preferred communication methods and organizational dynamics. Upon entering the analysis criteria, the program creates an interactive visualization of the social network.

Future plans for Big Data Analytics in Cisco are as follows:

1. Identify root causes for technical problems and issues: The knowledge that the support engineers acquire remains hidden in case notes. Mining of case notes for root causes can unlock the value of this unstructured data, thereby reducing time of problem resolution for other customers. It can also improve other systems processes and products.
2. By analyzing the service requests, it will be possible to generate insights into product usage, reliability and customer sentiment. Better product formulations can be enabled by better understanding of the customer feedback on the existing products. By mashing up the data from various service requests, quality data and service data, it will be possible to identify future product requirements and service requirements. This calls for the deployment of big data analytics techniques such as text analytics, entity extraction, correlation, sentiment analysis, alerting and machine learning.
3. No SQL databases provide better performance and scalability than traditional RDBMSs.

4. Expanding the number of nodes to 60 in future.

Case Study 15: JPMorgan Chase

JPMorgan Chase, the largest American banking and financial services company, is a big name in finance and banking, with more than 3.5 billion user accounts (amounting to 150 PB of data in 30,000 databases).

The financial services offered by JPMorgan Chase are wide ranging: commercial banking, private banking, investment banking, asset management, security services and treasury—all these services being offered in more than 100 countries, so as to serve the needs of individuals, financial institutions, governments, as well as the corporate sector as a whole.

Thanks to Big Data Analytics, JPMorgan Chase offers very successful and effective financial services to its customers, despite the huge large datasets which contain both structured data and unstructured data. The open-source Hadoop framework and Hadoop ecosystem are the infrastructure for its Big Data ecosystem. The Big Data Analytics implementation helps JPMorgan Chase to deliver the best financial services products to the customer at the right time through the right channel with the appropriate relevance to the context.

The huge amount of Big Data about the customers arrives from millions of credit card transactions in complex unstructured format which is processed by Apache Hadoop.

Hadoop is deployed to process massive amounts of information arriving from e-mails, social media postings, phone calls and any other sources or formats of unstructured information which cannot be processed and mined by traditional means such as conventional structured databases and by conventional data mining techniques on structured data. Apache Hadoop is deployed to process the customer data that is collected from thousands of banking products and a variety of different systems. While the conventional relational databases of the company continue to be there with full functionality, Apache Hadoop is being used for a variety of analytics functions including risk management and fraud detection.

JPMorgan started off with a small Hadoop cluster and then gradually expanded so as to be able to handle customer data that is collected from thousands of divergent sources as various banking products and systems. All the data, including the unstructured social media data, is aggregated and stored in Hadoop as a common platform for the use of a range of analytics tools for mining the customer data.

The application of Big Data Analytics in JPMorgan helps the sales of foreclosed properties; helps develop new marketing strategies, managing risks; and helps in credit assessment. Massive amounts of customer data are crunched to find out patterns in financial market or in customer behaviour that can help the bank identify any risks in market or identify possible opportunities to make money.

At a higher level, JPMorgan's big data analytics provides US policy makers with all the tools they need to revamp US Economy and thus help improve public good. The insights drawn will provide an opportunity for employers, policy makers and service providers to help people mitigate and manage any type of financial instability through groundbreaking tools, possible new programs or products of the bank.

As compared with the slow-moving surveys of the government, the big data analytics of JPMorgan data keeps up with a fast economy, providing an ability to gauge the fallout from natural disasters such as hurricanes or socioeconomic developments such as the impact of raising the minimum wage.

In fraud detection, JPMorgan made strides with the help of Big Data Analytics. By integrating conventional data mining with social media posts, e-mails and even phone calls, it is able to identify possible or probable fraudulent activities, which are otherwise impossible to detect. 'Palantio' software is deployed to keep track of employee communications to identify indications of internal fraud.

Further, JPMorgan Chase provides remittance information in lockboxes to customers and clients to provide deep insights into progress of their business. Customers and clients can effectively address any customer enquiries, cash forecasting and enhance the posting of financial transactions to their receivable systems. The customers are also enabled to benchmark their own information against that of their competitors.

In the context of credit market data also, conventionally, both the customers and banks were depending on manually prepared charts and graphs for visualizing the tables of statistical information. With such slow moving techniques, it was difficult for customers to read text reports and then viewing tables with large statistical information in order to get an understanding of the market trends. JPMorgan deploys a 'Credit map' application that uses 'Datawatch' platform to leverage the techniques of Big Data Analytics to provide clear-cut insights into market trends by providing accurate and fast real-time managed analytics. It enables its users to watch, view and analyze the information about color, size and easy user interface.

JPMorgan has adopted predictive analytics techniques for achieving effective cash management. The credit limits and credit line of the customers can be predicted by knowing customer cash requirements. The working capital needs of the customers are predicted. Forecasting mechanisms are provided for customer cash requirements and also working capital so as to support customers on their working capital cycle and capital protection. Such provisions of forecasting and predictive analytics techniques for cash management and working capital management also help the banks themselves in addition to their customers. Banks are enabled to rationalize, optimize their products and forecast cash flows among customer accounts. It will also help them gain insights into any loopholes in workflow of payments to addition better customer relationship management (CRM).

Thus, at JPMorgan, the adoption of the techniques Big Data Analytics, including the techniques of predictive analytics, became a very big differentiator for making smart and targeted investments and personalized customer experience that helps the growth of business and reduction of costs, in addition to helping mitigate risk by better detection of fraud.

At a broader level, JP Morgan's big data analytics provides US policy makers with all the tools they need to revamp US economy and thus help improve public good. The insights drawn will provide an opportunity for employees, policy makers and market providers to help people mitigate and manage any type of financial instability through groundbreaking analysis, possible new programs or products of the bank. As compared with the slow-moving surveys of the government, the big data analytics of JPMC is bound to keep up with a fast-moving economy providing an ability to gauge the fallout from natural disasters such as hurricanes or socioeconomic developments such as the impact of raising the minimum wages.

In fraud detection, JPMorgan made strides with the help of Big Data Analytics [illegible] data mining with social media posts, e-mails and even [illegible] possible [illegible] detect. [illegible]

Appendices

Appendix A: Statistics

Population

Population is the collection of objects. A population may be finite or infinite based on the number of objects in the population.

Example Group of students in a college is a finite population.

Example Batch and sample.

A finite subset of population is a batch or a sample.
A finite subset of a batch is a sample.
Batch size: Number of items in a batch is called batch size.
Sample size: Number of items in a sample is called sample size.
Types of sampling: Sampling is of four types:

(i) purposive sampling,
(ii) random sampling,
(iii) simple sampling and
(iv) stratified sampling.

In purposive sampling, sample elements are selected with a definite purpose in mind.

In random sampling, each element of it has an equal chance of being included in it. A random sampling is called simple sampling if each element of the population has an equal and independent chance of being included in the sample.

A stratified sampling is selecting elements from homogenous groups of a heterogeneous population.

C. S. R. Prabhu et al., *Big Data Analytics: Systems, Algorithms, Applications*,
https://doi.org/10.1007/978-981-15-0094-7

Measures of Central Tendency

One of the most important objectives of statistical analysis is to get one single value that describes the characteristics of the entire group. Such value is called 'the central value' or an 'average' or 'an expected value' of the variable.

Finding the average is called as measures of central tendency. Few important types of averages are as follows:

(i) arithmetic mean,
(ii) median,
(iii) mode,
(iv) GM and
(v) HM.

Arithmetic Mean

The most widely used measure of central tendency is 'arithmetic mean'. Its value is obtained by adding together all the items of the sample and dividing the sum by no. of items in the sample.

Arithmetic mean is of two types: (a) simple arithmetic mean and (b) weighted arithmetic mean.

(a) Simple arithmetic mean: Let x_1, x_2, x_3, …, x_n are individual observations or items of a sample. Then, the simple arithmetic mean of a sample represented by $\acute{x}$ is

$$\bar{x} = \frac{x_1 + x_2 + x_3 + \cdots + x_n}{n}$$

$$\Rightarrow \bar{x} = \frac{1}{n}\sum_{i=1}^{n} x_i$$

where n is the total no. of items in the sample.

(b) Weighted arithmetic mean: Let x_1, x_2, …, x_p are the individual observations with frequency f_1,f_2, …,f_p, respectively. Then, the weighted arithmetic mean $\bar{x}_w$ is obtained by the formula

$$\frac{x_1 * f_1 + x_2 * f_2 + \cdots + x_p * f_p}{(f_1 + f_2 + \cdots + f_p)}$$

$$\bar{x}_w = \frac{\sum_{i=1}^{n} f_i x_i}{\sum_{i=1}^{n} f_i}$$

Median

Median is a central value of a distribution. If a distribution is having odd number of observations, then the middle value of the sorted sample is called median. If a distribution is having even number of distributions, then the average of the two middle values of the sorted sample is called median.

Example Heights of five students are given as $5'4''$, $5'6''$, $5'2''$, $5'10''$, $5'5''$.

Heights in sorted order are $5'2''$, $5'4''$, $5'5''$, $5'6''$, $5'10''$.
The middle value of the above sample, i.e., $(5 + 1)/2 = 3$rd value, is the median.
The third value of the sample $= 5'5''$.
Therefore, the median of the given sample $= 5'5''$.

Example If heights of six students are given as $5'$, $4'$, $6'$, $4'$, $5'5''$, $4'6''$.

Sample elements in the sorted order $= 4'$, $4'$, $4'6''$, $5'$, $5'5''$, $6'$
Total number of elements $= 6$.
Then, the average of (6/2)th and $(6/2) + 1$th elements is called median.
$(6/2) = 3$ and $(6/2) + 1 = 4$.
Therefore, third and fourth elements of the sorted sample are $4'6''$ and $5'$.
Average of the two elements is $\frac{4'6''+5'}{2} = \frac{9'6''}{2} = 4'9''$.
Therefore, the median of the given sample is $4'9''$.

Mode

The mode or the model value is that value in a series of observations which occurs with the greatest frequency.

Example Mode of the sample 14, 12, 11, 13, 14, 12, 14, 13 is 14. Since this number (14) has highest frequency than any other number in the sample.

Number	Frequency
11	1
12	2
13	2
14	3 ← Mode

Geometric Mean

Geometric mean is defined as the nth root of the product of n items or values. Let $x_1, x_2, \ldots, x_n$ are the n observations of a sample, and then, the geometric mean of the given sample is

$$\mathrm{GM} = \sqrt[n]{x_1.x_2 \ldots x_n}$$

Example The geometric mean of 4, 8, 16 is

$$\mathrm{GM} = \sqrt[3]{4 * 8 * 16} = \sqrt[3]{_2 9} = 2^3 = 8.$$

Harmonic Mean

It is defined as the reciprocal of the arithmetic mean of the reciprocal of the individual observations. Thus by definition, if there are n elements in a sample, then

$$\mathrm{HM} = \frac{N}{1/x_1 + 1/x_2 + \cdots + 1/x_n}$$
$$\Rightarrow \mathrm{HM} = \frac{N}{\sum(1/x)}$$

Measures of Dispersion

Measures of dispersion give us one single figure that represents the entire data. But the average alone cannot adequately describe a set of observations, unless all the observations are the same. It is necessary to describe the variability or dispersion of the observation. There are different methods of studying variation. Important methods are as follows:

i. range,
ii. interquartile range and quartile deviation,
iii. mean deviation/average deviation and
iv. the standard deviation.

Range

This is the simplest method of studying dispersion. It is defined as the difference between the value of the smallest item and the value of the largest item included in the distribution.

$$\text{Range} = L - S,$$

where L is largest item and S is the smallest item.

Coefficient of Range

The relative measure corresponding to range, called the coefficient of range, is obtained by applying the following formula.

$$\text{Coefficient of Range} = \frac{L - S}{L + S}$$

Example Gold rates in seven days are given as 2500, 2450, 2460, 2600, 2550, 2490, 2440.

Solution

$$\text{Range} = L - S = 2600 - 2440 = 160,$$

where the largest value **L** = 2600 and the smallest value, **S** = 2440

$$\text{Coefficient of Range} = \frac{L - S}{L + S} = 160/5040 = 0.032$$

The Interquartile Range of the Quartile Deviation

Range has certain limitations as it is measured on two extreme values. For this reason, another measure called interquartile range developed. Interquartile range represents the difference between the third quartile and first quartile.

Interquartile range = $Q_3 - Q_1$
Semi-interquartile range/quartile deviation = $\frac{Q_3 - Q_1}{2}$
Coefficient of quartile deviation = $\frac{Q_3 - Q_1}{Q_3 + 1}$

Example Gold rates in seven days are given as 2500, 2450, 2460, 2600, 2550, 2490, 2440.

Solution Rates are arranged in ascending order: 2440, 2450, 2550, 2490, 2500, 2550, 2600.

First quartile Q_1 = Size of $\frac{N+1}{M}$ item = $\frac{7+1}{M}$ = 2nd item.
Size of second item is 2450; thus, $Q_1 = 2450$.

Third quartile Q_3 = Size of $\left(\frac{N+1}{M}\right) * 3$th item = Size of $\left(\frac{7+1}{M}\right) * 3$ = Size of sixth item.
Size of sixth item is 2550; thus, $Q_3 = 2550$.

Interquartile range $= Q_3 - Q_1 = 2550 - 2450 = 100$.
Quartile deviation $= (Q_3 - Q_1)/2 = (2550 - 2450)/2 = 50$.
Coefficient of quartile deviation $= (Q_3 - Q_1)/(Q_3 + Q_1) = 100/5000 = 0.02$.

The Mean Deviation or Average Deviation

The Mean Deviation is the average deviation or difference between the items is a distribution and the mean of that series.

If $x_1, x_2, x_3, \ldots, x_n$ are the N given observations, then the deviation about an average A is given by

$$\begin{aligned} \text{MD} &= \frac{1}{N}\sum_{i-1}^{N}|x_i - \bar{x}|, \\ &= 1/N\sum|X - A| = 1/N\sum|D| \end{aligned}$$

Example Calculate mean deviation from the following series.

Item: x :	20	25	30	35
Frequency: f :	2	4	3	2

Solution

Calculation of Arithmetic Mean

$$\begin{aligned} \text{A} &= \frac{\sum f \cdot x}{N} = \frac{\sum f \cdot x}{\sum f} = \frac{(20 * 2) + (25 * 4) + (30 * 3) + (35 * 2)}{11} \\ &= (40 + 100 + 90 + 105)/11 = 335/11 = 30.45 \approx 30.5 \end{aligned}$$

Calculation of Mean Deviation

X	f	$D = \|X - A\|$	fD
20	2	$(20 - 30.5) = 10.5$	21
25	4	$(25 - 30.5) = 0.5$	22
30	3	$(30 - 30.5) = 0.5$	1.5
35	2	$(35 - 30.5) = 4.5$	9.0
$\sum f = 11$		$\sum fD = 53.5$	

Mean deviation $= \sum fD / \sum f = \frac{53.5}{11} = 4.863$
Therefore, MD $= 4.863$.

Standard Deviation

This is the most important and widely used measure of studying dispersion. This is also known as root mean square deviation. Standard deviation is the square root of the mean of the squared deviation from the arithmetic mean. It is denoted by the Greek letter σ.

In case of individual observations, SD may be computed by applying any of the following two methods:

(a) by taking deviation of the items from the actual mean and
(b) by taking deviations of the items from assumed mean.

Deviation Taken from Actual Mean

When deviation is taken from actual mean, the following formula is applied.

$$\text{SD} = \sqrt{\frac{\sum d^2}{N} - \left(\frac{\sum d}{N}\right)^2},$$

where $d = (X - A)$.

Example Calculation of SD by actual mean method

X	$d = (X - A)$	d^2
120	(120 − 127.5) = −75	56.25
125	(125 − 127.5) = −2.5	6.25
130	(130 − 127.5) = 2.5	6.25
135	(135 − 127.5) = 7.5	56.25
$\sum x = 510$	$\sum d = 0$	125.0

Actual mean, A $= \frac{\sum x}{N} = 510/4 = 127.5$

$$\text{SD} = \sqrt{\frac{\sum d^2}{N} - \left(\frac{\sum d}{N}\right)^2} = \sqrt{\frac{125}{4}} - 0 = \sqrt{31.25} = 5.59$$

Deviation Taken from Assumed Mean

When deviations are taken from assumed mean, the following formula is applied

$$\sigma = \sqrt{\frac{\sum d^2}{N} - \left(\frac{\sum d}{N}\right)^2}$$

Example Calculation of SD by the assumed mean method

$$\text{SD} = \sqrt{\frac{\sum d^2}{N} - \left(\frac{\sum d}{N}\right)^2}$$

Actual mean, $A = \frac{\sum x}{N} = 510/4 = 127.5$
But assumed mean $= 127$

X	$d = (X - 127)$	d^2
120	−7	49
125	−2	4
130	3	9
135	8	64
$\sum x = 510$	$\sum d = 2$	$\sum d^2 = 126$

$$\text{SD, } \sigma = \sqrt{\frac{\sum d^2}{N} - \left(\frac{\sum d}{N}\right)^2} = \sqrt{\frac{126}{4} - \left(\frac{2}{4}\right)^2}$$

$$= \sqrt{31.5 - \frac{1}{4}} = \sqrt{31.5 - 0.25} = \sqrt{31.5} = 5.59$$

Variance

The variance or depression of a random variable x is given by

$$\sigma^2 = \frac{\sum d}{N} - \left(\frac{\sum d}{n}\right)^2$$

Coefficient of Variation

The variance or dispersion of a random variable x is given by

$$\text{CV} = \frac{\sigma}{\acute{x}} * 100$$

If the coefficient of variation is greater, the series is less consistent. If the coefficient of variation is less, the series is more consistent.

Standard deviation is an absolute measure of dispersion.

The corresponding relative measure is known as the coefficient of variation.

$$\text{Coefficient of variation} = \frac{\sigma}{\acute{x}} * 100 = \frac{5.59}{127.5} * 100$$

$$\text{CV} = 4.385\%$$

Correlation

If two quantities vary in such a way, that movements in one are accompanied by movements in other.

Example Price of commodity and amount demanded.

The degree of relationship between the variables under consideration is measured through the correlation analysis. The measure of correlation is called the correlation coefficient. The correlation analysis refers to the techniques used in measuring the closeness of the relationship between the variables. Correlation is a statistical device which helps us in analyzing the co-variation of two or more variables.

Types of Correlation

1. Positive or negative,
2. Simple, partial and multiple and
3. Linear and nonlinear.

Positive or Negative Correlation

Whether correlation is +ve (direct) or −ve (inverse) would depend upon the direction of change of the variables.

If both the variables are changing direction, i.e., if one variable is increasing, the other variable also increases. If one variable decreases, then the other variable also decreases. Then, correlation is said to be **positive**. If one variable is increasing and other is decreasing, or vice versa, then correlation is said to be **negative**.

Simple, Partial and Multiple Correlations

The difference between simple, partial and multiple correlations is based upon the number of variables studied.

When only two variables are studied, it is a problem of **simple correlation**.

In **partial correlation**, we recognize more two variables, but consider only two variables. To be influencing each other, the effect of other influencing variables is being kept constant.

Linear and Nonlinear Correlation

Linear Correlation If the amount of change in the variable tends to bear constant ratio to the amount of change in the other variable, then the correlation is said to be **linear**.

Example

$$X : 10\ 20\ 30\ 40\ 50$$
$$Y : 5\ \ 10\ 15\ 20\ 25$$

The variation between X and Y is a straight line.

Nonlinear Correlation (Curve Linear Correlation) If the amount of change in one variable does not bear a constant ratio to the amount change in the other variable.

Example If rainfall is doubled, the production of rice would not necessarily double.

Methods of Studying Correlation

Some of the methods used for studying correlation are as follows:

i. Scatter diagram method,
ii. Graphic method,
iii. Karl Pearson's coefficient of correlation,
iv. Rank method,
v. Concurrent deviation method and
vi. Method of least squares.

Scatter Diagram Method

For the bivariate distribution, if the values of the variables x and y are plotted in the xy-plane, the diagram of dots obtained is called scatter diagram. The greater the scatter of the plotted points on the diagram, the lesser is the relationship between the two variables.

(a) If all points lie on the straight line falling from the upper left-hand corner, correlation is said to be perfectly positive ($r = +1$).
(b) If all the points lie on a straight line rising from the upper left-hand corner to the lower right-hand corner, correlation is said to be perfectly negative ($r = -1$).

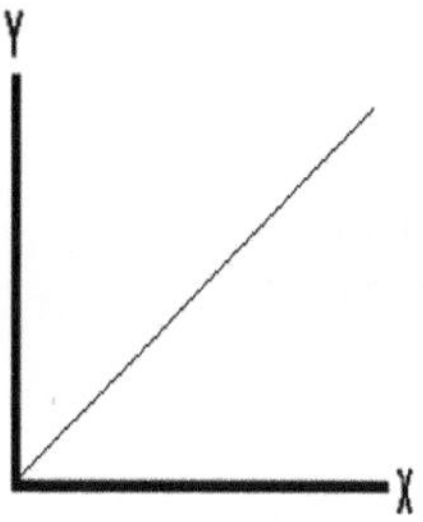

Fig. **a** ($r = +1$)

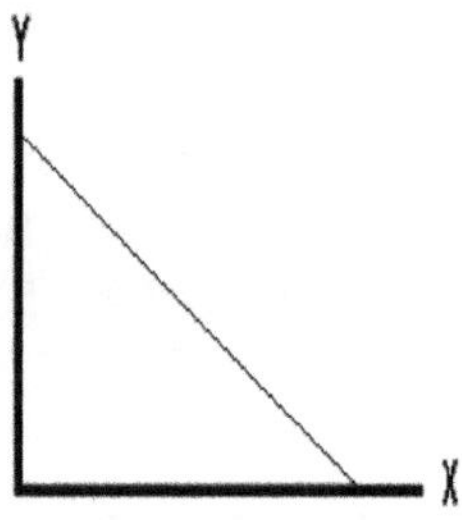

Fig. **b** ($r = -1$)

(c) If the plotted points fall in a narrowband and they show arising tendency from the lower left-hand corner to the upper right-hand corner, there would be a light degree of positive correlation.

(d) If the plotted points fall in a narrowband and they show a declining tendency from the upper left-hand corner to the lower right-hand corner, there would be high degree of negative correlation.

Fig. **c** ($r = +1$)

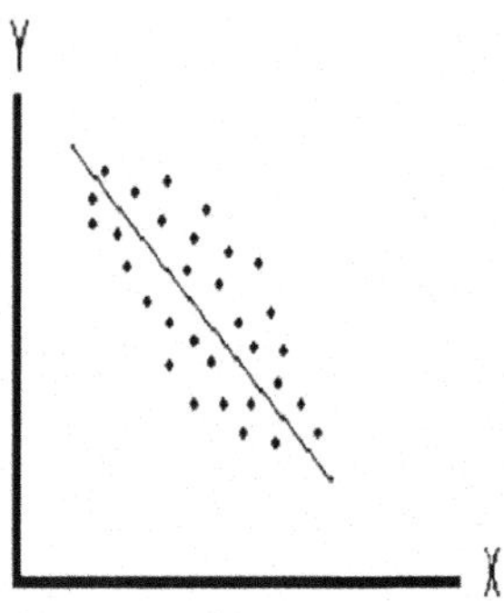

Fig. **d** ($r = -1$)

(e) If the points are widely scattered over the diagram and the points are rising from the lower left-hand corner to the upper right-hand corner, there would be low degree of positive correlation.

(f) If the points are widely scattered and the points are running from the upper left-hand side to be lower right-hand side, there would be low degree of negative correlation.

(g) If the plotted points lie on a straight line parallel to x-axis or in a haphazard manner, it shows the absence of correlation between two variables.

Fig. **e** ($r = +1$)

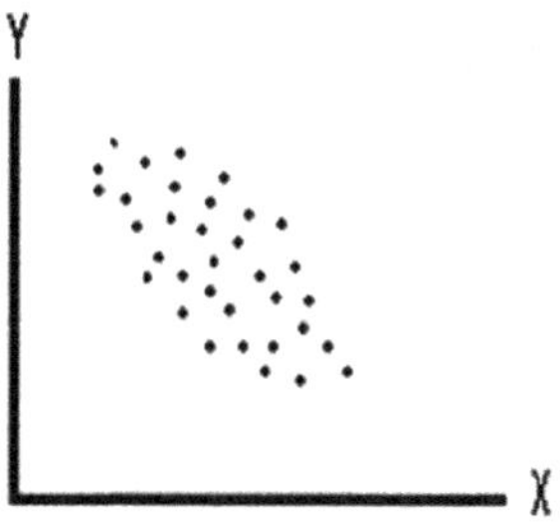

Fig. **f** ($r = -1$)

Graphic Method

In this method, the individual values of two variables are plotted on the graph paper, one curve for x variable and another for y-variable. If both the curves on the graph are changing in the same direction (either upward or downward), correlation is said to be positive. On the other hand, if the curves are moving in the opposite direction, correlation is said to be negative.

Karl Pearson Coefficient of Correlation

As a measure of degree of linear relationship between two variables, Karl Pearson developed a formula called correlation coefficient. The correlation coefficient between two random variables x and y usually denoted by γ_{xy} is a measure of linear relationship between them and is defined as

$$\gamma_{xy} = \frac{E(xy) - E(x)E(y)}{\sigma_x \sigma_y}$$

γ_{xy} Value lies always between -1 and $+1$.

Rank Correlation

Let us suppose that a group of n individuals is arranged in order of merit or proficiency in possession of two characteristics A and B. These ranks in the two characteristics will, in general, be different. For example, if we consider the relation between intelligence and beauty, it is not necessary that a beautiful individual is intelligent also.

If (X_i, Y_i) $I = 1, 2, 3, \ldots, n$ be the ranks of the individuals in two characteristics A and B, respectively. Then, the rank correlation coefficient is given by

$$\rho = 1 - \frac{6\sum d_i^2}{n\left(n^2 - 1\right)}.$$

Regression

Regression stands for stepping back toward the average. We can predict the value of one variable when the value of the other variable is given, using regression. Regression measures the relationship between the value of dependent variable and the corresponding values of series of other variables. It is a statistical tool for analyzing the mathematical relationship among variables. Regression involves techniques for analyzing several variables having dependent variable and one or more independent variables. The line described in the average relationship between two variables is known as line of regression. Generally, it is used to solve prediction problem.

Many methods of regression analysis are in use, but the most familiar methods such as linear regression and ordinary least square method are popular. Regression equation is an algebraic expression of regression line. The simplest form of regression involves two variables (bivariate) which represents the relationship between one independent variable (X) and a dependent variable (Y). We refer to this relationship as the regression curve of Y on X. This is a linear regression curve where for any given value of X, the mean of the distribution of Y is given by $Y = \alpha + \beta X + \epsilon$, and here, ϵ is a random variable. We can choose α so that the mean of the distribution of this random variable is equal to zero; here, α and β are constants. The magnitude and direction of that relation are given by the parameter β and Y-intercept α. β indicates the slope of the regression line and gives a measure of change of Y for a unit change in X. We also refer it as regression coefficient of Y on X. We can calculate the value of Y for any given value of X if we know the values of α, β.

Regression line fitted to the data by the method of least square gives the best possible mean value of one variable to specific value of other. There are always two lines of regression: One is Y on X; i.e., Y is dependent variable, and X is an independent variable. Another one is X on Y where we can predict X for any given value of Y. The two regression lines are not reversible or interchangeable because the basis and assumptions for equations are quite different. Regression function of Y on X is the conditional mean $E(Y/X)$ for a continuous distribution and is known as regression curve of Y on X. Similarly, the regression function of X on Y is $E(X/Y)$ and its graph is regression curve of X on Y.

If the regression curve is a straight line, then the corresponding regression is linear. If one is linear, then it does not imply that the other is linear. Correlation is different form regression analysis in a sense that both variables that are involved are random variables. In correlation, we are calculating the strength of the relationship between variables, while regression gives the type of relationship between variables. For example, there exists a relationship between blood pressure Y of a person and the age X of the person. Here, Y is dependent variable, and X is independent variable. We get the regression line of Y on X.

Types of Variables

Categorical Variables

Such variables include anything sort of measure that is "qualitative" or otherwise not amenable to actual quantification. There are a few subclasses of such variables.

i. Dummy variables take only two possible values, 0 and 1. They signify conceptual opposites:
   e.g., (a) war versus peace and (b) fixed exchange rate versus floating exchange rate.
ii. Nominal variables can range over any number of nonnegative integers.
   They signify conceptual categories that have no inherent relationship to one another,
   e.g., (a) red versus green versus black and (b) Christian versus Jewish versus Muslim.
iii. Ordinal variables are like nominal variables, and only, there is an ordered relationship among them,
   e.g., (a) no versus maybe versus yes.

Numerical Variables

Such variables describe data that can be readily quantified. Like categorical variables, there are a few relevant subclasses of numerical variables.

(a) Continuous variables can appear as fractions; in reality, they can have an infinite number of values. Examples include temperature, GDP, etc.
(b) Discrete variables can only take the form of whole numbers. Most often, these appear as count variables, signifying the number of times that something occurred: the number of firms invested in a country, the number of hate crimes committed in a county, etc.

Linear Regression Model (LRM)

The simple (or bivariate) LRM model is designed to study the relationship between a *pair* of variables that appear in a dataset.

The multiple LRM is designed to study the relationship between one variable and several of other variables.

In both cases, the sample is considered a random sample from some population. The two variables, *X* and *Y*, are two measured outcomes for each observation in the dataset.

χ^2 *Test (Chi-Square Test)*

Chi-square test: This is one of the simplest and most widely used nonparametric tests in statistical work. The symbol χ^2 is the Greek letter **chi**.

The quantity χ^2 describes the magnitude of the discrepancy between theory and observation. It is defined as

$$\chi^2 = \sum \frac{(O - E)^2}{E}$$

where *O* refers to the observed frequencies and *E* refers to the expected frequencies.

Procedure to Determine χ^2

(i) Calculate the expected frequency *E*, using the formula

$$E = \frac{\text{RT} * \text{CT}}{N}$$

where

RT the row total of the row containing the cell,
CT the column total for the column containing the cell,
N the total number of observations.

(ii) Calculate the square of the difference between observed and expected frequencies; i.e., obtain the value of $(O - E)^2$.
(iii) Calculate χ^2 using the formula

$$\chi^2 = \sum \frac{(O - E)^2}{E}$$

The value of χ^2 can range from **0** to **infinity**.
$\chi^2 = 0$ represents that observed and expected frequencies completely coincide.

The greater the discrepancy between the observed and expected frequencies, the greater shall be the value of χ^2. The value of χ is always +ve.

Chi-Square Distribution Curve

For large sample size, the (probability) sampling distribution of χ^2 can be closely approximated by a continuous curve known as **chi-square** distribution.

The probability function of χ^2 distribution is given by

$$f(\chi^2) = C(\chi^2)^{\frac{v}{2}-1} - \frac{\chi^2}{2}$$

where

e 2.71828,
v number of degrees of freedom,
C a constant, depending only on v.

For very small number of degrees of freedom, chi-square distribution is severely skewed to the right. As the number of degrees of freedom increases, the curve rapidly becomes more symmetrical.

For large values of v, the chi-square distribution is closely approximated by the normal curve.

Alternative Method of Applying the Value of χ

Marginal totals are as below

A	b	$(a+b)$
C	d	$(c+d)$
$(a+c)$	$(b+d)$	N

N is the total frequency and **ad** is the larger cross product. The value of χ^2 can easily be obtained by the following formula.

$$\chi^2 = \frac{(ad - bc)^2 N}{(a+c)(b+d)(c+d)(a+b)} \quad \text{or}$$

with Yate's corrections,

$$\chi^2 = \frac{(ad - bc - N/2)^2 N}{(a+c)(b+d)(c+d)(a+b)}$$

Conditions for Applying χ^2 Test

The following conditions should be satisfied before applying the χ^2 test.

1. In the first place, N must be reasonably large to ensure the similarity between theoretically correct distribution and our sampling distribution of χ^2, the chi-square static.

   The general rule is χ^2 test should not be used when N is less than 50.

2. No theoretical cell frequency should be small when the expected frequencies are too small. The value of χ^2 will be over-estimated and will result in too many rejections of null hypothesis. To avoid making incorrect inferences, a general rule is followed that expected frequency of less than 5 in one cell of a contingency table is too small to use.

When the table contains more than one cell with the expected frequency of less than 5, we **pool** the frequencies which are less than 5 with the preceding or succeeding frequency so that the resulting sum is 5 or more. However, in doing so, we reduce the number of categories of data and will gain less information from contingency table.

3. The constraints on the cell frequencies if any should be linear; i.e., they should not involve square and higher powers of the frequencies such as $\sum O = \sum E = N$.

Use of χ^2 Test

(a) χ^2 test as a test of independence. With the help of χ^2 test, we can find out whether two or more attributes are associated or not.
(b) χ^2 test as a goodness of fit: χ^2 test is very popularly known as test of goodness of fit for the reason that it enables us to ascertain how approximately the theoretical distributions such as binomial, Poisson, normal fit empirical distributions i.e. these obtained from sample data.
(c) χ^2 test as a test of homogeneity: It is an extension of chi-square test of independence. Tests of homogeneity are designed to determine whether two or more independent random samples are drawn from the same population, or from different populations.

Example From the data given below about the treatment of 250 patients suffering from a disease, state whether the new treatment is superior to the conventional treatment.

Treatment	No. of patients		Total
	Favorable	Not favorable	
New	140	30	170
Conventional	60	20	80
Total	200	50	250

Given for degree of freedom = 1, chi-square 5% = 3.84.

Solution

Let us take the hypothesis that there is no significant difference between the new and conventional statement, applying χ^2 test.

$$\text{Expectation of } (AB) = 200 * 170/250 = 136.$$

The table of expected frequencies shall be as follows:

New	136	34	170
Conventional	64	16	80
Total	200	50	250

O	E	$(O-E)^2$	$(O-E)^2/E$
140	136	16	0.118
60	64	16	0.250
30	34	16	0.471
20	16	16	1.000

$$\sum[(O-E)^{2/}/E] = 1.839$$

$$\chi^2 = \sum[(O-E)^2/E] = 1.839$$

$$v = (x-1)\cdot(c-1) = (2-1)(2-1) = 1$$

$$v = 1,\ \chi^2_{0.05} = 3.84$$

The calculated values of χ^2 are less than the table value. The hypothesis is accepted. Hence, there is no significant difference between the new and conventional treatment.

Estimations

Parameters Quantities appearing in distributions, such as p in the binomial distribution and μ and σ in the normal distribution, are called **parameters**.

Estimate An estimate is a statement made to find an unknown population parameter.

Estimator The procedure or rule to determine an unknown population parameter is called an estimator.

Types of Estimations

Basically, there are two kinds of estimates to determine the statistics of population parameters, namely

(a) Point Estimation
(b) Interval Estimation

Point Estimator

A point estimate is a single real number which is used as an estimate of n observations; say $x_1, x_2, \ldots, x_n$ are selected from a population $f(x;\theta)$, and then, some preconceived method is used to arrive from these observations, a real number $\hat{\theta}$.

Interval Estimates

An interval estimate of population parameter is a statement of two values between which it is estimated that the parameter lies. Thus, interval estimation refers to the estimation of a parameter by a random interval (called confidence interval) whose end points (called confidence limits or confidence coefficients) are satisfied.

Example If the weight of a student is measured as 50 kg, then the measurement gives point estimation. But if the weight is given as (50 ± 2) kg, then the weight lies b/w 148 and 152 kg. And the measurement gives the interval estimation.

Statistical Inference

The process by which we draw a conclusion about some measures of population is based on a sample value. The measure might be a variable, such as the mean, SD.

The purpose of sampling is to estimate some characteristics for the population from which the sample is selected.

There are two types of problems under statistical inference.

(i) Hypothesis testing and
(ii) Estimation.

Hypothesis Testing

To test some hypothesis about parent population from which the sample is drawn.

Estimations

To use the statistics obtained from the sample to estimate of the unknown parameter of the population from which sample is drawn.

Bayesian Estimation

The new concept introduced in Bayesian method is personal or subjective probability. Also, parameters are considered as random variable in Bayesian method.

Bayesian estimation is used to obtain mean and variance of distribution of a population.

If the prior distribution parameters, mean μ_0 and variance σ_0^2 of a population, are known and when the direct sample statistics, say mean $\bar{x}$, then it is possible to estimate the posterior distribution parameters of a given population. This is called **Bayesian estimation.**

Let μ_0 be the prior mean.

σ_0^2 be the prior variance

$\bar{x}$ be the sample mean

n be the sample size and

σ^2 be the sample variance.

Then, the posterior distribution parameters can well be approximated by normal distribution.

$$\text{The mean of the posteriors distribution} = \mu_1 = \frac{n\bar{x}\sigma^2 + \mu_0\sigma_0^2}{n\sigma^2 + \sigma_0^2}$$

$$\text{The variance of posterior distribution} = \sigma_1^2 = \frac{\sigma^2\sigma_0^2}{n\sigma^2 + \sigma_0^2}$$

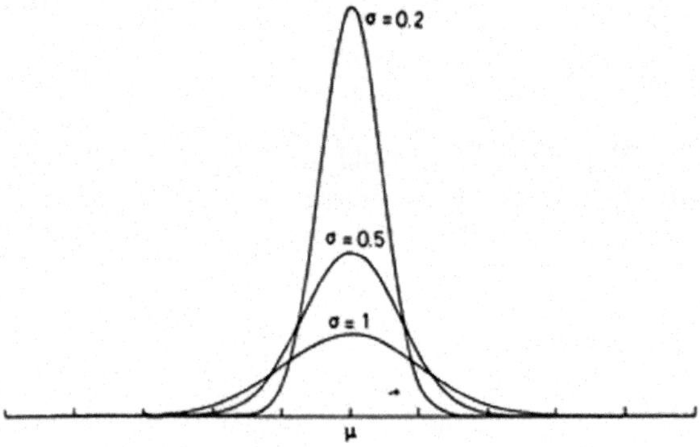

Fig. A.1 Gaussian distribution for various σ. The standard deviation determines the width of the distribution

The Gaussian or Normal Distribution

The Gaussian or normal distribution plays a central role in all of statistics and is the most ubiquitous distribution in all the sciences. Measurement errors, and in particular, instrumental errors, are generally described by this probability distribution. Moreover, even in cases where its application is not strictly correct, the Gaussian often provides a good approximation to the true governing distribution.

The Gaussian is a continuous, symmetric distribution whose density is given by

$$P(x) = \frac{1}{\sigma\sqrt{2\pi}} \exp\left(-\frac{(x-\mu)^2}{2\sigma^2}\right). \tag{A.1}$$

The two parameters μ and σ^2 correspond to the mean and variance of the distribution.

The shape of the Gaussian is shown in Fig. A.1 which illustrates this distribution for various sigmas. The significance of sigma as a measure of the distribution width is clearly seen. As can be calculated from (19), the standard deviation corresponds to the half width of the peak at about 60% of the full height. In some applications, however, the full width at half maximum (FWHM) is often used instead. This is somewhat larger than sigma and can easily be shown to be

$$\text{FWHM} = 2\sigma\sqrt{2\ln 2} = 2.35\sigma. \tag{A.2}$$

This is illustrated in Fig. A.2. In such cases, care should be taken to be clear about which parameter is being used. Another width parameter which is also seen in the literature is the full width at one-tenth maximum (FWTM).

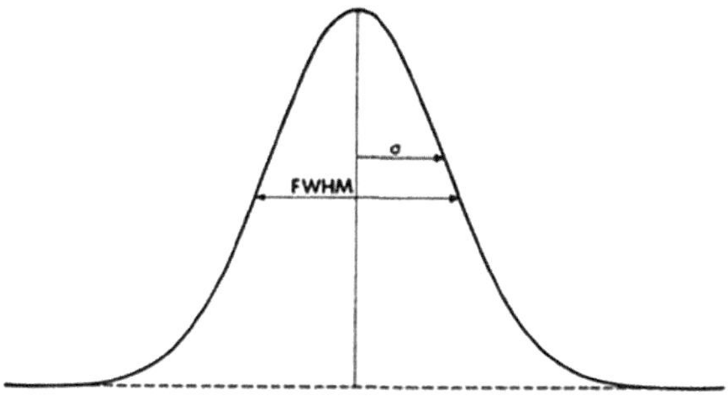

Fig. A.2 Relation between the standard deviation and the full width at half maximum (FWHM)

Fig. A.3 Area under Guassian

The integral distribution for the Gaussian density, unfortunately, cannot be calculated analytically so that one must resort to numerical integration. Tables of integral values are readily found as well. These are tabulated in terms of a reduced Gaussian distribution with $\mu = 0$ and $\sigma^2 = 1$. All Gaussian distributions may be transformed to this reduced form by making the variable transformation

$$z = \frac{x - \mu}{\sigma}, \tag{A.3}$$

where μ and σ are the mean and standard deviation of the original distribution. It is a trivial matter then to verify that z is distributed as a reduced Gaussian.

An important practical note is the area under the Gaussian between integral intervals of σ. This is shown in Fig. A.3. These values should be kept in mind when interpreting measurement errors. The presentation of a result as $x \pm \sigma$ signifies, in fact, that the true value has approximately 68% probability of lying between the limits $x - \sigma$ and $x + \sigma$ or a 95% probability of lying between $x - 2\sigma$ and $x + 2\sigma$, etc. Note that for a 1σ interval, there is almost a 1/3 probability that the true value is outside these limits! If two standard deviations are taken, then the probability of being outside is only approx 5%, etc.

Appendix B: Probability

Random Variables and Mathematical Expectation

Introduction

In a random experiment, the sample space 'S' consists of all outcomes (results) of that experiment. When the elements of the sample space are non-numeric, they can be quantified by assigning a real number to every event of the sample space.

This assignment is known as the random variable.

By a random variable, we mean a real number X associated with the outcome of a random experiment.

Example Suppose two coins are tossed simultaneously, then the sample space is $S = \{\text{HH, HT, TH, TT}\}$. We will consider the random variable, which is the number of heads (0, 1, 2).

Outcome	HH	HT	TH	TT
Value of X	2	1	1	0

Thus to each outcome 'S', there corresponds a real number $X(s)$.

Note: The real number is denoted by $X(s)$, and it is defined for each $s \in S$.

Definition of random variables

A random variable X on a sample space "S" is a function from S to the set of real numbers R, which assign a real number $X(s)$ to each outcome "s" of S.

The function is given as $X: S \rightarrow R$ random variable is also known as stochastic variable or variable.

Example

(1) If a coin is tossed, then sample space is $S = \{\text{H, T}\}$

Here, we consider the random variable 1 if $s = \text{H}$ $X(s) = \{0\}$ if $s = \text{T}$.

Types of random variables

There are two types of random variables:

(1) Discrete random variable and
(2) continuous random variable.

Discrete Random Variable (Def) A random variable X is said to be discrete random variable if its set of all possible outcomes (sample space) is countable (finite or an un-ending sequence with as many elements as there in whole numbers).

Continuous Random Variables (Def) A random variable X is said to be continuous random variable if the sample space contains infinite numbers equal to the number of points on a line segment.

Probability Distribution

Probability distribution of a random variable X is the description of the set of possible values which X can take along with the probability associated with the possible values of X.

For example, if X is a random variable which can take the values $x_1, x_2, \ldots$ such that

$$\begin{aligned} P(X = x_i) &= p_i \\ &= f(x_i)(i = 1, 2, 3, \ldots) \end{aligned}$$

Probability distribution always satisfies the following conditions

$$f(x) \geq 0 \text{ for all } x \sum_x f(x) = 1$$

Probability mass function

$$f(x) \geq 0$$
$$\sum_x f(x) = 1$$

Cumulative Probability Distribution function or distribution function

Cumulative distribution function or simply distribution function of a discrete random variable X is $F(x)$ and is defined by

$$F(x) = P[X \leq x] = \sum_{x_i \leq x} f(x_i); \quad -\infty < x < \infty.$$

Probability Densities

Introduction

Continuous sample spaces and continuous random variables arise when we deal with quantities that are measured on a continuous scale—for instance, when we measure the speed of a car, the amount of alcohol in a person's blood, the efficiency of a solar collector or the tensile strength of a new alloy.

Continuous random variables

A random variable X is said to be continuous random variable if the sample space contains infinite numbers equal to the numbers of points in a line segment.

OR

A random variable X is said to be continuous if it can assume all possible values between certain limits.

For example, weight, height.

Continuous Probability Distribution

Probability Density Function

For a continuous random variable X, the function $f(x)$ satisfying the following conditions is known as probability density function or simply density function.

(1) $f(x) \geq 0$ for all x.
(2) $\int_{-\infty}^{\infty} f(x)\mathrm{d}x = 1$.
(3) $P[a < x < b] = \int_a^b f(x)\mathrm{d}x =$ Area under $f(x)$ between ordinates $x = a$ and $x = b$.

Cumulative Distribution (Distribution Function)

For a continuous random variable X, with a probability density function $f(x)$, the cumulative distribution $F(x)$ is defined as

$$F(x) = P[X \leq x] = \int_{-\infty}^{X} f(x)\mathrm{d}x \quad -\infty < x < \infty.$$

It follows that

(1) $F(-\infty) = 0, F(\infty) = 1, 0 \leq F(x) \leq 1$ for $-\infty < x < \infty$
(2) $P[a < x < b] = F(b) - F(a)$

Relation between probability density function and distribution function

$$f(x) = \frac{\mathrm{d}}{\mathrm{d}x}(F(x)) = F'(x) \geq 0$$

Joint Distribution—discrete and continuous

Discrete variables

For two discrete random variables X_1 and X_2, we write the probability that X_1 will take the value x_1 and X_2 will take the value x_2 as $P(X_1 = x_1, X_2 = x_2)$.

The probability distribution $f_1(x_1)$ of X_1 appears in the lower margin of this table.

The probability distribution $f_2(x_2)$ of X_2 appears in the right-hand margin of the table. Consequently, the individual distributions are called marginal probability distribution.

For each fixed value of x_1, the marginal probability distribution is given by $P(X_1 = x_1) = f_1(x_1) = \sum_{x_2} f(x_1, x_2)$ where the sum is over all possible values of the second variable.

When A is the event $X_1 = x_1$ and B is the event $X_2 = x_2$, the conditional probability distribution of X_1 given $X_2 = x_2$ is defined as

$$f_1(x_1/x_2) = \frac{f(x_1, x_2)}{f_2(x_2)} \quad \text{for all } x_1 \text{ provided}$$

$$f_2(x_2) \neq 0$$

If $f_1(x_1/x_2) = f_1(x_1)$ for all x_1 and x_2 so the conditional probability distribution is free of x_2 equivalently if $f(x_1, x_2) = f_1(x_1) f_2(x_2)$ for all $x_{1,}x_2$, the two random variables are independent.

Suppose that we are concerned with 'k' random variables $X_1, X_2, \ldots, X_k$.

Let x_1 be a possible value for the first random variable X_1, x_2 be a possible value for the second random variable X_2 and so on with x_k a possible value of the kth random variable.

The function $f(x_1, x_2, \ldots, x_k) = P[X_1 = x_1, X_2 = x_2, \ldots, X_k = x_k]$ as the probability distribution of $X_1, X_2, \ldots, X_k$.

The probability distribution $f_i(x_i)$ of the individual variable X_i is called the marginal probability distribution of the ith random variable.

$$f_i(x_i) = \sum_{x_1} \cdots \sum_{x_{i-1}} \sum_{x_{i+1}} \cdots \sum_{x_k} f(x_{1,}x_2, \ldots, x_k)$$

where the summation is over all possible k-tuples where the ith component is held fixed at the specified value x_i.

Joint Probability Density

Let X_1, X_2, X_k are k continuous random variables, then $f(x_1, x_2, x_3, \ldots, x_k)$ is the joint probability density of these random variables if

(1)
$$P(a_1 \le X_1 \le b_1, a_2 \le X_2 \le b_2, \ldots, a_k \le X_k \le b_k) = \int_{a_1}^{b_1} \int_{a_2}^{b_2} \cdots \int_{a_k}^{b_k} f(x_1, x_2, \ldots, x_k) \mathrm{d}x_1 \mathrm{d}x_2 \ldots \mathrm{d}x_k$$

(2) $f(x_1, x_2, \ldots, x_k) \ge 0$ for all values of $x_1, x_2, \ldots, x_k$

(3) $\int_{-\infty}^{\infty} \int_{-\infty}^{\infty} \cdots \int_{-\infty}^{\infty} f(x_1, x_2, \ldots, x_k) \mathrm{d}x_1 \mathrm{d}x_2 \mathrm{d}x_3 \ldots \mathrm{d}x_k = 1$

Joint Cumulative Distribution Function

$$F(x_1, x_2, x_3, \ldots, x_k) = \int_{-\infty}^{x_1} \int_{-\infty}^{x_2} \cdots \int_{-\infty}^{x_k} f(x_1, x_2, \ldots, x_k) \mathrm{d}x_1 \mathrm{d}x_2 \mathrm{d}x_3 \ldots \mathrm{d}x_k$$

Independent random variables

k random variable $X_1, X_2, \ldots, X_k$ are independent iff $F(x_1, x_2, x_3, \ldots, x_k) = F_1(x_1)$ $F_2(x_2) \ldots F_k(x_k)$ for all values $x_1, x_2, x_3, \ldots, x_k$ of these random variables where $F(x_1, x_2, \ldots, x_k)$ is the joint distribution function of the k random variables. While $F_i(x_i)$ for $i = 1, 2, \ldots, k$ are the corresponding individual distribution functions of the respective random variables.

Conditional Probability Density

Given two continuous random variables X_1 and X_2, then the conditional probability density of the first given that the second takes on the value x_2.

As $f_1(x_1/x_2) = \frac{f(x_1,x_2)}{f_2(x_2)}$ provided $f_2(x_2) \neq 0$ where $f(x_1, x_2)$ and $f_2(x_2)$ are the joint density of the two random variables and the marginal density of the second.

Properties of Expectation

Consider a random variable X with probability density function $f(x)$.

Then, the mean or expectation of X is given by

$$E(X) = \int_{-\infty}^{\infty} x f(x) \mathrm{d}x.$$

In discrete case where X has probability distribution f, then

$$E(x) = \sum_{x_i} x f(x_i)$$

where x_i is a possible value for X.

Result

If $Y = aX + b$ then 1 $E(y)$

$$\begin{aligned} E(y) = E(ax + b) &= \int_{-\infty}^{\infty} (ax + b)\, f(x)\mathrm{d}x \\ &= a \int_{-\infty}^{\infty} x f(x)\mathrm{d}x + b \int_{-\infty}^{\infty} f(x)\mathrm{d}x \\ &= aE(x) + b \end{aligned}$$

Variance

The variance of probability distribution is given

$$\sigma^2 = \sum_{\text{all}\, x} (x - \mu)^2 f(x)$$

where $\mu = E(X)$.

$\mathrm{Var}(X) = E(X^2) - (E(X))^2$

The standard deviation is defined as

$$\sigma = \sqrt{\sum_{\text{all}\, x} (x - \mu)^2 f(x)}$$

Results

Var($aX+b$) = $a^2V(X)$
$V(b) = 0$
$V(X+b) = V(X)$
Variance written as σ^2.

Variance of a probability density

$$\sigma^2 = \int_{-\infty}^{\infty} (x-\mu)^2 f(x)\mathrm{d}x$$
$$= \int_{-\infty}^{\infty} x^2 f(x)\mathrm{d}x - \mu^2$$
$$= E\left(x^2\right) - (E(x))^2$$

σ is known as the standard deviation.

Covariance and variance of sums of Random variables

The covariance of two random variable X and Y, written Cov(X, Y), is defined by

$$\mathrm{CoV}(X, Y) = E[(X-\mu_x)(Y-\mu_y)]$$

where $\mu_x = E[X]$ and $\mu_y = E[Y]$.

Results

$$\mathrm{CoV}(X, Y) = \mathrm{CoV}(Y, X)$$
$$\mathrm{CoV}(X, X) = \mathrm{Var}(X)$$
$$\mathrm{CoV}(aX, Y) = a\mathrm{Cov}(X, Y)$$
$$\mathrm{CoV}(X+Z, Y) = \mathrm{Cov}(X, Y) + \mathrm{Cov}(Z, Y)$$

$$\mathrm{Cov}\left(\sum_{i=1}^{n} X_i, Y\right) = \sum_{i=1}^{n} \mathrm{Cov}(X_i, Y)$$
$$\mathrm{Cov}\left(\sum_{i=1}^{n} X_i, \sum_{j=1}^{m} Y_j\right) = \sum_{i=1}^{n}\sum_{j=1}^{m} \mathrm{Cov}(X_i, Y_j)$$
$$\mathrm{Var}\left(\sum_{i=}^{n} X_i\right) = \sum_{i=1}^{n} \mathrm{Var}(X_i) + \sum_{i=1}^{n}\sum_{j=1, j^1 i}^{n} \mathrm{Cov}(X_i, X_j)$$
$$\mathrm{Var}(X+Y) = \mathrm{Var}(X) + \mathrm{Var}(Y) + 2\mathrm{Cov}(X, Y)$$

Cov(X, Y) = 0, if X and Y are independent variables.

For independent variables $X_1, X_2, \ldots, X_n$

$$\mathrm{Var}\left(\sum_{i=}^{n} X_i\right) = \sum_{i=1}^{n} \mathrm{Var}(X_i)$$

$$\mathrm{Cov}(X, Y) = \frac{\mathrm{Cov}(X, Y)}{\sqrt{\mathrm{Var}(X)\mathrm{Var}(Y)}}$$

Moment-Generating Functions

The moment-generating function $\phi(t)$ of the random variable X is defined for all values 't' by

$$\phi(t) = E[\mathrm{e}^{tX}] \left\{ \begin{array}{ll} \sum_{x} \mathrm{e}^{tx} p(x) & \text{if } X \text{ is Discrete} \\ \int_{-\infty}^{\infty} \mathrm{e}^{tx} f(x)\mathrm{d}x & \text{if } X \text{ is continuos} \end{array} \right\}$$

Binomial Distribution

Binomial distribution is a discrete probability distribution.

Conditions for Binomial Distribution

Binomial distribution will be applied under the following experimental conditions.

(1) The number of trails (n) is finite.
(2) The trails are independent of each other.
(3) The probability of success p is constant for each trail.
(4) Each trail results in two mutually exclusive events known as success and failure.

Definition
A random variable X is said to follow binomial distribution if it assumes only non-negative values and its probability mass function is given by

$$P[X = x] = f(x) = nC_X p^X q^{n-x}; \quad x = 0, 1, 2, \ldots, n$$

n and p are two parameters of the distribution and $q = 1 - p$.

The Poisson Approximation to the Binomial Distribution

Poisson distribution is the discrete probability distribution of a discrete random variable X, which has no upper bound. It is defined for nonnegative values of x as follows:

$$f(x; \lambda) = \frac{\lambda^x \mathrm{e}^{-\lambda}}{x!} \text{ for } x = 0, 1, 2, \ldots, \infty; \quad \lambda > 0$$

λ is called the parameter of the distribution.

Poisson distribution is suitable for rare events for which the probability of occurrence 'p' is very small and the number of trials 'n' is very large.

Also, binomial distribution can be approximated by poisson distribution when $n \to \infty$ and $p \to 0$ such that $\lambda = np =$ constant variance of Poisson distribution = mean of Poisson distribution.

Standard deviation is $\sqrt{\lambda}$.

A random variable X with mean μ and variance σ^2 and following the normal probability density is expressed by

$$X \sim N\left(\mu, \sigma^2\right)$$

The normal probability density's graph is given below.

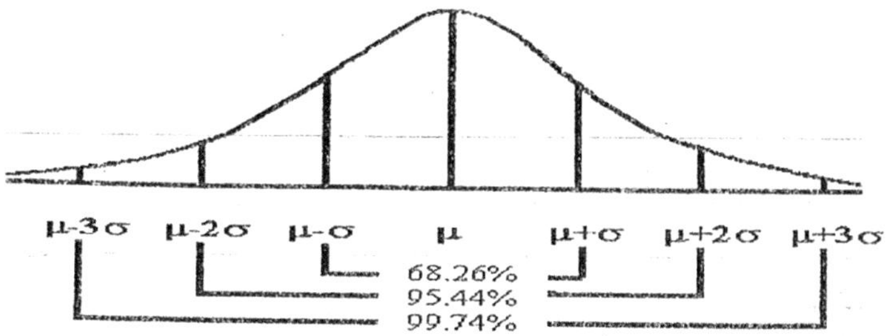

Arithmetic Mean of Normal Distribution

The AM of a continuous distribution $f(x)$ is given by

$$\text{Mean} = E(x) = \int_{-\infty}^{\infty} x f(x) \mathrm{d}x$$

$$= \int_{-\infty}^{\infty} x \frac{1}{\sigma\sqrt{2\pi}} \mathrm{e}^{\frac{-(x-\mu)^2}{2\sigma^2}} \mathrm{d}x = \mu$$

Variance of normal distribution

$$\text{Variance} = \sigma^2 \text{ Sampling Distribution}$$

Population and Samples

Population

Population is a collection of objects. Population may be finite or infinite depending upon size of the population. Here, size refers to total numbers of objects in the population and it is denoted by N.

Sample: A finite subset of the population is known as sample.
Sample size: The number of items in a sample is called as sample size.
Sample: A finite subset of the population is known as sample.
Sample size: The number of items in a sample is called as sample size.

Random sample (finite population)
A set of observations $X_1, X_2, \ldots, X_n$ constitutes a random sample of size 'n' from a finite population of size N, if its values are chosen so that each subset of n of the N elements of the population has the same probability of being selected.
Random sample (infinite population)
A set of observations $X_1, X_2, \ldots, X_n$ constitutes a random sample of size n from the infinite population $f(x)$ if

1. Each X_i is a random variable whose distribution is given by $f(x)$.
2. These n random variables are Parameters and statistics.

The measures of population, namely mean (μ), variance (σ^2), standard deviation (σ), are known as population parameters or parameters.

The measures computed from the sample observations, namely mean ($\bar{x}$), variance (s^2) standard deviation (s), are known as sample statistics or statistics independent.

Sampling Distribution

The probability distribution of a statistic calculated on the basis of a random sample.

Theorem
If a random sample of size 'n' is taken from a population having the mean μ and the variance σ^2, then $\overline{X}$ is random variable whose distribution has the mean $\mu_{\bar{x}} = \mu$.

For samples from infinite population, the variance of this distribution

$$\sigma_{\bar{x}}^2 = \frac{\sigma^2}{n}.$$

For samples from a finite population of size N, the variance is

$$\sigma_{\bar{x}}^2 = \frac{\sigma^2}{n} \cdot \frac{N-n}{N-1}$$

Note $\frac{N-n}{N-1}$ is often called the finite population correction factor.

Standard Error

The standard deviation of the sampling distribution of a statistic is known as standard error.

$$\text{Standard error of the mean } (\sigma_{\bar{x}}) = \sqrt{\sigma_{\bar{x}}^2} = \frac{\sigma}{\sqrt{n}}$$

Central Limit Theorem

If $\overline{X}$ is the mean of a sample of size 'n' taken from a population having the mean μ and finite variance σ^2, then $Z = \frac{\overline{X}-\mu}{\frac{\sigma}{\sqrt{n}}}$ is a random variable whose distribution function approaches that of the standard normal distribution as $n \to \infty$.

A Random Variable Having the *t*-Distribution

Let $\overline{X}$ be the mean of a random sample of size 'n' drawn from a normal population with μ and variance $\sigma^2 =$ then $t = \frac{\overline{X}-\mu}{\frac{s}{\sqrt{n}}}$ is a random variable having t-distribution with the parameter $\upsilon = (n-1)$ degree of freedom where $S^2 = \frac{\sum(xi-\bar{x})^2}{n-1}$ (degrees of freedom $\upsilon = n - k$ is the difference between 'n' the sample size and 'k' is the number of population parameters which are calculated using the sample data).

The t-distribution curve is symmetric about the mean zero, unimodal, bell shaped and asymptotic on both sides of t-axis.

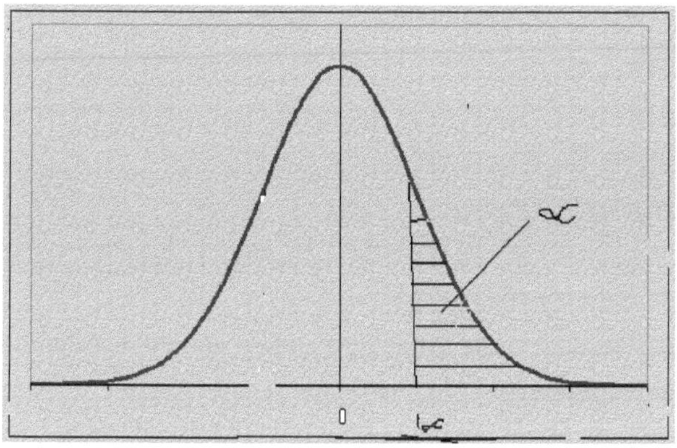

t_α represents the area under the curve which is similar to normal curve while the variance for normal distribution is more than 1, since it depends upon the parameter 'υ'.

As $n \to \alpha$ variance of t-distribution approaches 1. Since t-distribution is symmetric,

$$t_{1-\alpha} = -t_{\alpha}$$
$$t_{0.95} = -t_{0.05}$$
$$t_{1-0.05} = -t_{0.05}$$

Inferences concerning means

Theory of statistical inference is divided into two major areas:

(1) estimation and
(2) tests of hypothesis.

Estimation

Procedure of estimation of a population by using sample information is referred as estimation.

Estimation procedures are divided into two types:

1. point estimation and
2. interval estimation.

Point Estimate

If we use the value of a sample statistic to estimate a population parameter, this value is called the point estimate of the parameter.

Point Estimator

The statistic, whose value is used as the point estimate of a parameter, is called a point estimator.

Maximum Error of Estimate

To examine the error, between the sample mean $\overline{X}$ and population mean μ, $(\overline{X} - \mu)$, let us make use of the fact that for large $n \geq 30$.

$Z = \frac{\overline{X}-\mu}{\frac{\sigma}{\sqrt{n}}}$ is a random variable having standard normal distribution with mean 0 and unit standard deviation 'σ'. If $Z_{\frac{\alpha}{2}}$ is a value of z such that the area under the normal curve between $Z_{\frac{\alpha}{2}}$ and ∞ is $\frac{\alpha}{2}$ or area from $z = 0$ to $z = Z_{\frac{\alpha}{2}}$ is $\frac{1}{2}(1 - \alpha)$. Then, the probability that z lies between $-Z_{\frac{\alpha}{2}}$ to $Z_{\frac{\alpha}{2}}$ is $(1 - \alpha)$ or $P\left(-Z_{\frac{\alpha}{2}} \leq \frac{\overline{X}-\mu}{\frac{\sigma}{\sqrt{n}}} \leq Z_{\frac{\alpha}{2}}\right) = (1 - \alpha)$.

The equality $-Z_{\frac{\alpha}{2}} \leq \frac{\overline{X}-\mu}{\frac{\sigma}{\sqrt{n}}} \leq Z_{\frac{\alpha}{2}}$ will be satisfied that $\frac{|\overline{X}-\mu|}{\frac{\sigma}{\sqrt{n}}} \leq Z_{\frac{\alpha}{2}}$. If we now let E stand for the maximum of the values of $\left|\overline{X} - \mu\right|$, the maximum error of estimate, we have that the error $\left|\overline{X} - \mu\right|$ will be less than Maximum error estimate $E = Z_{\frac{\alpha}{2}} \frac{\sigma}{\sqrt{n}}$ with probability $(1 - \alpha)$.

The most widely used values for $(1 - \alpha)$ are 0.95 and 0.99, and the corresponding values of $Z_{\frac{\alpha}{2}}$ are $Z_{0.025} = 1.96$ and $Z_{0.005} = 2.575$.

Sample Size

We know that maximum error $E = Z_{\frac{\alpha}{2}} \frac{\sigma}{\sqrt{n}}$

$$\text{Sample size } n = \left(\frac{Z_{\frac{\alpha}{2}}}{E}\sigma\right)^2$$

σ is the standard deviation of population
E is the maximum error
$Z_{\frac{\alpha}{2}}$ is the ordinate.

Maximum error estimate for normal population σ unknown (small sample)
We know that $t = \frac{\overline{X}-\mu}{\frac{s}{\sqrt{n}}}$ is a random variable having the t-distribution with $(n - 1)$ degrees of freedom.

Error estimate for small sample is

$$E = t_{\frac{\alpha}{2}} \frac{s}{\sqrt{n}}$$

s standard deviation of sample
n sample size.

Interval estimation
In general, a point estimator does not coincide with a true value of parameter. So it is preferred to obtain an interval, in which the parameter value lies, interval for mean (large sample).

To illustrate the construction of such an interval, suppose that we have a large ($n \geq 30$) random sample from a population with the unknown mean μ and unknown variance σ^2. We have already seen the inequality $-Z_{\frac{\alpha}{2}} < \frac{\overline{X}-\mu}{\frac{\sigma}{\sqrt{n}}} < Z_{\frac{\alpha}{2}}$ will be satisfied with probability $1 - \alpha$. The above inequality is same as

$$\overline{X} - Z_{\frac{\alpha}{2}} \frac{\sigma}{\sqrt{n}} < \mu < \overline{X} + Z_{\frac{\alpha}{2}} \frac{\sigma}{\sqrt{n}}$$

Thus when a sample has been obtained and the value of $\overline{X}$ has been calculated, we can claim with 100% confidence that the interval from $\overline{X} - Z_{\frac{\alpha}{2}} \frac{\sigma}{\sqrt{n}}$ to $\overline{X} + Z_{\frac{\alpha}{2}} \frac{\sigma}{\sqrt{n}}$ contains μ. An interval of this kind is referred as large sample confidence interval for μ having the degrees of confidence $1 - \alpha$ or $1 - \alpha$ 100% and to its end points $\overline{X} - Z_{\frac{\alpha}{2}} \frac{\sigma}{\sqrt{n}}$ and $\overline{X} - Z_{\frac{\alpha}{2}} \frac{\sigma}{\sqrt{n}}$ are called confidence limits.

Confidence interval for mean when σ unknown (small sample)

$$\left(\overline{X} - t_{\frac{\alpha}{2}} \frac{s}{\sqrt{n}}, \overline{X} + t_{\frac{\alpha}{2}} \frac{s}{\sqrt{n}}\right)$$

Tests of Hypothesis

There are many problems in which, rather than estimate the value of a parameter, we must decide whether a statement concerning a parameter is true or false; that is, we must test a hypothesis about a parameter.

There are two types of hypothesis: null hypothesis and alternate hypothesis.

Null hypothesis: It is the hypothesis which is tested for possible rejection under the assumption that it is true. It is denoted by H_0.

Alternate hypothesis: It is the hypothesis other than the null hypothesis. It is denoted by H_1 test of hypothesis which is a procedure to decide whether to accept or reject the null hypothesis. When null hypothesis is accepted, the result is said to be significant, and when it is rejected, the result is said to be insignificant.

Type I error and Type II error: To accept or reject, the hypothesis always gives rise to same error or risk. There are two types of error in testing the hypothesis.

Type error I: It involves the rejection of null hypothesis when it is there.

Type error II: It involves acceptance of null hypothesis when it is false and should be rejected, i.e., P (rejecting H_0 when it is true) $= \alpha P$ (accepting H_0 when it is wrong) $= \beta$.

Level of Significance: The maximum probability of committing type I error is called the level of significance and is denoted by α. Level of significance generally taken as 0.05 or 0.01 level of significance is also expressed as percentage. Thus, level of significance $\alpha = 5\%$ means there are 5 chances in 100 that null hypothesis is rejected when it is true or one is 95% confident that a right decision is made.

Critical Region: The area under probability curve is divided into two regions: Region of rejection (where N.H is rejected)

1. Region of acceptance (where N.H is accepted)
   Critical region is the region of rejection of N.H. The area of critical region equals to the level of significance α.
   Simple hypothesis:
   Simple hypothesis is a statistical hypothesis which completely specifies a parameter.
   Null hypothesis is always a simple hypothesis stated as an equality specifying an exact value of the parameter.
   Simple hypothesis:
   Simple hypothesis is a statistical hypothesis which completely specifies a parameter.
   Null hypothesis is always a simple hypothesis stated as an equality specifying an exact value of the parameter.
   Composite hypothesis:
   Composite hypothesis is stated in terms of several possible values, i.e., by an inequality.
   Alternate hypothesis is a composite hypothesis involving statements expressed as inequalities such as < or > or

Example

1. $H_1\colon \mu > \mu_0$
2. $H_1\colon \mu < \mu_0$
3. $H_1\colon \mu \neq \mu_0$

One-tailed and two-tailed tests:
For a test, the critical region is represented at only one side (left or right) of the sampling distribution of statistic, and then, it is called one-tailed test. To indicate the critical region, left or right < or > symbol is used in alternate hypothesis.
One-tailed and two-tailed tests:
For a test, the critical region is represented at only one side (left or right) of the sampling distribution of statistic, and then, it is called one-tailed test. To indicate the critical region, left or right < or > symbol is used in alternate hypothesis.

If A.H. is of the not equal to type i.e.; $H_1\colon \mu \neq \mu_0$ then the critical region lies on both sides of the right and left tails of the curve such that the critical region of area $\frac{\alpha}{2}$ lies on the right tail and critical region of area $\frac{\alpha}{2}$ lies on the left tail.

If A.H. is of the not equal to type i.e.; $H_1\colon \mu \neq \mu_0$ then the critical region lies on both sides of the right and left tails of the curve such that the critical region of area $\frac{\alpha}{2}$ lies on the right tail and critical region of area $\frac{\alpha}{2}$ lies on the left tail as shown in the figure below.

Hypothesis concerning one mean (large sampleσ known)
To test whether the population mean μ equals to a specified constant μ_0 or not, then formulate the test hypothesis as follows.

Test Procedure

1. Null hypothesis

$$H_0\colon \mu = \mu_0$$

2. Alternate hypothesis

$$H_1\colon \mu \neq \mu_0 \text{ (Two tailed test)}$$

or

$$H_1\colon \mu > \mu_0 \text{ (Right tailed test)}$$

or

$$H_1\colon \mu > \mu_0 \text{ (Left tailed test)}$$

3. Level of significance: α

4. Critical region:

If the alternate hypothesis is not equal to type, a two-tailed test is concerned. For given α, critical values $-Z_{\frac{\alpha}{2}}$ and $Z_{\frac{\alpha}{2}}$ are determined from normal table, since normal distribution is assumed.

Computation:
The test statistic Z, denoted by

$$Z_{\text{cal}} \text{ or } Z \text{ by } Z = \frac{\overline{X} - \mu}{\frac{\sigma}{\sqrt{n}}}$$

where $\overline{X}$, the mean of the sample of size *n*, is calculated from the sample data.

Example
For $\alpha = 5\%$ or $\alpha = 0.05$ from normal table $Z_{\frac{\alpha}{2}} = Z_{0.025} = 1.96$.

Thus, the critical region consists of two shaded regions in the figure; i.e., reject null hypothesis H_0 if $Z < -Z_{\frac{\alpha}{2}}$ or $Z > Z_{\frac{\alpha}{2}}$; otherwise, accept H_0.

If the alternate hypothesis is of greater than type, a right-tailed test or is concerned. For a given α, critical value $Z\alpha$ is determined from normal table.

Example
For $\alpha = 5\%\alpha = 0.05$ from normal table $Z_{0.05} = 1.645$, thus the critical region consists of right shaded region in the figure; i.e., reject null hypothesis H_0 than type, and a left-tailed test is concerned. For a given α, critical value $-Z\alpha$ is determined from normal table.

Example
For $\alpha = 5\%$, $\alpha = 0.05$ from normal table $Z_{0.05} = 1.645$. Thus, the critical region in the figure; i.e., reject null hypothesis H_0 if $Z < -Z\alpha$; otherwise, accept H_0.

Conclusion
For two-tailed test, reject H_0 if Z_{cal} or Z falls in the critical region ($Z < -Z_{\frac{\alpha}{2}}$ or $Z > Z_{\frac{\alpha}{2}}$); otherwise, accept H_0. For right-tailed test, reject H_0 if $Z > Z_\alpha$; otherwise, accept H_0.

For left-tailed test, reject H_0 if $Z < -Z_\alpha$; otherwise, accept H_0.

Level of Significance

	$1\% = 0.01$	$5\% = 0.05$	$10\% = 0.1$
Two-tailed test	$\left\lvert Z_{\frac{\alpha}{2}} \right\rvert = 2.58$	$\left\lvert Z_{\frac{\alpha}{2}} \right\rvert = 1.96$	$\left\lvert Z_{\frac{\alpha}{2}} \right\rvert = 1.645$
Right-tailed test	$z_\alpha = 2.33$	$z_\alpha = 1.645$	$z_\alpha = 1.28$
Left-tailed test	$-z_\alpha = -2.33$	$-z_\alpha = -1.645$	$-z_\alpha = -1.28$

Hypothesis concerning two mean (large sample)
Difference of mean

Procedure

1. Null hypothesis

$$H_0\colon \mu_1 = \mu_2$$

Alternate hypothesis

$$H_1\colon \mu_1 \neq \mu_2 \text{ (Two tailed test)}$$

or

$$H_1\colon \mu_1 > \mu_2 \text{ (Right Tailed Test)}$$

or

$$H_1\colon \mu_1 < \mu_2 \text{ (Left tailed test)}$$

Level of significance α
Critical region

For two-tailed test
Reject null hypothesis if $|z| > Z_{\frac{\alpha}{2}}$

For right-tailed test
Reject null hypothesis if $Z > Z_\alpha$

For left-tailed test
Reject null hypothesis if $Z < -Z_\alpha$

Computation

$$\text{Test statistic } Z = \frac{\overline{X}_1 - \overline{X}_2}{\sqrt{\frac{\sigma_1^2}{n_1} + \frac{\sigma_2^2}{n_2}}}$$

Conclusion
Refer critical region.

Confidence interval for two means (large sample)

$$(\overline{X}_1 - \overline{X}_2) \pm \left(Z_{\frac{\alpha}{2}} \sqrt{\frac{\sigma_1^2}{n_1} + \frac{\sigma_2^2}{n_2}} \right)$$

$$\text{Maximum error of estimate} = \left(Z_{\frac{\alpha}{2}} \sqrt{\frac{\sigma_1^2}{n_1} + \frac{\sigma_2^2}{n_2}} \right)$$

Test of significance for single mean (small sample)

Procedure

Null hypothesis

$$H_0: \mu = \mu_0$$

Alternate hypothesis

$$H_1: \mu \neq \mu_0 \text{ (Two tailed test)}$$

or

$$H_1: \mu > \mu_0 \text{ (Right Tailed Test)}$$

or

$$H_1: \mu < \mu_0 \text{ (Left tailed test)}$$

Level of significance α

Critical region

For two-tailed test

Reject null hypothesis if $|t| > t_{\frac{\alpha}{2}}$ with $(n - 1)$ degrees of freedom

For right-tailed test

Reject null hypothesis if $t > t_\alpha$ with $(n - 1)$ degrees of freedom.

For left-tailed test

Reject null hypothesis if $t < -t_\alpha$ with $(n - 1)$ degrees of freedom

Computation

$$\text{Test statistic } t = \left(\frac{\bar{x} - \mu}{\frac{s}{\sqrt{n}}}\right)$$

where

$$s^2 = \frac{\sum (x_i - \bar{x})^2}{n - 1}.$$

Conclusion

Refer critical region.

Hypothesis concerning two mean (small sample)

Difference of mean

Procedure

Null hypothesis

$$H_0: \mu_1 - \mu_2 = 0$$

Alternate hypothesis

$$H_1: \mu_1 - \mu_2 \neq 0 \text{ (Two tailed test)}$$

or

$$H_1: \mu_1 > \mu_2 > 0 \text{ (Right Tailed Test)}$$

or

$$H_1: \mu_1 - \mu_2 < 0 \text{ (Left tailed test)}$$

Level of significance α
Critical region

For two-tailed test
Reject null hypothesis if $|t| > t_{\frac{\alpha}{2}}$ with $n_1 + n_2 - 2$ degrees of freedom

For right-tailed test
Reject null hypothesis if $t > t_\alpha$ with $n_1 + n_2 - 2$ degrees of freedom

For left-tailed test
Reject null hypothesis if $t < -t_\alpha$ with $n_1 + n_2 - 2$ degrees of freedom

Computation

$$\text{Test statistic } t = \frac{(\bar{x}_1 - \bar{x}_2)}{\left(s\sqrt{\frac{1}{n_1} + \frac{1}{n_2}}\right)}$$

with $n_1 + n_2 - 2$ degrees of freedom
where

$$s = \sqrt{\frac{s_1^2(n_1 - 1) + s_2^2(n_2 - 1)}{n_1 + n_2 - 1}}$$

where

$$s^2 = \frac{\sum (x_i - \bar{x})^2}{n - 1}.$$

Conclusion
Refer critical region.
Confidence interval for two means (difference of mean) small sample

$$(\bar{x}_1 - \bar{x}_2) \pm \left(t_{\frac{\alpha}{2}} s \sqrt{\frac{1}{n_1} + \frac{1}{n_2}}\right)$$

with $n_1 + n_2 - 2$ degrees of freedom.

The paired sample$t$-test

Not two samples but a pair of values before and after test will be given, take the difference of the values. Here, the variables are not independent. We can consider

$$d_i = \bar{x}(\text{say})$$

where d_i = difference of the values after and before work

$$t = \left(\frac{\bar{x} - \mu}{\left(\frac{s}{\sqrt{n}}\right)}\right)$$

where $\bar{x}$ and s, mean and standard deviation of the difference s, d_i's, μ = mean of the population differences.

Procedure

Null hypothesis

$$\mu = 0$$

Alternative hypothesis

$$\mu > 0$$
$$\mu < 0$$

or

Level of significance α

Critical region

For right-tailed test

Reject null hypothesis if $t > t_\alpha$ with $(n - 1)$ degrees of freedom

For left-tailed test

Reject null hypothesis if $t < -t_\alpha$ with $(n - 1)$ degrees of freedom

Computation

$$\text{Test statistic} t = \left(\frac{\bar{x} - \mu}{\left(\frac{s}{\sqrt{n}}\right)}\right)$$

Conclusion

Refer critical region.

Sampling distribution of the variance

A random variable has chi-square distribution (χ^2-distribution).

This is related to gamma distribution.

Theorem If S^2 is the variance of a random sample of size n taken from a normal population having the variance σ^2, then

$$\chi^2 = \frac{(n-1)s^2}{\sigma^2} = \frac{\sum_{i=1}^{n}(X_i - \overline{X})^2}{\sigma^2}$$

is a random variable having the chi-square distribution with the parameter $\upsilon = n - 1$ (degree of freedom).

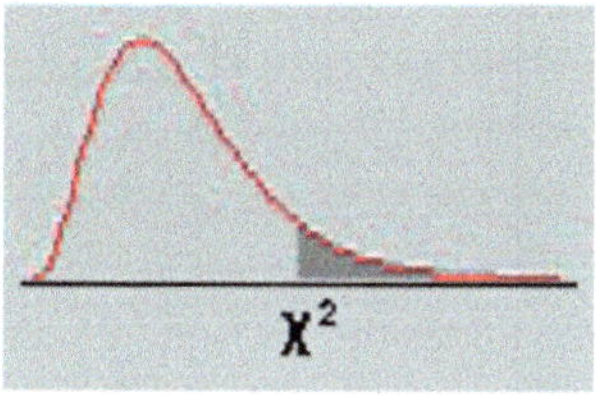

Lower critical values of chi-square distribution with νdegrees of freedom

Probability of exceeding the critical value

ν	0.90	0.95	0.975	0.99	0.999
1.	0.016	0.004	0.001	0.000	0.000
2.	0.211	0.103	0.051	0.020	0.002
3.	0.584	0.352	0.216	0.115	0.024
4.	1.064	0.711	0.484	0.297	0.091
5.	1.610	1.145	0.831	0.554	0.210
6.	2.204	1.635	1.237	0.872	0.381
7.	2.833	2.167	1.690	1.239	0.598
8.	3.490	2.733	2.180	1.646	0.857
9.	4.168	3.325	2.700	2.088	1.152
10.	4.865	3.940	3.247	2.558	1.479
11.	5.578	4.575	3.816	3.053	1.834

Lower critical values of chi-square distribution with νdegrees of freedom

Probability of exceeding the critical value

ν	0.90	0.95	0.975	0.99	0.999
1.	0.016	0.004	0.001	0.000	0.000
2.	0.211	0.103	0.051	0.020	0.002
3.	0.584	0.352	0.216	0.115	0.024
4.	1.064	0.711	0.484	0.297	0.091
5.	1.610	1.145	0.831	0.554	0.210
6.	2.204	1.635	1.237	0.872	0.381
7.	2.833	2.167	1.690	1.239	0.598
8.	3.490	2.733	2.180	1.646	0.857
9.	4.168	3.325	2.700	2.088	1.152
10.	4.865	3.940	3.247	2.558	1.479
11.	5.578	4.575	3.816	3.053	1.834

In the table, we can see values of χ^2_α for various values of υ (degrees of freedom) where χ^2_α is such that the area under the chi-square distribution to its right is equal to α.

The chi-square distribution is not symmetrical.

A random variable having the *F*-distribution

Theorem 6.5

If S_1^2 and S_2^2 are the variances of independent random samples of size n_1 and n_2, respectively, taken from two normal populations having the same variance, then $F = \frac{s_1^2}{s_2^2}$ is a random variable having the F-distribution with the parameters $\upsilon_1 = n_1 - 1$ and $\upsilon_2 = n_2 - 1$.

F-distribution determines whether the ratio of two sample variance S_1 and S_2 too small or too large. The F-distribution is related to the beta distribution, and its two parameters υ_1 and υ_2 are called the numerator and denominator degrees of freedom.

$F_{0.05}$ and $F_{0.01}$ for the various combinations of values of υ_1 and υ_2 are given in the F-distribution table.

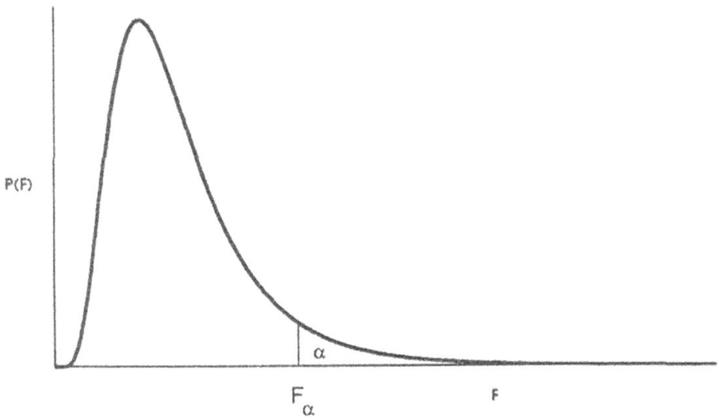

Figure K.1: The F distribution

F values for ® = 0:05

*d*1	*d*2								
	1	2	3	4	5	6	7	8	9
1	161.4	199.5	215.7	224.6	230.2	234.0	236.8	238.9	240.5
2	18.51	19.00	19.16	19.25	19.3	19.33	19.35	19.37	19.38
3	10.13	9.55	9.28	9.12	9.01	8.94	8.89	8.85	8.81
4	7.71	6.94	6.59	6.39	6.26	6.16	6.09	6.04	6.00
5	6.61	5.79	5.41	5.19	5.05	4.95	4.88	4.82	4.77
6	5.99	5.14	4.76	4.53	4.39	4.28	4.21	4.15	4.10
7	5.59	4.74	4.35	4.12	3.97	3.87	3.79	3.73	3.68
8	5.32	4.46	4.07	3.84	3.69	3.58	3.50	3.44	3.39

$F\alpha$ (υ_1, υ_2) is the value of F with υ_1 and υ_2 DOF such that the area under the F-distribution curve to the right of $F\alpha$ is α.

$$F_{1-\alpha}(\upsilon_1, \upsilon_2) = \frac{1}{F_\alpha(\upsilon_2, \upsilon_1)}$$

(1) F-distribution is always positive.
(2) The F-distribution curve lies entirely in first quadrant, and it is unimodel.
(3) Testing for the equality of variances of two normal population.
(4) F-test is used to determine whether two independent estimates of the population variance differ significantly or whether the two samples may be regarded as drawn from the normal population having the same variance.
(5) $(\sigma_A)^2 = (\sigma_B)^2 = \sigma^2$
(6) $F = \frac{(S_A)^2}{(S_B)^2}$

$$S_A^2 = \frac{\sum (x_i - \bar{x})^2}{n_1 - 1}$$
$$S_B^2 = \frac{\sum (y_i - \bar{y})^2}{n_2 - 1}$$

The degrees of freedom are $\upsilon_1 = n_1 - 1$, $\upsilon_2 = n_2 - 2$.
The numerator variance must be always greater than the denominator variance. That is $S_A^2 > S_B^2$.

Hypothesis concerning the variance of a normal population

Suppose we want to test a random sample $X_i(i = 1, 2, 3, \ldots)$ has been drawn from a normal population with a specified variance σ^2.

Test statistic $\chi^2 = \frac{ns^2}{\sigma^2}$ forms a chi-square distribution with $(n - 1)$ degree of freedom.

R Language

There are many in-built functions for statistical analysis in R. Most of them are part of R package. The in-built functions take R vector and other arguments as an input for giving the result. The in-built functions that we will discuss now are mean, median and mode.

Mean

It is calculated by taking the summation of all the values and dividing with the number of total number of values in a data series.
Syntax—mean (A, trim = 0, na.rm = FALSE, ...)

Following is the description of the parameters used

- A is the input vector.
- trim is used to drop some observations from both end of the sorted vector.
- na.rm is used to remove the missing values from the input vector.
- Example

```
y <- c(1,2,3,4,5)
result.mean <- mean(y)
print(result.mean)
[1] 3
```

With trim option

After applying trim parameter, all the values in the vector get sorted, and then, the required numbers of observations are dropped from calculating the mean.

When trim = 0.1, 1, values from each end will be dropped from the calculations to find mean. In this case, the sorted vector is (1,2,3,4,5,6) and the values removed from the vector for calculating mean are (1) from left and (6) from right.

Example

```
> x <- c(6,5,4,3,2,1)
> result.mean <- mean(x,trim = 0.1)
> print(result.mean)
[1] 3.5
```

With NA Option

The mean function returns NA if there are any missing values. To remove the missing values from the calculation, use na.rm = TRUE.

Example

```
> x <- c(6,5,4,3,2,1,NA)
> result.mean <- mean(x)
> print(result.mean)
[1] NA
```

Example

```
> x <- c(6,5,4,3,2,1,NA)
> result.mean <- mean(x,na.rm = TRUE)
> print(result.mean)
[1] 3.5
```

Median

The 'median' is the 'middle' value in the set of numbers. To find the median, your numbers have to be sorted first and then find the middle number.
Syntax—median (A, na.rm = FALSE)

With an even amount of numbers, we find the middle number in different way. In that case, we find the middle pair of numbers, by adding them together and dividing by two.

Following is the description of the parameters used.

- A is the input vector.
- na.rm is used to remove the missing values from the input vector.

Example 1

```
> x <- c(1,2,3,4,5)
> median(x)
[1] 3
```

Example 2

```
> x <- c(1,2,3,4,5,6)
> median(x)
[1] 3.5
```

Mode

The mode is the value that has maximum number of occurrences in a set of data. Mode can have both character data and numeric values. R does not have a standard in-built function to calculate mode. So we create a user function to calculate mode of a dataset in R.

Example 1

```
getmode <- function(p) {
A <- unique(p)
A[which.max(tabulate(match(p,A)))]
}
v <- c(1,2,3,1,1,4,5)
result <- getmode(v)
print(result)
[1] 1
```

Example 2

```
v <- c("o","it","prabhu","prabhu","raj")
result <- getmode(v)
print(result)
[1] "prabhu"
```

R-Linear Regression

It is a statistical tool which establishes relationship between two variables. One of the variables is called input variable whose values are given in the input dataset that is gathered through many experiments, and the other is output variable whose value is derived from input variable.

In linear regression, both the variables are related by an equation. The equation for linear regression is given by

$$y = mx + c$$

where

y is the output variable;
x is the input variable;
m, c are constants also known as coefficients.

Procedure to establish regression

A simple example of regression is given height of person as an input then predicts his weight. To predict the weight of person, we need to establish relationship between height and weight.

Steps involved to create the relationship:

- Gather a dataset of height of person and his corresponding weight.
- Use lm() built-in function of R-language to create relationship between input and output variable.
- Take the coefficients from the model and create the regression equation.
- Use predict function in R for predicting the weight of new person.

Input Data

Below is the dataset representing the observations

Height = 150 Weight = 62
Height = 173 Weight = 80
Height = 137 Weight = 55
Height = 185 Weight = 90
Height = 127 Weight = 46
Height = 135 Weight = 56
Height = 178 Weight = 75
Height = 162 Weight = 71

Height = 151 Weight = 61

Height = 130 Weight = 47

---

lm() Function

This function creates the relationship model between the input and output variables. Syntax:

lm (formula, inputdata)
Formula represents relationship between input and output variables.
Inputdata—It is a vector on which the formula is applied.

Create relationship and get the coefficients

```
x <- c(150,173,137,185,127,135,178,162,151,130)
y <- c(62,80,55,90,46,56,75,71,61,47)
relation <- lm(y ~ x)
print(relation)
Call:
lm(formula = y ~ x)
Coefficients:
 (Intercept)     x
 −38.7805 0.6746
```

Get the summary of relationship

```
print(summary(relation))
Call:
lm(formula = y ~ x)
Residuals:
   Min       1Q    Median   3Q     Max
−6.3002 −1.6629 0.0412 1.8944 3.9775
Coefficients:
Estimate Std. Error t value Pr(>|t|)
(Intercept) −38.78048 7.99754 −4.849   0.00127**
     x         0.67461   0.05191  12.997  1.16e − 06***
----
Signif. codes: 0 '***' 0.001 '**' 0.01 '*' 0.05 '.' 0.1 ' ' 1
Residual standard error: 3.253 on 8 degrees of freedom
Multiple R-squared: 0.9548, adjusted R-squared: 0.9491
F-statistic: 168.9 on 1 and 8 DF, p-value: 1.164e−06
```

Predict() Function

Syntax:

Predict (formula, newdataset)

formula is the formula which is already created by lm() function.
newdataset—It contains value for input variable.

Predict weight of new person

```
a <- data.frame (x = 170)
result <- predict (relation,a)
print(result)
    1
75.9033
```

Visualize Graphically

```
# Give the chart file a name.
png(file = "linearsion.png")
```

 # Plot the chart.

```
plot(y,x,col = "blue",main = "Height Weight Regression",
abline(lm(x~y)),cex = 1.3,pch = 16,xlab = "Weight in Kg",ylab = "Height in cm")
```

 # Save the file.

```
dev.off()
```

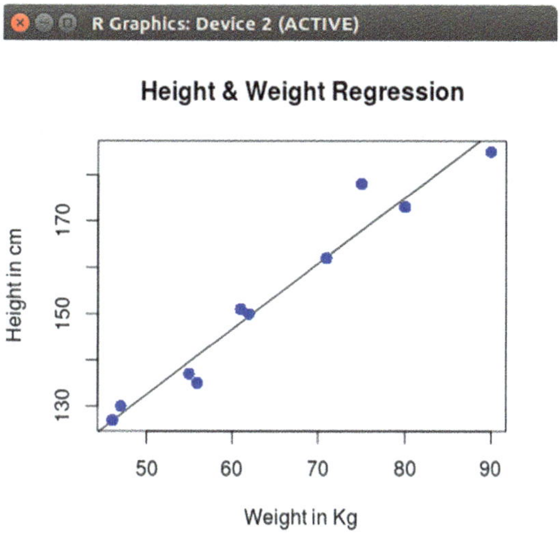

R—Multiple Regressions

Multiple regressions are a similar to linear regression, but there is a relationship between more than two variables. In simple linear relation, we have one input and one output variable, but in multiple regressions we have more than one input variable and one output variable.

The general mathematical equation for multiple regressions is

$$y = c + d_1x_1 + d_2x_2 + \cdots d_nx_n$$

Following is the description of the parameters used

- y is the output variable.
- $c, d_1, d_2, \ldots, d_n$ are the coefficients.
- $x_1, x_2, \ldots, x_n$ are the input variables.

We create the regression model using the lm() function in R. The model determines the value of the coefficients using the input data. Next, we can predict the value of the output variable for a given set of input variables using these coefficients.

lm() Function

This function creates the relationship model between the input and the output variables.
Syntax:

lm (formula, inputdata)
Formula represents relationship between input and output variables.
Eg: a ~ b1 + b2 + b3
Inputdata—It is a vector on which the formula is applied.

Example
Consider the dataset 'mtcars' available in the R environment. It gives a comparison between different car models in terms of quarter miletime (qsec), horse power('hp'), weight of the car('wt') and some more parameters.

The goal of the model is to establish the relationship between 'qsec' as a output variable with 'hp' and 'wt' as input variables. We create a subset of these variables from the mtcars dataset for this purpose.

```
input <- mtcars[,c("qsec","hp","wt")]
print(head(input))
```

	qsec	hp	wt
Mazda RX4	16.46	110	2.620
Mazda RX4 Wag	17.02	110	2.875
Datsun 710	18.61	93	2.320
Hornet 4 Drive	19.44	110	3.215
Hornet Sportabout	17.02	175	3.440
Valiant	20.22	105	3.460

Create model and get coefficients

```
model <- lm(qsec~hp+wt, data = input)
print(model)
```

Call:

```
lm(formula = qsec ~ hp + wt, data = input)
```

Coefficients:

```
(Intercept)        hp          wt
  18.82559   -0.02731   0.94153
```

```
m <- coef(model)[1]
print(m)
(Intercept)
18.82559
```

```
Xhp <- coef(model)[2]
print(Xhp)

   hp

-0.02730962
```

```
Xwt <- coef(model)[3]
print(Xwt)

   wt

0.9415324
```

Create Equation for Regression Model

Based on the above intercept and coefficient values, we create the mathematical equation.

$$y = m + Xhp * x1 + Xwt * x2$$
$$y = 18.82559 + (-0.02730962 * x1) + (0.9415324 * x2)$$

Appendix D: R Scripts

```
#install.packages("e1071")
library(e1071)
model <- naiveBayes(Species ~ ., data = iris)
class(model)
```

```
print(model)
#model1 <- naiveBayes(Species ~ iris$Sepal.Length,iris$Petal.Length,data =iris)
#install.packages("rminer")
#library("rminer")
#require("rminer")

set.seed(100) # setting seed to reproduce results of random sampling
trainingRowIndex <- sample(1:nrow(iris), 0.8*nrow(iris)) # row indices for training
data
trainingData <- iris[trainingRowIndex,] # model training data
testData <- iris[-trainingRowIndex,]
model1 <- naiveBayes(Species ~ .,data=trainingData)
preds <- predict(model1,newdata = testData)
conf_matrix <- table(preds, testData$Species)
mmetric(testData$Species,preds,c("ACC"))
## load RWeka
library(RWeka)
library(party)
## look for a package providing id3
WPM("refresh-cache")
WPM("list-packages", "available") ## look for id3
## install package providing id3
WPM("install-package", "simpleEducationalLearningSchemes")
## load the package
WPM("load-package", "simpleEducationalLearningSchemes")
## make classifier
ID3 <- make_Weka_classifier("weka/classifiers/trees/Id3")
## test it out.
DF2 <- read.arff(system.file("arff", "iris.arff", package = "RWeka"))
View(DF2)
exam <-ID3('class' ~ ., data = DF2)
summary(exam)
print(exam)

library(datasets)
head(iris)

library(ggplot2)
ggplot(iris, aes(Petal.Length, Petal.Width, color = Species)) + geom_point()

set.seed(20)
irisCluster <- kmeans(iris[, 3:4], 3, nstart = 20)
irisCluster

table(irisCluster$cluster, iris$Species)
irisCluster$cluster <- as.factor(irisCluster$cluster)
```

```
ggplot(iris,aes(Petal.Length,Petal.Width,color = irisCluster$cluster)) +
geom_point()
#points(irisCluster$centers)
plot(iris$Petal.Length,iris$Petal.Width,col=irisCluster$cluster)
points(irisCluster$centers, pch=11)

library(cluster)
clusplot(iris, irisCluster$cluster, color=TRUE, shade=TRUE,
labels=2, lines=0)
## load RWeka
library(RWeka)
library(party)
## look for a package providing id3
WPM("refresh-cache")
WPM("list-packages", "available") ## look for id3
## install package providing id3
WPM("install-package", "simpleEducationalLearningSchemes")
## load the package
WPM("load-package", "simpleEducationalLearningSchemes")
## make classifier
ID3 <- make_Weka_classifier("weka/classifiers/trees/Id3")
## test it out.
DF2 <- read.arff(system.file("arff", "iris.arff", package = "RWeka"))
View(DF2)
exam <-ID3(‘class’ ~ ., data = DF2)
summary(exam)
print(exam)
library(RWeka)
library(party)
library(partykit)

str(iris)
View(iris)
m1 <- J48(Species~., data = iris)
if(require("party", quietly = TRUE)) plot(m1)
summary(m1)
print(m1)
library(FSelector)
information.gain(Species~., data = iris)

library(ggvis)

# Iris scatter plot
iris %>% ggvis(~Sepal.Length, ~Sepal.Width, fill = ~Species) %>% layer_points()
iris %>% ggvis(~Petal.Length, ~Petal.Width, fill = ~Species) %>% layer_points()

# Overall correlation ‘Petal.Length’ and ‘Petal.Width’
```

```
cor(iris$Petal.Length, iris$Petal.Width)

# Return values of 'iris' levels
x=levels(iris$Species)

# Print Setosa correlation matrix
print(x[1])
cor(iris[iris$Species==x[1],1:4])

# Print Versicolor correlation matrix
print(x[2])
cor(iris[iris$Species==x[2],1:4])

# Print Virginica correlation matrix
print(x[3])
cor(iris[iris$Species==x[3],1:4])

library("class")

# Build your own 'normalize()' function
normalize <- function(x) {

  num <- x - min(x)
  denom <- max(x) - min(x)
  return (num/denom)

}

# Normalize the 'iris' data
iris_norm <- as.data.frame(lapply(iris[1:4], normalize))

# Summarize 'iris_norm'
summary(iris_norm)

set.seed(1234)
ind <- sample(2, nrow(iris), replace=TRUE, prob=c(0.67, 0.33))

# Compose training set
iris.training <- iris[ind==1, 1:4]

# Inspect training set
head(iris.training)

# Compose test set
iris.test <- iris[ind==2, 1:4]

# Inspect test set
head(iris.test)

# Compose 'iris' training labels
iris.trainLabels <- iris[ind==1,5]
```

```
# Inspect result
print(iris.trainLabels)

# Compose 'iris' test labels
iris.testLabels <- iris[ind==2, 5]

# Inspect result
print(iris.testLabels)

iris_pred <- knn(train = iris.training, test = iris.test, cl = iris.trainLabels, k=3)

# Inspect 'iris_pred'
iris_pred
# Put 'iris.testLabels' in a data frame
irisTestLabels <- data.frame(iris.testLabels)

# Merge 'iris_pred' and 'iris.testLabels'
merge <- data.frame(iris_pred, iris.testLabels)

# Specify column names for 'merge'
names(merge) <- c("Predicted Species", "Observed Species")

# Inspect 'merge'
merge

library(gmodels)
CrossTable(x = iris.testLabels, y = iris_pred, prop.chisq=FALSE)

###########################################

# Create index to split based on labels
index <- createDataPartition(iris$Species, p=0.75, list=FALSE)

# Subset training set with index
iris.training <- iris[.......,]

# Subset test set with index
iris.test <- iris[-.........,]

# Overview of algos supported by caret
names(getModelInfo())

# Train a model
model_knn <- train(iris.training[, 1:4], iris.training[, 5], method='knn')

# Predict the labels of the test set
predictions<-predict(object=model_knn,iris.test[,1:4])

# Evaluate the predictions
table(predictions)
```

```
# Confusion matrix
confusionMatrix(predictions,iris.test[,5])

# Train the model with preprocessing
model_knn <- train(iris.training[, 1:4], iris.training[, 5], method='knn', preProcess=c("center", "scale"))

# Predict values
predictions<-predict.train(object=model_knn,iris.test[,1:4], type="raw")

# Confusion matrix
confusionMatrix(predictions,iris.test[,5])
```

Printed by Printforce, the Netherlands